Lecture Notes in Computer Science　　7515

Commenced Publication in 1973
Founding and Former Series Editors:
Gerhard Goos, Juris Hartmanis, and Jan van Leeuwen

Gordon Fraser   Jerffeson Teixeira de Souza (Eds.)

Search Based
Software Engineering

4th International Symposium, SSBSE 2012
Riva del Garda, Italy, September 28-30, 2012
Proceedings

 Springer

Volume Editors

Gordon Fraser
The University of Sheffield
Department of Computer Science
Regent Court
211 Portobello
Sheffield S1 4DP, UK
E-mail: gordon.fraser@sheffield.ac.uk

Jerffeson Teixeira de Souza
State University of Ceara
Department of Research and Graduate Studies
Av. Paranjana 1700
60714-903 Fortaleza, CE, Brazil
E-mail: jeff@larces.uece.br

ISSN 0302-9743   e-ISSN 1611-3349
ISBN 978-3-642-33118-3   e-ISBN 978-3-642-33119-0
DOI 10.1007/978-3-642-33119-0
Springer Heidelberg Dordrecht London New York

Library of Congress Control Number: 2012945433

CR Subject Classification (1998): D.2, D.4, D.1, F.1

LNCS Sublibrary: SL 2 – Programming and Software Engineering

Typesetting: Camera-ready by author, data conversion by Scientific Publishing Services, Chennai, India

Printed on acid-free paper

Springer is part of Springer Science+Business Media (www.springer.com)

Preface

Message from the SSBSE 2012 General Chair

Welcome to the proceedings of the 4th Symposium on Search-Based Software Engineering, SSBSE 2012, held in Riva del Garda, in the Province of Trento, in Italy. Riva del Garda is a small city located in the north-western corner of the Garda Lake, in the middle of the Alps, surrounded by Mediterranean vegetation with olive and lemon trees. The symposium was co-located with the 28th IEEE International Conference on Software Maintenance, the premiere international venue in software maintenance and evolution.

The field of search-based software engineering is experiencing a growing interest from the software engineering community. SBSE is more and more influencing all phases of the software engineering process. From requirements engineering, where several works proposing the conjoint use of search-based techniques with other requirements engineering techniques have been presented in previous editions of the RE conference, to the more traditional application areas of search-based techniques such as software testing and maintenance phases, as witnessed by several works presented in the major conferences in these areas. This transversal exploitation of SBSE in the whole software engineering lifecycle is also reflected in the papers presented in this volume.

Many people contributed in different ways to the organization of SSBSE 2012 and to the preparation of the proceedings. Thus, there are many people to thank. First of all, the authors of the papers and their enthusiastic interest in search-based software engineering. The Program Chairs, Gordon Fraser and Jerffeson Teixeira de Souza, and the Graduate Students Track Chair, Shin Yoo, that organized a rich and high-quality scientific program; it was a great pleasure for me to work with them sharing the organizational decisions in the last year. The members of the Program Committee, as well as other reviewers, who reviewed the papers with efficacy and accuracy giving valuable feedback to the authors and selecting a pool of high-quality papers. The Steering Committee, chaired by Mark Harman, and the General Chair of SSBSE 2011, Phil McMinn, who provided me with several precious suggestions during the preparation of the event.

I would like to thank the members of the Organizing Committee: the Local Chair, Alessandro Marchetto, and the webmaster, Matthieu Vergne. They were fundamental for the success of the symposium. I would also like to give my appreciation to the Fondazione Bruno Kessler in Trento, Italy, for the encouragement and support in the organization of the symposium, and in particular to Moira Osti and Silvia Malesardi of the "eventi" office of the Fondazione for their help, dedication, and patience.

I am delighted with the two outstanding keynote speakers, Kalyanmoy Deb and Massimiliano di Penta, who focused their attention to multi-objective optimization and to new opportunities for using search-based techniques in software maintenance, respectively. It is also a pleasure to thank the speakers of the tutorials, and the participants of the panel who accepted to share their valuable experiences with all the participants.

Thanks to the sponsors of the symposium: UCL CREST department, Berner & Mattner, and IBM. I would also like to thank the FITTEST project, and in particular Tanja Vos, for her support in the acquisition of the sponsorships for the event. I am grateful to Alfred Hofmann, at Springer, who accepted to publish the proceedings of SSBSE.

My final thanks go to Paolo Tonella, General Chair of the International Conference on Software Maintenance, ICSM 2012, who accepted to host SSBSE 2012 in conjunction with this major conference, giving an occasion to reinforce the cross-fertilization between the two research communities.

I hope this symposium was a great opportunity to create new synergies between research groups from around the world, as well as between different disciplines and cultures. For those who have not personally attended the symposium, I hope that the spirit of the event can be captured thanks to this proceedings volume. See you next year at SSBSE 2013 in St. Petersburg, in Russia.

July 2012   Angelo Susi

Message from the SSBSE 2012 Program Chairs

On behalf of the SSBSE 2012 Program Committee, it is our pleasure to present the proceedings of the 4th International Symposium on Search-Based Software Engineering, held in beautiful Riva del Garda, Trento, Italy. SSBSE 2012 continued a recent tradition of bringing together the international SBSE community in an annual event to discuss and to celebrate progress in the field.

This year, we had 38 papers submitted to the conference (34 to the Research Track and 4 to the Student Track), with authors from 20 different countries (Austria, Brazil, Canada, China, Czech Republic, France, Germany, India, Ireland, Israel, Italy, Luxembourg, New Zealand, Norway, Saudi Arabia, Spain, Sweden, UK, and USA). At the end of the review process, where each submitted paper was refereed by at least three SBSE researchers, 15 papers were accepted as full papers and three were accepted as posters. In the Student Track, two papers were accepted.

We would like to thank the members of the SSBSE 2012 Program Committee. Their continuing support was essential to further improving the quality of accepted submissions and the resulting success of the conference. We also wish to especially thank the General Chair, Angelo Susi, who led a group of wonderful people that managed the organization of every single aspect in order to make the conference special to all of us. In addition, we want to thank Shin Yoo, SSBSE 2012 Student Track Chair, for managing the submissions of the bright young minds who will be responsible for the future of the SBSE field.

Maintaining a successful tradition, SSBSE 2012 attendees had the opportunity to learn from experts both from the research fields of search as well as software engineering, in two outstanding keynote talks. This year, we had the honor of receiving a keynote from Kalyanmoy Deb, who, with his distinguished contributions on evolutionary multi-objective optimization, has influenced many researchers in the SBSE field. Furthermore, we received a keynote from Massimiliano Di Penta, who enlightened us about the state of the art in software maintenance, laying down several research opportunities for SBSE researchers. Finally, we received a tutorial by Simon Poulding on the recently popular topic of exploiting graphics processing units (GPGPU) for search. Finally, we would like to thank all the authors who submitted papers to SSBSE 2012, regardless of acceptance or rejection, and everyone who attended the conference. We hope that with these proceedings, anybody who did not have the chance to be at Riva del Garda will have the opportunity to feel the liveliness of the SBSE community.

July 2012   Gordon Fraser
                                                                      Jerffeson Souza

Conference Organization

General Chair

Angelo Susi   Fondazione Bruno Kessler, Italy

Program Chairs

Gordon Fraser   University of Sheffield, UK
Jerffeson Souza   Universidade Estadual do Ceara, Brazil

Graduate Students Track Chair

Shin Yoo   University College London, UK

Organizing Committee

Alessandro Marchetto   Fondazione Bruno Kessler, Italy
Matthieu Vergne   Fondazione Bruno Kessler, Italy
Moira Osti   Fondazione Bruno Kessler, Italy

Program Committee

Enrique Alba   University of Málaga, Spain
Giulio Antoniol   Ecole Polytechnique de Montreal, Canada
Andrea Arcuri   Schlumberger and Simula Research Laboratory,
                               Norway
Marcio Barros   Universidade Federal do Estado do Rio
                               de Janeiro, Brazil
Leonardo Bottaci   University of Hull, UK
Lionel Briand   University of Luxembourg, Luxembourg
Francisco Chicano   University of Málaga, Spain
John Clark   University of York, UK
Myra Cohen   University of Nebraska-Lincoln, USA
Massimiliano Di Penta   RCOST - University of Sannio, Italy
Robert Feldt   University of Blekinge, Chalmers University
                               of Technology, Sweden
Mark Harman   University College London, UK
Rob Hierons   Brunel University, UK
Colin Johnson   University of Kent, UK
Gregory Kapfhammer   Allegheny College, UK

Yvan Labiche	Carleton University, Canada
Spiros Mancoridis	Drexel University, USA
Phil McMinn	University of Sheffield, UK
Mel Ó Cinnéide	University College Dublin, Ireland
Pasqualina Potena	Università degli Studi di Bergamo, Italy
Simon Poulding	University of York, UK
Xiao Qu	ABB Corporate Research, USA
Marek Reformat	University of Alberta, Canada
Marc Roper	University of Strathclyde, UK
Guenther Ruhe	University of Calgary, Canada
Paolo Tonella	Fondazione Bruno Kessler, Italy
Silvia Vergilio	Universidade Federal do Paraná, Brazil
Tanja Vos	Universidad Politecnica de Valencia, Spain
Westley Weimer	University of Virginia, USA
Yuanyuan Zhang	University College London, UK

External Reviewers

Nadia Alshahwan	University of Luxembourg, Luxembourg
Nesa Asoudeh	Carleton University, Canada
Arthur Baars	Universidad Politecnica de Valencia, Spain
S. M. Didar-Al-Alam	University of Calgary, Canada
Iman Hemati Moghadam	University College Dublin, Ireland
Reza Karimpour	University of Calgary, Canada
Sebastian Maurice	University of Calgary, Canada
Shiva Nejati	Simula Research Laboratory, Norway
Arash Niknafs	University of Calgary, Canada
Mehrdad Sabetzadeh	Simula Research Laboratory, Norway
Pingyu Zhang	University of Nebraska-Lincoln, USA

Steering Committee

Mark Harman (Chair)	University College London, UK
Giulio Antoniol	Ecole Polétechnique de Montreal, Canada
Lionel Briand	Universite du Luxembourg, Luxembourg
Myra Cohen	University of Nebraska Lincoln, USA
Massimiliano Di Penta	University of Sannio, Italy
Phil McMinn	University of Sheffield, UK
Mel Ó Cinnéide	University College Dublin, Ireland
Jerffeson Souza	Universidade Estadual do Ceara, Brazil
Joachim Wegener	Berner and Mattner, Germany

Sponsors

"The IBM logo is a registered trademark of International Business Machines Corporation (IBM) in the United States and other countries"

Table of Contents

Short Papers

Graduate Track Papers

Advances in Evolutionary Multi-objective Optimization

Kalyanmoy Deb*

Koenig Endowed Chair Professor,
Department of Electrical and Computer Engineering,
Michigan State University,
East Lansing, MI 48824, USA
kdeb@egr.msu.edu
http://www.iitk.ac.in/kangal/deb.htm

Abstract. Started during 1993-95 with three different algorithms, evolutionary multi-objective optimization (EMO) has come a long way in a quick time to establish itself as a useful field of research and application. Till to date, there exist numerous textbooks and edited books, commercial softwares dedicated to EMO algorithms, freely downloadable codes in most-used computer languages, a biannual conference series (called EMO conference series) running successfully since 2001, and special sessions and workshops held in almost all major evolutionary computing conferences. In this paper, we discuss briefly the principles of EMO through an illustration of one specific algorithm.Thereafter, we focus on mentioning a few recent research and application developments of EMO. Specifically, we discuss EMO's use with multiple criterion decision making (MCDM) procedures and EMO's applicability in handling of a large number of objectives. Besides, the concept of multi-objectivization and *innovization* – which are practically motivated, is discussed next. A few other key advancements are also highlighted. The development and application of EMO to multi-objective optimization problems and their continued extensions to solve other related problems have elevated the EMO research to a level which may now undoubtedly be termed as an active field of research with a wide range of theoretical and practical research and application opportunities. EMO concepts are ready to be applied to search based software engineering (SBSE) problems.

Keywords: Evolutionary optimization, Multi-objective optimization, Evolutionary multi-objective optimization, EMO.

1 Introduction

Evolutionary multi-objective optimization (EMO) has become a popular and useful field of research and application over the past decade. In a recent survey

---

* Also at Department of Mechanical Engineering, IIT Kanpur, India (deb@iitk.ac.in) and Department of Information and Service Economy, Aalto University School of Economics, Helsinki, Finland.

G. Fraser (Ed.): SSBSE 2012, LNCS 7515, pp. 1–26, 2012.

announced during the World Congress on Computational Intelligence (WCCI) in Vancouver 2006, EMO has been judged as one of the three fastest growing fields of research and application among all computational intelligence topics. Evolutionary optimization (EO) algorithms for solving single-objective optimization problems use a population based approach in which more than one solution participates in an iteration and evolves a new population of solutions in each iteration. The reasons for their popularity are many: (i) EOs do not require any derivative information (ii) EOs are relatively simple to implement and (iii) EOs are flexible and have a wide-spread applicability. For solving single-objective optimization problems or in other tasks focusing on finding a single optimal solution, the use of a population of solutions in each iteration may at first seem like an *overkill* (but they help provide an implicit parallel search ability, thereby making EOs computationally efficient [60,55]), in solving multi-objective optimization problems an EO procedure becomes a perfect match [20].

Multi-objective optimization problems, by nature, give rise to a set of Pareto-optimal solutions which need a further processing to arrive at a single preferred solution. To achieve the initial task of finding multiple trade-off solutions, it becomes quite a natural proposition to use an EO, because the use of a population in an iteration helps an EO to simultaneously find multiple trade-off solutions in a single run of the algorithm. This fact alone caused a surge of research and application of EMO over the years.

EMO concepts are applied only to a limited number of scenarios in search based software engineering (SBSE) [100,85,57]. Most applications have been concentrated in bi-objective problems and specifically applying to software clustering problems of minimizing the number of intra-cluster relations and maximizing the number of intra-cluster relations [40,75,79,76]. Recent developments of EMO for many-objective problems, EMO with multi-criterion decision-making, multi-objectivization, and EMOs to gather knowledge about problem – that are discussed in this paper – are all relevant topics that will have an immediate application to SBSE. A multi-objective genetic programming (MOGP) technique is suggested for SBSE [47], which shows how other multi-objective evolutionary algorithms can be useful to SBSE. Parameter tuning strategies used in SBSE [1] usually involve more than one criteria (maximizing robustness of an algorithm and minimizing computational time, for example) and an EMO procedure can be applied for the purpose as well.

In this paper, we begin with a brief description of the principles of an EMO in solving multi-objective optimization problems and then illustrate its working through a specific EMO procedure, which has been popularly and extensively used over the past 10 years. Besides this specific algorithm, there exist a number of other equally efficient EMO algorithms which we do not describe here for brevity. Instead, in this paper, we discuss a number of recent advancements of EMO research and application which are driving the researchers and practitioners ahead. The diversity of EMO's research is bringing researchers and practitioners together with different backgrounds including computer scientists, mathematicians, economists, engineers. The topics we discuss here amply demonstrate why and how EMO researchers from different backgrounds must and should collaborate in solving complex problem-solving tasks which have become the need of the hour in most branches of science, engineering, and commerce today.

2   Introduction to Multi-objective Optimization (MO)

A multi-objective optimization problem involves a number of objective functions which are to be either minimized or maximized subject to a number of constraints and variable bounds:

$$\left.\begin{array}{ll} \text{Minimize/Maximize} & f_m(\mathbf{x}), & m = 1, 2, \ldots, M; \\ \text{subject to} & g_j(\mathbf{x}) \geq 0, & j = 1, 2, \ldots, J; \\ & h_k(\mathbf{x}) = 0, & k = 1, 2, \ldots, K; \\ & x_i^{(L)} \leq x_i \leq x_i^{(U)}, & i = 1, 2, \ldots, n. \end{array}\right\} \quad (1)$$

A solution $\mathbf{x} \in \mathbf{R}^n$ is a vector of n decision variables: $\mathbf{x} = (x_1, x_2, \ldots, x_n)^T$. The solutions satisfying the constraints and variable bounds constitute a *feasible set* S in the decision variable space $\mathbf{R}^n$. One of the striking differences between single-objective and multi-objective optimization is that in multi-objective optimization the objective function vectors belong to a multi-dimensional objective space $\mathbf{R}^M$. The objective function vectors constitute a feasible set Z in the objective space. For each solution $\mathbf{x}$ in S, there exists a point $\mathbf{z} \in Z$, denoted by $\mathbf{f}(\mathbf{x}) = \mathbf{z} = (z_1, z_2, \ldots, z_M)^T$. To make the descriptions clear, we refer a decision variable vector as a solution and the corresponding objective vector as a point.

The optimal solutions in multi-objective optimization can be defined from a mathematical concept of *partial ordering*. In the parlance of multi-objective optimization, the term *domination* is used for this purpose. In this section, we restrict ourselves to discuss unconstrained (without any equality, inequality or bound constraints) optimization problems. The domination between two solutions is defined as follows [20,78]:

Definition 1. *A solution* $\mathbf{x}^{(1)}$ *is said to dominate the another solution* $\mathbf{x}^{(2)}$, *if both the following conditions are true:*

1. *The solution* $\mathbf{x}^{(1)}$ *is no worse than* $\mathbf{x}^{(2)}$ *in all objectives. Thus, the solutions are compared based on their objective function values (or location of the corresponding points ($\mathbf{z}^{(1)}$ and $\mathbf{z}^{(2)}$) in the objective function set Z).*
2. *The solution* $\mathbf{x}^{(1)}$ *is strictly better than* $\mathbf{x}^{(2)}$ *in at least one objective.*

For a given set of solutions (or corresponding points in the objective function set Z, for example, those shown in Figure 1(a)), a pair-wise comparison can be made using the above definition and whether one point dominates another point can be established. All points which are not dominated by any other member of the set are called non-dominated points of class one, or simply the non-dominated points. For the set of six points shown in the figure, they are points 3, 5, and 6. One property of any two such points is that a gain in an objective from one point to the other happens only due to a sacrifice in at least one other

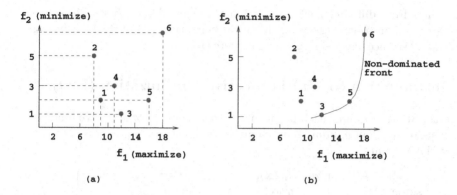

Fig. 1. (a) A set of points in a two-objective space and (b) Non-dominated front of level one

objective. This *trade-off* property between the non-dominated points makes the practitioners interested in finding a wide variety of them before making a final choice. These points make up a front (albeit disconnected in some scenarios) when viewed together on the objective space; hence the non-dominated points are often visualized to represent a *non-dominated front*.

The theoretical computational effort needed to select the points of the non-dominated front from a set of N points is $O(N \log N)$ for 2 and 3 objectives, and $O(N \log^{M-2} N)$ for $M > 3$ objectives [69], but for a moderate number of objectives, the procedure need not be particularly computationally effective in practice.

With the above concept, now it is easier to define the *Pareto-optimal solutions* in a multi-objective optimization problem. If the given set of points for the above task contain all points in the feasible decision variable space, the points lying on the non-domination front, by definition, do not get dominated by any other point in the objective space, hence are Pareto-optimal points (together they constitute the Pareto-optimal front) and the corresponding pre-images (decision variable vectors) are called Pareto-optimal solutions. However, more mathematically elegant definitions of Pareto-optimality (including the ones for continuous search space problems) exist in the multi-objective optimization literature [78,61,42].

2.1   Principles of Multi-objective Optimization

In the context of multi-objective optimization, the extremist principle of finding the optimum solution cannot be applied to one objective alone, when the rest of the objectives are also important. This clearly suggests two ideal goals of multi-objective optimization:

Convergence: Find a (finite) set of solutions which lie on the Pareto-optimal front, and

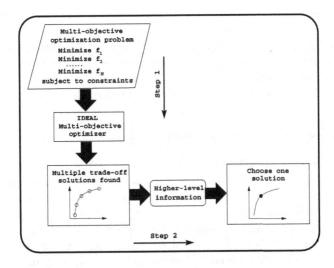

Fig. 2. Schematic of a two-step multi-criteria optimization and decision-making procedure

Diversity: Find a set of solutions which are diverse enough to represent the entire range of the Pareto-optimal front.

2.2 Principles of Evolutionary Multi-objective Optimization (EMO)

EMO algorithms attempt to follow both the above principles, similar to a posteriori MCDM method. Figure 2 shows schematically the principles followed in an EMO procedure. Since EMO procedures are heuristic based, they may not guarantee finding exact Pareto-optimal points, as a theoretically provable optimization method would do for tractable (for example, linear or convex) problems. But EMO procedures have essential operators to constantly improve the evolving non-dominated points (from the point of view of convergence and diversity mentioned above) similar to the way most natural and artificial evolving systems continuously improve their solutions. To this effect, a recent study [39] has demonstrated that a particular EMO procedure, starting from random non-optimal solutions, can progress towards theoretical Karush-Kuhn-Tucker (KKT) points with iterations in real-valued multi-objective optimization problems. The main difference and advantage of using an EMO compared to a posteriori MCDM procedures is that multiple trade-off solutions can be found in a single run of an EMO algorithm, whereas most a posteriori MCDM methodologies would require multiple independent runs.

In Step 1 of the EMO-based multi-objective optimization and decision-making procedure (the task shown vertically downwards in Figure 2), multiple trade-off, non-dominated points are found. Thereafter, in Step 2 (the task shown horizontally, towards the right), higher-level information is used to choose one of the obtained trade-off points.

Recent advances in EMO suggest a diagonal possibility. Both optimization and decision-making tasks can be combined together in an interactive manner and a more efficient multi-objective optimization task is possible [36]. We discuss this aspect in Section 4.4 under interactive EMO algorithms.

2.3   A Posteriori MCDM Methods and EMO

In the 'a posteriori' MCDM approaches (also known as 'generating MCDM methods'), the task of finding multiple Pareto-optimal solutions is achieved by executing many independent single-objective optimizations, each time finding a single Pareto-optimal solution [78]. A parametric scalarizing approach (such as the weighted-sum approach, ϵ-constraint approach, and others discussed in MCDM textbooks [78,10]) can be used to convert multiple objectives into a parametric single-objective objective function. By simply varying the parameters (weight vector or ϵ-vector) and optimizing the scalarized function, different Pareto-optimal solutions can be found. In contrast, in an EMO, multiple Pareto-optimal solutions are attempted to be found in a single run of the algorithm by emphasizing multiple non-dominated and isolated solutions in each iteration of the algorithm and without the use of any scalarization of objectives.

Consider Figure 3, in which we sketch how multiple independent parametric single-objective optimizations (through a posteriori MCDM method) may find different Pareto-optimal solutions. It is worth highlighting here that the Pareto-optimal front corresponds to global optimal solutions of several problems

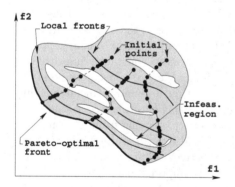

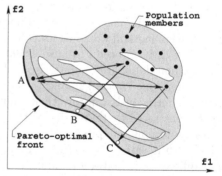

Fig. 3. A posteriori MCDM methodology employing independent single-objective optimizations

Fig. 4. EMO's parallel search ability brings multiple solutions (B and C) close to the Pareto-optimal front quickly from a population member A through its recombination operator

each formed with a different scalarization of objectives. During the course of an optimization task, algorithms must overcome a number of difficulties, such as infeasible regions, local optimal solutions, flat or non-improving regions of objective landscapes, isolation of optimum, etc., to finally converge to the global optimal solution. Moreover, due to practical limitations, an optimization task must also be completed in a reasonable computational time. All these difficulties in a problem require that an optimization algorithm strikes a good balance between exploring new search directions and exploiting the extent of search in currently-best search direction. When multiple runs of an algorithm need to be performed independently to find a set of Pareto-optimal solutions, the above balancing act must have to performed in every single run. Since runs are performed independently from one another, no information about the success or failure of previous runs is utilized to speed up the overall process. In difficult multi-objective optimization problems, such memory-less, a posteriori methods may demand a large overall computational overhead to find a set of Pareto-optimal solutions [89]. Moreover, despite the issue of global convergence, independent runs may not guarantee achieving a good distribution among obtained points by an easy variation of scalarization parameters.

EMO, as mentioned earlier, constitutes an inherent parallel search. When a particular population member overcomes certain difficulties and makes a progress towards the Pareto-optimal front, its variable values and their combination must reflect this fact. When a recombination takes place between this solution and another population member, such valuable information of variable value combinations gets shared through variable exchanges and blending, thereby making the overall task of finding multiple trade-off solutions a parallelly processed task (depicted in Figure 4).

3   Elitist EMO: NSGA-II

The NSGA-II procedure [22] is one of the popularly used EMO procedures which attempt to find multiple Pareto-optimal solutions in a multi-objective optimization problem and has the following three features:

1. It uses an elitist principle,
2. it uses an explicit diversity preserving mechanism, and
3. it emphasizes non-dominated solutions.

At any generation t, the offspring population (say, Q_t) is first created by using the parent population (say, P_t) and the usual genetic operators. Thereafter, the two populations are combined together to form a new population (say, R_t) of size $2N$. Then, the population R_t is classified into different non-dominated classes. Thereafter, the new population is filled by points of different non-dominated fronts, one at a time. The filling starts with the first non-dominated front

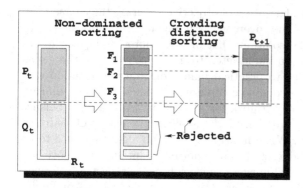
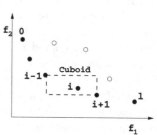

Fig. 5. Schematic of the NSGA-II procedure

Fig. 6. The crowding distance calculation

(of class one) and continues with points of the second non-dominated front, and so on. Since the overall population size of R_t is $2N$, not all fronts can be accommodated in N slots available for the new population. All fronts which could not be accommodated are deleted. When the last allowed front is being considered, there may exist more points in the front than the remaining slots in the new population. This scenario is illustrated in Figure 5. Instead of arbitrarily discarding some members from the last front, the points which will make the diversity of the selected points the highest are chosen.

The crowded-sorting of the points of the last front which could not be accommodated fully is achieved in the descending order of their *crowding distance values* and points from the top of the ordered list are chosen. The crowding distance d_i of point i is a measure of the objective space around i which is not occupied by any other solution in the population. Here, we simply calculate this quantity d_i by estimating the perimeter of the cuboid (Figure 6) formed by using the nearest neighbors in the objective space as the vertices (we call this the *crowding distance*).

3.1 Sample Results

Here, we show results from several runs of the NSGA-II algorithm on two test problems. The first problem (ZDT2) is two-objective, 30-variable problem with a concave Pareto-optimal front:

$$\text{ZDT2}: \begin{cases} \text{Minimize } f_1(\mathbf{x}) = x_1, \\ \text{Minimize } f_2(\mathbf{x}) = s(\mathbf{x})\left[1 - (f_1(\mathbf{x})/s(\mathbf{x}))^2\right], \\ \text{where} \quad s(\mathbf{x}) = 1 + \frac{9}{29}\sum_{i=2}^{30} x_i \\ \quad\quad\quad 0 \le x_1 \le 1, \\ \quad\quad\quad -1 \le x_i \le 1, \quad i = 2, 3, \ldots, 30. \end{cases} \tag{2}$$

The second problem (KUR), with three variables, has a disconnected Pareto-optimal front:

$$\text{KUR}: \begin{cases} \text{Minimize } f_1(\mathbf{x}) = \sum_{i=1}^{2}\left[-10\exp(-0.2\sqrt{x_i^2 + x_{i+1}^2})\right], \\ \text{Minimize } f_2(\mathbf{x}) = \sum_{i=1}^{3}\left[\text{abs}(x_i)^{0.8} + 5\sin(x_i^3)\right], \\ -5 \leq x_i \leq 5, \quad i = 1, 2, 3. \end{cases} \tag{3}$$

NSGA-II is run with a population size of 100 and for 250 generations. The variables are used as real numbers and an SBX recombination operator [21] with $p_c = 0.9$ and distribution index of $\eta_c = 10$ and a polynomial mutation operator [20] with $p_m = 1/n$ (n is the number of variables) and distribution index of $\eta_m = 20$ are used. Figures 7 and 8 show that NSGA-II converges to the Pareto-optimal front and maintains a good spread of solutions on both test problems.

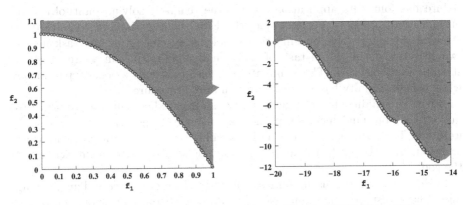

Fig. 7. NSGA-II on ZDT2   **Fig. 8.** NSGA-II on KUR

There also exist other competent EMOs, such as strength Pareto evolutionary algorithm (SPEA) and its improved version SPEA2 [103], Pareto archived evolution strategy (PAES) and its improved versions PESA and PESA2 [17], multi-objective messy GA (MOMGA) [97], multi-objective micro-GA [15], neighborhood constraint GA [73], ARMOGA [86] and others. Besides, there exists other EA based methodologies, such as particle swarm EMO [14,80], ant-based EMO [77,56], and differential evolution based EMO [2].

The constraint handling method modifies the binary tournament selection, where two solutions are picked from the population and the better solution is chosen. In the presence of constraints, each solution can be either feasible or infeasible. Thus, there may be at most three situations: (i) both solutions are feasible, (ii) one is feasible and other is not, and (iii) both are infeasible. We consider each case by simply redefining the domination principle. Details can be found in [20].

Since the early development of EMO algorithms in 1993, they have been applied to many challenging real-world optimization problems. Descriptions of some of these studies can be found in books [20,16,13,84], dedicated conference proceedings [101,49,12,83,43,95], and domain-specific books, journals and proceedings. Interested readers can refer to them for more detail.

4   Recent Developments of EMO

As soon as basic EMO algorithms were developed, researchers realized that there are plethora of further developments that could be made. In this section, we describe a number of such salient recent developments of EMO.

4.1   Multi-objectivization

Interestingly, the act of finding multiple trade-off solutions using an EMO procedure has found its application outside the realm of solving multi-objective optimization problems. For example, the EMO concept is used to solve constrained single-objective optimization problems by converting the task into a two-objective optimization task of additionally minimizing an aggregate constraint violation [11,24]. This eliminates the need to specify a penalty parameter while using a penalty based constraint handling procedure.

A well-known difficulty in genetic programming studies, called *bloating*, arises due to the continual increase in the size of evolved 'genetic programs' with iteration. The reduction of bloating by minimizing the size of a program as an additional objective helped find high-performing solutions with a smaller size of the code [6,19].

In clustering algorithms, minimizing the intra-cluster distance and maximizing inter-cluster distance simultaneously in a bi-objective formulation of a clustering problem is found to yield better solutions than the usual single-objective minimization of the ratio of the intra-cluster distance to the inter-cluster distance [58]. In SBSE, such bi-objective optimization is already used [79,75,76] and their advantage vis-a-vis a single-objective minimization of the ratio of number of intra-cluster relations to inter-cluster relations is shown. An EMO is also used to solve minimum spanning tree problem better than a single-objective EA [82].

A recent edited book [66] describes many such interesting applications in which EMO methodologies have helped solve problems which are otherwise (or traditionally) not treated as multi-objective optimization problems.

4.2   *Innovization*: Knowledge Extraction through EMO

One striking difference between a single and a multi-objective optimization is the cardinality of the solution set. In latter, multiple solutions are the outcome and each solution is theoretically an optimal solution corresponding to a particular trade-off among the objectives. Thus, if an EMO procedure can find solutions close to the true Pareto-optimal set, what we have in our hand are a number of high-performing solutions trading-off the conflicting objectives considered in the study. Since they are all near optimal, these solutions can be analyzed for finding properties which are common to them. Such a procedure can then become a systematic approach in deciphering important and hidden properties which optimal and high-performing solutions must have for that problem. In a number of practical problem-solving tasks, the so-called *innovization* procedure is shown to find important knowledge about high-performing solutions [37].

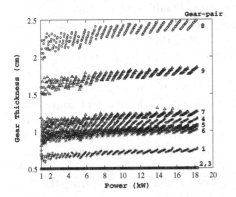

Fig. 9. Gear thickness does not vary much from one trade-off solution to another

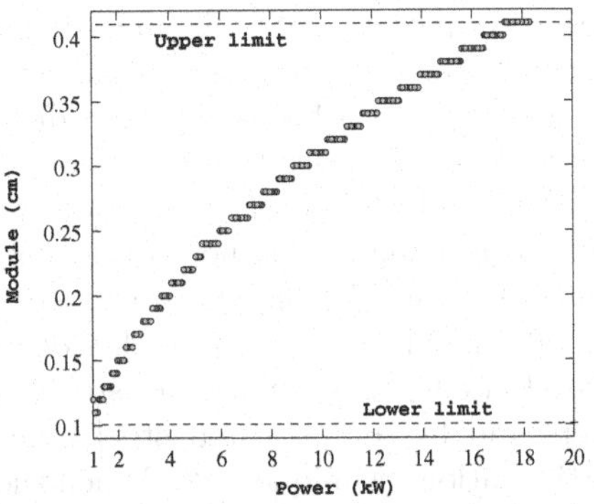

Fig. 10. Module of the gear system show a definite relationship among trade-off solutions

For example, consider a gear-box design problem involving 14 gear pairs. There are 29 design variables of which 14 are gear thickness values, 14 are number of teeth in each of 14 gear pairs and one is the module indicating the tooth profile of the gears that is same for all 14 gear pairs [30]. 200 trade-off solutions are obtained using NSGA-II procedure and variable values are shown with one of the objective values in Figures 9 and 10. As evident, the all trade-off solutions have more or less similar gear thickness values (a similar observation is made with number of gear teeth as well); the only way they vary significantly is with their module values (Figure 10). In fact, a regression fit indicates the following property among module (m) and power delivered (P):

$$m \propto \sqrt{P}. \tag{4}$$

This relationship that is common to most near Pareto-optimal solution reveals that the way to design a gear-box with more delivered power is to increase the size of the gears by increasing module as a squared-root of the desired delivered power. An increase in module will increase the diameter of gears and will hence increase the size of the gear-box which is then capable of delivering an increased power. By increasing the gear diameters as squared-root of required power is the *optimal* way of modifying the design, as dictated by the above *innovization* analysis.

Such useful properties that are hidden in a problem are valuable to designers and practitioners. The *innovization* procedure is one way to unveil such important information. The current efforts to automate the knowledge extraction procedure through a sophisticated data-mining task should make the overall approach more appealing to and useful in practice [4,3,23].

4.3 Handling Uncertainty in EMO

A major surge in EMO research has taken place in handling uncertainties among decision variables and problem parameters in multi-objective optimization. Practice is full of uncertainties and almost no parameter, dimension, or property can be guaranteed to be fixed at a value it is aimed at. Either a probability distribution of the uncertainty is known beforehand (alleatory uncertainty) or an imprecise information is known (epsitemic uncertainty). In such scenarios, evaluation of a solution is not precise, and the resulting objective and constraint function values becomes probabilistic quantities. Optimization algorithms are usually designed to handle such stochastiticies by using crude methods, such as the Monte Carlo simulation of stochasticities in uncertain variables and parameters and by sophisticated stochastic programming methods involving nested optimization techniques [28]. When these effects are taken care of during the optimization process, the resulting solution is usually different from the optimum solution of the problem and is known as a 'robust' solution. Such an optimization procedure will then find a solution which may not be the true global optimum solution, but one which is less sensitive to uncertainties in decision variables and problem parameters. In the context of multi-objective optimization, a consideration of uncertainties for multiple objective functions will result in a robust frontier which may be different from the globally Pareto-optimal front. Each and every point on the robust frontier is then guaranteed to be less sensitive to uncertainties in decision variables and problem parameters. Some such studies in EMO are [26,5].

When the evaluation of constraints under uncertainties in decision variables and problem parameters are considered, deterministic constraints become stochastic (they are also known as 'chance constraints') and involves a *reliability index* (R) to handle the constraints. A constraint $g(\mathbf{x}) \geq 0$ then becomes $\mathrm{Prob}(g(\mathbf{x}) \geq 0) \geq R$. In order to find left side of the above chance constraint, a separate optimization methodology [18], is needed, thereby making the overall algorithm a bi-level optimization procedure. Approximate single-loop algorithms exist [41] and recently one such methodology has been integrated with an EMO [28] and shown to find a 'reliable' frontier corresponding a specified reliability index, instead of the Pareto-optimal frontier, in problems having uncertainty in decision variables and problem parameters. Recently, two different types of epsitemic uncertainties are handled using modified NSGA-II procedures [93,94]. More such investigations are needed, as uncertainties is an integral part of practical problem-solving and multi-objective optimization researchers must look for better and faster algorithms to handle them.

4.4   EMO and Decision Making

Searching for a set of Pareto-optimal solutions by using an EMO fulfills only one aspect of multi-objective optimization, as choosing a particular solution for an implementation is the remaining decision-making task which is equally important. For many years, EMO researchers have postponed the decision-making aspect and concentrated on developing efficient algorithms for finding multiple trade-off solutions. Having pursued that part somewhat, now for the past couple of years or so, EMO researchers are putting efforts to design combined algorithms for optimization and decision making. In the view of the author, the decision-making task can be considered from two main considerations in an EMO framework:

1. **Generic consideration:** There are some aspects which most practical users would like to use in narrowing down their choice. We have discussed above the importance of finding robust and reliable solutions in the presence of uncertainties in decision variables and/or problem parameters. In such scenarios, an EMO methodology can straightway find a robust or a reliable frontier [26,28] and no subjective preference from any decision maker may be necessary. Similarly, if a problem resorts to a Pareto-optimal front having *knee* points, such points are often the choice of decision makers. Knee points demands a large sacrifice in at least one objective to achieve a small gain in another thereby making it discouraging to move out from a knee point [27,8].

2. **Subjective consideration:** In this category, any problem-specific information can be used to narrow down the choices and the process may even lead to a single preferred solution at the end. Most decision-making procedures use some preference information (utility functions, reference points [98], reference directions [67], marginal rate of return and a host of other considerations [78]) to select a subset of Pareto-optimal solutions. A recent book [9] is dedicated to the discussion of many such multi-criteria decision analysis (MCDA) tools and collaborative suggestions of using EMO with such MCDA tools. Some hybrid EMO and MCDA algorithms are suggested in the recent past [38,32,31,96,74].

Subjective consideration can again be pursued in three different ways:

1. **A priori method:** The preference information is provided *before* any optimization task is executed. This will usually result in converting a multi-objective optimization problem into a single-objective optimization problem. This is how most classical multi-objective optimization methodologies work [10,78].

2. **A posteriori method:** The preference information is provided *after* a multi-objective optimization task is executed and a set of trade-off solutions are found. This is the principles with which most EMO methodologies work. However, such a method requires a large number of trade-off solutions for many-objective (> 3) optimization problems and most EMO methods become inefficient in handling such problems.

3. **Interactive method:** The preference information is provided *during* the multi-objective optimization process. One such methodology is recently suggested elsewhere [36]. After an EMO runs for a few generations and a set of trade-off solutions are found, a few representative ones can be shown to the decision-maker for any preference information. Decision-maker's choice on pair-wise comparison of solution are then used to construct a mathematical *utility* function. From then on, the derived utility function is used to steer the EMO search, thereby preferring the information provided by the decision-maker.

An interactive method is clearly a viable approach and provides best flexibility in combining multi-objective optimization and multi-criterion decision making. Other MCDM approaches for eliciting preference information and integrating them with NSGA-II or other EMOs can also be tried.

Many other generic and subjective considerations are needed and it is interesting that EMO and MCDM researchers are collaborating on developing such complete algorithms for multi-objective optimization [9].

4.5   EMO for Handling a Large Number of Objectives

Soon after the development of efficient EMO methodologies, researchers were interested in exploring whether existing EMO methodologies are adequate to handle a large number of objectives (say, 10 or more). An earlier study [63] with eight objectives revealed somewhat negative results. But the author in his book [20] and recent other studies [64] have clearly explained the reason for the poor performance of EMO algorithms to many-objective problems (having more than three objectives).

EMO methodologies work by emphasizing non-dominated solutions in a population. Unfortunately, as the number of objectives increase, most population members in a randomly created population tend to become non-dominated to each other. For example, in a three-objective scenario, about 10% members in a population of size 200 are non-dominated, whereas in a 10-objective problem scenario, as high as 90% members in a population of size 200 are non-dominated. Thus, in a large-objective problem, an EMO algorithm runs out of room to introduce new population members into a generation, thereby causing a stagnation in the performance of an EMO algorithm. It has been argued that to make EMO procedures efficient, an exponentially large population size (with respect to number of objectives) is needed. This makes an EMO procedure slow and computationally less attractive. There are other difficulties related to visualization and adequate working of a recombination operator in a high-dimensional problem.

Earlier thoughts on handling many-objective optimization problems using an EMO were two-pronged:

1. **Solve structured problems:** In such problems, objective functions tend to correlate to each other as the solutions come close to the Pareto-optimal

front. In such cases, even though the original problem is many-dimensional (say, 10-objective), since the Pareto-optimal front is lower-dimensional (say three-dimensional and other seven objectives get correlated with these three objective), the EMO methodologies may work, as long as the Pareto-optimal front is two or three-dimensional. Several studies [35,88] have suggested a PCA-based NSGA-II to handle a 10 or 20-dimensional problem and find the complete range of its Pareto-optimal front which is two or three-dimensional.

2. **Use preference information:** If a priori preference information is known beforehand for a many-objective optimization problem, such knowledge can be used to focus EMO's search in the preferred part of the Pareto-optimal front. Although a large-dimensional front is still the target, the focused nature of the search does not require the EMO to establish a large diversity among solutions. Recent a priori based NSGA-II studies [38,31,32] have amply demonstrated NSGA-II's use in solving 10 and 15-objective problems, but focused on a small part of the entire Pareto-optimal front.

Recently, the above earlier thoughts have been challenged and a few many-objective evolutionary methods have been suggested to handle a large-objective problem. A recent study [29] showed that by specifying a set of reference points [29] or reference directions [99], an efficient EMO can be designed to find a representative set of points on the entire Pareto-optimal front in 10 to 15-objective optimization problems. The results from a three-objective and 15-objective DTLZ2 problems using M-NSGA-II algorithm [29] are shown in Figures 11 and 12. These are interesting developments and require further detail investigations for making them applicable to practice.

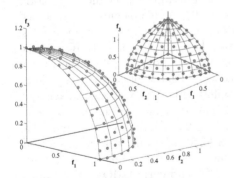

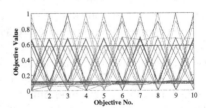

Fig. 11. M-NSGA-II finds a nicely distributed set of solutions on the entire three-dimensional Pareto-optimal front for the DTLZ4 problem

Fig. 12. M-NSGA-II finds a nicely distributed set of solutions on the entire 10-dimensional Pareto-optimal front for the DTLZ4 problem

4.6   Dynamic EMO

Dynamic optimization involves objectives, constraints, or problem parameters which change over time. Often, in such dynamic optimization problems, an algorithm is usually not expected to find the optimum, instead it is best expected to track the changing optimum with iteration. The performance of a dynamic optimizer then depends on how close it is able to track the true optimum (which is changing with iteration or time). A recent study [34] proposed the following procedure for dynamic optimization involving single or multiple objectives. In the case of dynamic multi-objective problem solving tasks, there is an additional difficulty which is worth mentioning here. Not only does an EMO algorithm needs to find or track the changing Pareto-optimal fronts, in a real-world implementation, it must also make an immediate decision about which solution to implement from the current front before the problem changes to a new one. If dynamic EMO is to be applied in practice, *automated* procedures for making decisions must be developed. This remains as a fertile area for research.

4.7   Hybrid EMO Algorithms

The search operators used in EMO are heuristic based. Thus, these methodologies are not guaranteed to find Pareto-optimal solutions with a finite number of solution evaluations in an arbitrary problem. In single-objective EA research, hybridization of EAs is common for ensuring convergence to an optimal solution, it is not surprising that studies on developing hybrid EMOs are now being pursued to ensure finding true Pareto-optimal solutions by hybridizing them with mathematically convergent ideas.

EMO methodologies provide adequate emphasis to currently non-dominated and isolated solutions so that population members progress towards the Pareto-optimal front iteratively. To make the overall procedure faster and to perform the task with a more theoretical emphasis, EMO methodologies are combined with mathematical optimization techniques having local convergence properties. A simple-minded approach would be to start the process with an EMO and the solutions obtained from EMO can be improved by optimizing a composite objective derived from multiple objectives to ensure a good spread by using a local search technique [25]. Another approach would be to use a local search technique as a mutation-like operator in an EMO so that all population members are at least guaranteed to be local optimal solutions [25,92]. To save computational time, instead of performing the local search for every solution in a generation, a mutation can be performed only after a few generations. Some recent studies [92,62,87] have demonstrated the usefulness of such hybrid EMOs for a guaranteed convergence.

Although these studies have concentrated on ensuring convergence to the Pareto-optimal front, some emphasis should now be placed in providing an adequate diversity among obtained solutions, particularly when a continuous Pareto-optimal front is represented by a finite set of points. Some ideas of maximizing hypervolume measure [48] or maintenance of uniform distance between points are proposed for this purpose, but how such diversity-maintenance techniques would be integrated with convergence-ensuring principles in a synergistic way would be interesting and useful future research. Some relevant studies in this direction exist [62,70,7].

4.8   Quality Estimates for EMO

When algorithms are developed and test problems with known Pareto-optimal fronts are available, an important task is to have performance measures with which the EMO algorithms can be evaluated. Thus, a major focus of EMO research has been spent to develop different performance measures. Since the focus in an EMO task is multi-faceted – convergence to the Pareto-optimal front and diversity of solutions along the entire front, it is also expected that one performance measure to evaluate EMO algorithms will be unsatisfactory. In the early years of EMO research, three different sets of performance measures were used:

1. Metrics evaluating convergence to the known Pareto-optimal front (such as error ratio, distance from reference set, etc.),
2. Metrics evaluating spread of solutions on the known Pareto-optimal front (such as spread, spacing, etc.), and
3. Metrics evaluating certain combinations of convergence and spread of solutions (such as hypervolume, coverage, R-metric, etc.).

Some of these metrics are described in texts [16,20]. A detailed study [65] comparing most existing performance metrics based on out-performance relations has recommended the use of the S-metric (or the hypervolume metric) and R-metric suggested by [59]. A recent study has argued that a single unary performance measure or any finite combination of them (for example, any of the first two metrics described above in the enumerated list or both together) cannot adequately determine whether one set is better than another [104]. That study also concluded that binary performance metrics (indicating usually two different values when a set of solutions A is compared against B and B is compared against A), such as epsilon-indicator, binary hypervolume indicator, utility indicators R1 to R3, etc., are better measures for multi-objective optimization. The flip side is that the chosen binary metric must be computed $K(K-1)$ times when comparing K different sets to make a fair comparison, thereby making the use of binary metrics computationally expensive in practice. Importantly, these performance measures have allowed researchers to use them directly as fitness measures within indicator based EAs (IBEAs) [102]. In addition, attainment indicators of [50,51] provide further information about location and inter-dependencies among obtained solutions.

4.9   Exact EMO with Run-Time Analysis

Since the suggestion of efficient EMO algorithms, they have been increasingly applied in a wide variety of problem domains to obtain trade-off frontiers. Simultaneously, some researchers have also devoted their efforts in developing exact

EMO algorithms with a theoretical complexity estimate in solving certain discrete multi-objective optimization problems. The first such study [72] suggested a pseudo-Boolean multi-objective optimization problem – a two-objective LOTZ (Leading Ones Trailing Zeroes) – and a couple of EMO methodologies – a simple evolutionary multi-objective optimizer (SEMO) and an improved version fair evolutionary multi-objective optimizer (FEMO). The study then estimated the worst-case computational effort needed to find all Pareto-optimal solutions of the problem LOTZ. This study spurred a number of improved EMO algorithms with run-time estimates and resulted in many other interesting test problems [53,71,54,68]. Although these test problems may not resemble common practical problems, the working principles of suggested EMO algorithms to handle specific problem structures bring in a plethora of insights about the working of multi-objective optimization, particularly in comprehensively finding all (not just one, or a few) Pareto-optimal solutions.

4.10   EMO with Meta-models and a Budget of Fixed Number of Evaluations

The practice of optimization algorithms is often limited by the computational overheads associated with evaluating solutions. Certain problems involving expensive computations, such as numerical solution of partial differential equations describing the physics of the problem, finite difference computations involving an analysis of a solution, computational fluid dynamics simulation to study the performance of a solution over a changing environment etc. In some such problems, evaluation of each solution to compute constraints and objective functions may take a few hours to a complete day or two. In such scenarios, even if an optimization algorithm needs one hundred solutions to get anywhere close to a good and feasible solution, the application needs an easy three to six months of continuous computational time. In most practical purposes, this is considered a 'luxury' in an industrial set-up. Optimization researchers are constantly at their toes in coming up with approximate yet faster algorithms.

A little thought brings out an interesting fact about how optimization algorithms work. The initial iterations deal with solutions which may not be close to optimal solutions. Therefore, these solutions need not be evaluated with high precision. Meta-models for objective functions and constraints have been developed for this purpose. Two different approaches are mostly followed. In one approach, a sample of solutions are used to generate a meta-model (approximate model of the original objectives and constraints) and then efforts have been made to find the optimum of the meta-model, assuming that the optimal solutions of both the meta-model and the original problem are similar to each other [44,52]. In the other approach, a successive meta-modeling approach is used in which the algorithm starts to solve the first meta-model obtained from a sample of the entire search space [81,33,45]. As the solutions start to focus near the optimum region of the meta-model, a new and more accurate meta-model is generated in the region dictated by the solutions of the previous optimization. A coarse-to-fine-grained meta-modeling technique based on artificial neural networks is shown to reduce the computational effort by about 30 to 80% on different problems [81]. Other successful meta-modeling implementations for multi-objective optimization based on Kriging and response surface methodologies exist [45,46].

Recent studies [91,90] shows the importance of fixing some finite number of predefined number of solution evaluations in the presence of replicates in an EMO study for handling practical problems. Efficient strategies for creating new solutions, allocating function evaluations to replicates and varying sizing of populations are all need to be enforced to achieve such a task. The above references just scratches the surface to this important practical optimization task and more such studies are needed to make EMOs efficient and quick for handling practical multi- and many-objective optimization problems.

5   Conclusions

The research and application in evolutionary multi-objective optimization (EMO) is now almost 20 years old and has resulted in a number of efficient algorithms for finding a set of well-diversified, near Pareto-optimal solutions. EMO algorithms are now regularly being applied to different problems involving most branches of science, engineering and commerce.

This paper has started with discussing principles of EMO and illustrated the principle by depicting one efficient and popularly used EMO algorithm. However, the highlight of this paper is the description of some of the current research and application activities involving EMO. One critical area of current research lies in collaborative EMO-MCDM algorithms for achieving a complete multi-objective optimization task of finding a set of trade-off solutions and finally arriving at a single preferred solution. Another direction taken by the researchers is to address guaranteed convergence and diversity of EMO algorithms through hybridizing them with mathematical and numerical optimization techniques as local search algorithms. Interestingly, EMO researchers have discovered its potential in solving traditionally hard optimization problems, but not necessarily multi-objective in nature, in a convenient manner using EMO algorithms. The so-called multi-objectivization studies are attracting researchers from various fields to develop and apply EMO algorithms in many innovative ways. A considerable research and application interest has also been put in addressing practical aspects into existing EMO algorithms. Towards this direction, handling uncertainty in decision variables and parameters, meeting an overall desired system reliability in obtained solutions, achieving a multi-objective optimization task with a limited budget of solution evaluations, handling dynamically changing problems (on-line optimization), handling a large number of objectives have been discussed in this paper.

It is clear that the field of EMO research and application, in a short span of about 20 years, now has efficient algorithms and numerous interesting and useful applications, and has been able to attract theoretically and practically oriented researchers to come together and make collaborative contributions. The practical importance of EMO's working principle, the flexibility of evolutionary optimization which lies at the core of EMO algorithms, and demonstrated diversification of EMO's principle to a wide variety of different problem-solving tasks are the main cornerstones for their success so far. The scope of research and application in EMO and using EMO are enormous and open-ended. This paper remains an open invitation to search-based software engineering (SBSE) researchers to take a look at what has been done in EMO and to explore ways take advantage of EMO's flexible problem solving ability.

Acknowledgments. The author acknowledges the support of the Academy of Finland (Grant # 133387). This paper is an updated version of a few recent articles by the author:

- Deb, K. (2010). *Recent Developments in Evolutionary Multi-Objective Optimization, In S. Greco, M. Ehrgott and J. R. Figueira (Eds.) "Trends in Multiple Criteria Decision Analysis Trends in Multiple Criteria Decision Analysis", International Series in Operations Research & Management Science, Volume 142, Springer, 339-368*
- Deb, K. (2006). Multi-objective optimization. In E. Burke, G. Kendall, *Search Methodologies: Introductory Tutorials in Optimization and Decision Support Techniques*, Springer, (pp. 273–316).

References

1. Arcuri, A., Fraser, G.: On Parameter Tuning in Search Based Software Engineering. In: Cohen, M.B., Ó Cinnéide, M. (eds.) SSBSE 2011. LNCS, vol. 6956, pp. 33–47. Springer, Heidelberg (2011)
2. Babu, B., Jehan, M.L.: Differential Evolution for Multi-Objective Optimization. In: Proceedings of the 2003 Congress on Evolutionary Computation (CEC 2003), vol. 4, pp. 2696–2703. IEEE Press, Canberra (2003)
3. Bandaru, S., Deb, K.: Automated discovery of vital knowledge from pareto-optimal solutions: First results from engineering design. In: World Congress on Computational Intelligence (WCCI 2010). IEEE Press (2010)
4. Bandaru, S., Deb, K.: Towards automating the discovery of certain innovative design principles through a clustering based optimization technique. Engineering Optimization 43(9), 911–941 (2011)
5. Basseur, M., Zitzler, E.: Handling uncertainty in indicator-based multiobjective optimization. International Journal of Computational Intelligence Research 2(3), 255–272 (2006)
6. Bleuler, S., Brack, M., Zitzler, E.: Multiobjective genetic programming: Reducing bloat using SPEA2. In: Proceedings of the 2001 Congress on Evolutionary Computation, pp. 536–543 (2001)
7. Bosman, P.A.N., Thierens, D.: The balance between proximity and diversity in multiobjective evolutionary algorithms. IEEE Transactions on Evolutionary Computation 7(2) (2003)
8. Branke, J., Deb, K., Dierolf, H., Osswald, M.: Finding Knees in Multi-objective Optimization. In: Yao, X., Burke, E.K., Lozano, J.A., Smith, J., Merelo-Guervós, J.J., Bullinaria, J.A., Rowe, J.E., Tiňo, P., Kabán, A., Schwefel, H.-P. (eds.) PPSN VIII. LNCS, vol. 3242, pp. 722–731. Springer, Heidelberg (2004)

9. Branke, J., Deb, K., Miettinen, K., Slowinski, R.: Multiobjective optimization: Interactive and evolutionary approaches. Springer, Berlin (2008)
10. Chankong, V., Haimes, Y.Y.: Multiobjective Decision Making Theory and Methodology. North-Holland, New York (1983)
11. Coello Coello, C.A.: Treating objectives as constraints for single objective optimization. Engineering Optimization 32(3), 275–308 (2000)
12. Coello Coello, C.A., Hernández Aguirre, A., Zitzler, E. (eds.): EMO 2005. LNCS, vol. 3410. Springer, Heidelberg (2005)
13. Coello Coello, C.A., Lamont, G.B.: Applications of Multi-Objective Evolutionary Algorithms. World Scientific (2004)
14. Coello Coello, C.A., Lechuga, M.S.: MOPSO: A Proposal for Multiple Objective Particle Swarm Optimization. In: Congress on Evolutionary Computation (CEC 2002), vol. 2, pp. 1051–1056. IEEE Service Center, Piscataway (2002)
15. Coello Coello, C.A., Toscano, G.: A micro-genetic algorithm for multi-objective optimization. Tech. Rep. Lania-RI-2000-06, Laboratoria Nacional de Informatica Avanzada, Xalapa, Veracruz, Mexico (2000)
16. Coello Coello, C.A., VanVeldhuizen, D.A., Lamont, G.: Evolutionary Algorithms for Solving Multi-Objective Problems. Kluwer, Boston (2002)
17. Corne, D.W., Knowles, J.D., Oates, M.: The Pareto Envelope-based Selection Algorithm for Multiobjective Optimization. In: Deb, K., Rudolph, G., Lutton, E., Merelo, J.J., Schoenauer, M., Schwefel, H.-P., Yao, X. (eds.) PPSN VI. LNCS, vol. 1917, pp. 839–848. Springer, Heidelberg (2000)
18. Cruse, T.R.: Reliability-based mechanical design. Marcel Dekker, New York (1997)
19. De Jong, E.D., Watson, R.A., Pollack, J.B.: Reducing bloat and promoting diversity using multi-objective methods. In: Proceedings of the Genetic and Evolutionary Computation Conference (GECCO 2001), pp. 11–18 (2001)
20. Deb, K.: Multi-objective optimization using evolutionary algorithms. Wiley, Chichester (2001)
21. Deb, K., Agrawal, R.B.: Simulated binary crossover for continuous search space. Complex Systems 9(2), 115–148 (1995)
22. Deb, K., Agrawal, S., Pratap, A., Meyarivan, T.: A fast and elitist multi-objective genetic algorithm: NSGA-II. IEEE Transactions on Evolutionary Computation 6(2), 182–197 (2002)
23. Deb, K., Bandaru, S., Celal Tutum, C.: Temporal Evolution of Design Principles in Engineering Systems: Analogies with Human Evolution. In: Pavone, M., Nicosia, G. (eds.) PPSN 2012, Part II. LNCS, vol. 7492, pp. 1–10. Springer, Heidelberg (2012)
24. Deb, K., Datta, R.: A fast and accurate solution of constrained optimization problems using a hybrid bi-objective and penalty function approach. In: Proceedings of the IEEE World Congress on Computational Intelligence (WCCI 2010), pp. 165–172 (2010)
25. Deb, K., Goel, T.: A Hybrid Multi-objective Evolutionary Approach to Engineering Shape Design. In: Zitzler, E., Deb, K., Thiele, L., Coello Coello, C.A., Corne, D.W. (eds.) EMO 2001. LNCS, vol. 1993, pp. 385–399. Springer, Heidelberg (2001)
26. Deb, K., Gupta, H.: Introducing robustness in multi-objective optimization. Evolutionary Computation Journal 14(4), 463–494 (2006)
27. Deb, K., Gupta, S.: Understanding knee points in bicriteria problems and their implications as preferred solution principles. Engineering Optimization 43(11), 1175–1204 (2011)

28. Deb, K., Gupta, S., Daum, D., Branke, J., Mall, A., Padmanabhan, D.: Reliability-based optimization using evolutionary algorithms. IEEE Trans. on Evolutionary Computation 13(5), 1054–1074 (2009)

29. Deb, K., Jain, H.: An improved NSGA-II procedure for many-objective optimization Part I: Problems with box constraints. Tech. Rep. KanGAL Report Number 2012009, Indian Institute of Technology Kanpur (2012)

30. Deb, K., Jain, S.: Multi-speed gearbox design using multi-objective evolutionary algorithms. ASME Transactions on Mechanical Design 125(3), 609–619 (2003)

31. Deb, K., Kumar, A.: Interactive evolutionary multi-objective optimization and decision-making using reference direction method. In: Proceedings of the Genetic and Evolutionary Computation Conference (GECCO 2007), pp. 781–788. The Association of Computing Machinery (ACM), New York (2007)

32. Deb, K., Kumar, A.: Light beam search based multi-objective optimization using evolutionary algorithms. In: Proceedings of the Congress on Evolutionary Computation (CEC 2007), pp. 2125–2132 (2007)

33. Deb, K., Nain, P.K.S.: An Evolutionary Multi-objective Adaptive Meta-modeling Procedure Using Artificial Neural Networks. In: Yang, S., Ong, Y.-S., Jin, Y. (eds.) Evolutionary Computation in Dynamic and Uncertain Environments. SCI, vol. 51, pp. 297–322. Springer, Heidelberg (2007)

34. Deb, K., Rao, U.B.N., Karthik, S.: Dynamic Multi-objective Optimization and Decision-Making Using Modified NSGA-II: A Case Study on Hydro-thermal Power Scheduling. In: Obayashi, S., Deb, K., Poloni, C., Hiroyasu, T., Murata, T. (eds.) EMO 2007. LNCS, vol. 4403, pp. 803–817. Springer, Heidelberg (2007)

35. Deb, K., Saxena, D.: Searching for Pareto-optimal solutions through dimensionality reduction for certain large-dimensional multi-objective optimization problems. In: Proceedings of the World Congress on Computational Intelligence (WCCI 2006), pp. 3352–3360 (2006)

36. Deb, K., Sinha, A., Korhonen, P., Wallenius, J.: An interactive evolutionary multi-objective optimization method based on progressively approximated value functions. IEEE Transactions on Evolutionary Computation 14(5), 723–739 (2010)

37. Deb, K., Srinivasan, A.: Innovization: Innovating design principles through optimization. In: Proceedings of the Genetic and Evolutionary Computation Conference (GECCO 2006), pp. 1629–1636. ACM, New York (2006)

38. Deb, K., Sundar, J., Uday, N., Chaudhuri, S.: Reference point based multi-objective optimization using evolutionary algorithms. International Journal of Computational Intelligence Research (IJCIR) 2(6), 273–286 (2006)

39. Deb, K., Tiwari, R., Dixit, M., Dutta, J.: Finding trade-off solutions close to KKT points using evolutionary multi-objective optimization. In: Proceedings of the Congress on Evolutionary Computation (CEC 2007), pp. 2109–2116 (2007)

40. Doval, D., Mancoridis, S., Mitchell, B.S.: Automatic clustering of software systems using a genetic algorithm. In: Proceedings of International Conference on Software Tools and Engineering Practice (STEP 1999), pp. 73–81. IEEE Press, Piscatway (1999)

41. Du, X., Chen, W.: Sequential optimization and reliability assessment method for efficient probabilistic design. ASME Transactions on Journal of Mechanical Design 126(2), 225–233 (2004)

42. Ehrgott, M.: Multicriteria Optimization, 2nd edn. Springer, Berlin (2005)

43. Ehrgott, M., Fonseca, C.M., Gandibleux, X., Hao, J.-K., Sevaux, M. (eds.): EMO 2009. LNCS, vol. 5467. Springer, Heidelberg (2009)

44. El-Beltagy, M.A., Nair, P.B., Keane, A.J.: Metamodelling techniques for evolutionary optimization of computationally expensive problems: promises and limitations. In: Proceedings of the Genetic and Evolutionary Computation Conference (GECCO1999), pp. 196–203. Morgan Kaufman, San Mateo (1999)

45. Emmerich, M., Giannakoglou, K.C., Naujoks, B.: Single and multiobjective evolutionary optimization assisted by gaussian random field metamodels. IEEE Transactions on Evolutionary Computation 10(4), 421–439 (2006)

46. Emmerich, M., Naujoks, B.: Metamodel-assisted multiobjective optimisation strategies and their application in airfoil design. In: Adaptive Computing in Design and Manufacture VI, pp. 249–260. Springer, London (2004)

47. Ferrucci, F., Gravino, C., Sarro, F.: How Multi-Objective Genetic Programming Is Effective for Software Development Effort Estimation? In: Cohen, M.B., Ó Cinnéide, M. (eds.) SSBSE 2011. LNCS, vol. 6956, pp. 274–275. Springer, Heidelberg (2011)

48. Fleischer, M.: The Measure of Pareto Optima: Applications to Multi-objective Optimization. In: Fonseca, C.M., Fleming, P.J., Zitzler, E., Deb, K., Thiele, L. (eds.) EMO 2003. LNCS, vol. 2632, pp. 519–533. Springer, Heidelberg (2003)

49. Fonseca, C.M., Fleming, P.J., Zitzler, E., Deb, K., Thiele, L. (eds.): EMO 2003. LNCS, vol. 2632. Springer, Heidelberg (2003)

50. Fonseca, C.M., Fleming, P.J.: On the Performance Assessment and Comparison of Stochastic Multiobjective Optimizers. In: Voigt, H.M., Ebeling, W., Rechenberg, I., Schwefel, H.P. (eds.) PPSN IV. LNCS, vol. 1141, pp. 584–593. Springer, Heidelberg (1996)

51. Fonseca, C.M., da Fonseca, V.G., Paquete, L.: Exploring the Performance of Stochastic Multiobjective Optimisers with the Second-Order Attainment Function. In: Coello Coello, C.A., Hernández Aguirre, A., Zitzler, E. (eds.) EMO 2005. LNCS, vol. 3410, pp. 250–264. Springer, Heidelberg (2005)

52. Giannakoglou, K.C.: Design of optimal aerodynamic shapes using stochastic optimization methods and computational intelligence. Progress in Aerospace Science 38(1), 43–76 (2002)

53. Giel, O.: Expected runtimes of a simple multi-objective evolutionary algorithm. In: Proceedings of Congress on Evolutionary Computation (CEC 2003). IEEE Press, Piscatway (2003)

54. Giel, O., Lehre, P.K.: On the effect of populations in evolutionary multi-objective optimization. In: Proceedings of the 8th Annual Genetic and Evolutionary Computation Conference (GECCO 2006), pp. 651–658. ACM Press, New York (2006)

55. Goldberg, D.E.: Genetic Algorithms for Search, Optimization, and Machine Learning. Addison-Wesley, Reading (1989)

56. Gravel, M., Price, W.L., Gagné, C.: Scheduling continuous casting of aluminum using a multiple objective ant colony optimization metaheuristic. European Journal of Operational Research 143(1), 218–229 (2002)

57. Gueorguiev, S., Harman, M., Antoniol, G.: Software project planning for robustness and completion time in the presence of uncertainty using multi objective search based software engineering. In: Proceedings of the Nineth Annual Conference on Genetic and Evolutionary Computation (GECCO 2009), pp. 1673–1680. ACM Press, New York (2009)

58. Handl, J., Knowles, J.D.: An evolutionary approach to multiobjective clustering. IEEE Transactions on Evolutionary Computation 11(1), 56–76 (2007)

59. Hansen, M.P., Jaskiewicz, A.: Evaluating the quality of approximations to the non-dominated set. Tech. Rep. IMM-REP-1998-7, Lyngby: Institute of Mathematical Modelling, Technical University of Denmark (1998)

60. Holland, J.H.: Adaptation in Natural and Artificial Systems. MIT Press, Ann Arbor (1975)
61. Jahn, J.: Vector optimization. Springer, Berlin (2004)
62. Jin, H., Wong, M.L.: Adaptive diversity maintenance and convergence guarantee in multiobjective evolutionary algorithms. In: Proceedings of the Congress on Evolutionary Computation (CEC 2003), pp. 2498–2505 (2003)
63. Khare, V., Yao, X., Deb, K.: Performance Scaling of Multi-objective Evolutionary Algorithms. In: Fonseca, C.M., Fleming, P.J., Zitzler, E., Deb, K., Thiele, L. (eds.) EMO 2003. LNCS, vol. 2632, pp. 376–390. Springer, Heidelberg (2003)
64. Knowles, J., Corne, D.: Quantifying the Effects of Objective Space Dimension in Evolutionary Multiobjective Optimization. In: Obayashi, S., Deb, K., Poloni, C., Hiroyasu, T., Murata, T. (eds.) EMO 2007. LNCS, vol. 4403, pp. 757–771. Springer, Heidelberg (2007)
65. Knowles, J.D., Corne, D.W.: On metrics for comparing nondominated sets. In: Congress on Evolutionary Computation (CEC 2002), pp. 711–716. IEEE Press, Piscataway (2002)
66. Knowles, J.D., Corne, D.W., Deb, K.: Multiobjective problem solving from nature. Natural Computing Series. Springer (2008)
67. Korhonen, P., Laakso, J.: A visual interactive method for solving the multiple criteria problem. European Journal of Operational Reseaech 24, 277–287 (1986)
68. Kumar, R., Banerjee, N.: Analysis of a multiobjective evolutionary algorithm on the 0-1 knapsack problem. Theoretical Computer Science 358(1), 104–120 (2006)
69. Kung, H.T., Luccio, F., Preparata, F.P.: On finding the maxima of a set of vectors. Journal of the Association for Computing Machinery 22(4), 469–476 (1975)
70. Laumanns, M., Thiele, L., Deb, K., Zitzler, E.: Combining convergence and diversity in evolutionary multi-objective optimization. Evolutionary Computation 10(3), 263–282 (2002)
71. Laumanns, M., Thiele, L., Zitzler, E.: Running Time Analysis of Multiobjective Evolutionary Algorithms on Pseudo-Boolean Functions. IEEE Transactions on Evolutionary Computation 8(2), 170–182 (2004)
72. Laumanns, M., Thiele, L., Zitzler, E., Welzl, E., Deb, K.: Running Time Analysis of Multi-objective Evolutionary Algorithms on a Simple Discrete Optimization Problem. In: Guervós, J.J.M., Adamidis, P.A., Beyer, H.-G., Fernández-Villacañas, J.-L., Schwefel, H.-P. (eds.) PPSN VII. LNCS, vol. 2439, pp. 44–53. Springer, Heidelberg (2002)
73. Loughlin, D.H., Ranjithan, S.: The neighborhood constraint method: A multi-objective optimization technique. In: Proceedings of the Seventh International Conference on Genetic Algorithms, pp. 666–673 (1997)
74. Luque, M., Miettinen, K., Eskelinen, P., Ruiz, F.: Incorporating preference information in interactive reference point based methods for multiobjective optimization. Omega 37(2), 450–462 (2009)
75. Mahdavi, K., Harman, M., Hierons, R.M.: A multiple hill climbing approach to software module clustering. In: Proceedings of the International Conference on Software Maintenance (ICSM 2003), pp. 315–324. IEEE Computer Society (2003)
76. Mancoridis, S., Mitchell, B.S., Chen, Y., Gansner, E.R.: Bunch: A clustering tool for the recoveryand maintenance of software system structures. In: Proceedings of the IEEE International Conference on Software Maintenance (ICSM 1999), pp. 50–59. IEEE Press, Piscatway (1999)
77. McMullen, P.R.: An ant colony optimization approach to addessing a JIT sequencing problem with multiple objectives. Artificial Intelligence in Engineering 15, 309–317 (2001)

78. Miettinen, K.: Nonlinear Multiobjective Optimization. Kluwer, Boston (1999)
79. Mitchell, B.S., Mancoridis, S., Traverso, M.: Using Interconnection Style Rules to Infer Software Architecture Relations. In: Deb, K., Tari, Z. (eds.) GECCO 2004. LNCS, vol. 3103, pp. 1375–1387. Springer, Heidelberg (2004)
80. Mostaghim, S., Teich, J.: Strategies for Finding Good Local Guides in Multi-objective Particle Swarm Optimization (MOPSO). In: 2003 IEEE Swarm Intelligence Symposium Proceedings, pp. 26–33. IEEE Service Center, Indianapolis (2003)
81. Nain, P.K.S., Deb, K.: Computationally effective search and optimization procedure using coarse to fine approximations. In: Proceedings of the Congress on Evolutionary Computation (CEC 2003), pp. 2081–2088 (2003)
82. Neumann, F., Wegener, I.: Minimum spanning trees made easier via multi-objective optimization. In: GECCO 2005: Proceedings of the 2005 Conference on Genetic and Evolutionary Computation, pp. 763–769. ACM, New York (2005)
83. Obayashi, S., Deb, K., Poloni, C., Hiroyasu, T., Murata, T. (eds.): EMO 2007. LNCS, vol. 4403. Springer, Heidelberg (2007)
84. Osyczka, A.: Evolutionary algorithms for single and multicriteria design optimization. Physica-Verlag, Heidelberg (2002)
85. Saliu, M.O., Ruhe, G.: Bi-objective release planning for evolving software. In: ESEC/SIGSOFT FSE, pp. 105–114. ACM press, New York (2007)
86. Sasaki, D., Morikawa, M., Obayashi, S., Nakahashi, K.: Aerodynamic Shape Optimization of Supersonic Wings by Adaptive Range Multiobjective Genetic Algorithms. In: Zitzler, E., Deb, K., Thiele, L., Coello Coello, C.A., Corne, D.W. (eds.) EMO 2001. LNCS, vol. 1993, pp. 639–652. Springer, Heidelberg (2001)
87. Zapotecas Martínez, S., Coello Coello, C.A.: A Proposal to Hybridize Multi-Objective Evolutionary Algorithms with Non-gradient Mathematical Programming Techniques. In: Rudolph, G., Jansen, T., Lucas, S., Poloni, C., Beume, N. (eds.) PPSN X. LNCS, vol. 5199, pp. 837–846. Springer, Heidelberg (2008)
88. Saxena, D., Deb, K.: Trading on infeasibility by exploiting constraint's criticality through multi-objectivization: A system design perspective. In: Proceedings of the Congress on Evolutionary Computation, CEC 2007 (2007) (in press)
89. Shukla, P., Deb, K.: On finding multiple pareto-optimal solutions using classical and evolutionary generating methods. European Journal of Operational Research (EJOR) 181(3), 1630–1652 (2007)
90. Siegmund, F., Bernedixen, J., Pehrsson, L., Ng, A.H.C., Deb, K.: Reference point-based evolutionary multi-objective optimization for industrial systems simulation. In: Proceedings of Winter Simulation Conference 2012, Berlin, Germany (2012)
91. Siegmund, F., Ng, A.H.C., Deb, K.: Finding a preferred diverse set of pareto-optimal solutions for a limited number of function calls. In: Proceedings of 2012 IEEE World Congress on Computational Intelligence, pp. 2417–2424 (2012)
92. Sindhya, K., Deb, K., Miettinen, K.: A Local Search Based Evolutionary Multi-objective Optimization Approach for Fast and Accurate Convergence. In: Rudolph, G., Jansen, T., Lucas, S., Poloni, C., Beume, N. (eds.) PPSN 2008. LNCS, vol. 5199, pp. 815–824. Springer, Heidelberg (2008)
93. Srivastava, R., Deb, K.: Bayesian Reliability Analysis under Incomplete Information Using Evolutionary Algorithms. In: Deb, K., Bhattacharya, A., Chakraborti, N., Chakroborty, P., Das, S., Dutta, J., Gupta, S.K., Jain, A., Aggarwal, V., Branke, J., Louis, S.J., Tan, K.C. (eds.) SEAL 2010. LNCS, vol. 6457, pp. 435–444. Springer, Heidelberg (2010)

94. Srivastava, R., Deb, K., Tulshyan, R.: An evolutionary algorithm based approach to design optimization using evidence theory. Tech. Rep. KanGAL Report No. 2011006, Indian Institite of Technology Kanpur, India (2011)

95. Takahashi, R.H.C., Deb, K., Wanner, E.F., Greco, S. (eds.): EMO 2011. LNCS, vol. 6576. Springer, Heidelberg (2011)

96. Thiele, L., Miettinen, K., Korhonen, P., Molina, J.: A preference-based interactive evolutionary algorithm for multiobjective optimization. Tech. Rep. Working Paper Number W-412, Helsingin School of Economics, Helsingin Kauppakorkeakoulu, Finland (2007)

97. Veldhuizen, D.V., Lamont, G.B.: Multiobjective evolutionary algorithms: Analyzing the state-of-the-art. Evolutionary Computation Journal 8(2), 125–148 (2000)

98. Wierzbicki, A.P.: The use of reference objectives in multiobjective optimization. In: Fandel, G., Gal, T. (eds.) Multiple Criteria Decision Making Theory and Applications, pp. 468–486. Springer, Berlin (1980)

99. Zhang, Q., Li, H.: MOEA/D: A multiobjective evolutionary algorithm based on decomposition. IEEE Transactions on Evolutionary Computation 11(6), 712–731 (2007)

100. Zhang, Y., Harman, M., Mansouri, S.A.: The multi-objective next release problem. In: Proceedings of the Nineth Annual Conference on Genetic and Evolutionary Computation (GECCO 2007), pp. 1129–1137. ACM Press, New York (2007)

101. Zitzler, E., Deb, K., Thiele, L., Coello Coello, C.A., Corne, D.W. (eds.): EMO 2001. LNCS, vol. 1993. Springer, Heidelberg (2001)

102. Zitzler, E., Künzli, S.: Indicator-Based Selection in Multiobjective Search. In: Yao, X., Burke, E.K., Lozano, J.A., Smith, J., Merelo-Guervós, J.J., Bullinaria, J.A., Rowe, J.E., Tiňo, P., Kabán, A., Schwefel, H.-P. (eds.) PPSN VIII. LNCS, vol. 3242, pp. 832–842. Springer, Heidelberg (2004)

103. Zitzler, E., Laumanns, M., Thiele, L.: SPEA2: Improving the strength pareto evolutionary algorithm for multiobjective optimization. In: Giannakoglou, K.C., Tsahalis, D.T., Périaux, J., Papailiou, K.D., Fogarty, T. (eds.) Evolutionary Methods for Design Optimization and Control with Applications to Industrial Problems, pp. 95–100. International Center for Numerical Methods in Engineering (CIMNE), Athens (2001)

104. Zitzler, E., Thiele, L., Laumanns, M., Fonseca, C.M., Fonseca, V.G.: Performance assessment of multiobjective optimizers: An analysis and review. IEEE Transactions on Evolutionary Computation 7(2), 117–132 (2003)

SBSE Meets Software Maintenance: Achievements and Open Problems

Massimiliano Di Penta

Dept. of Engineering, University of Sannio,
Palazzo ex Poste, Via Traiano, 82100 Benevento, Italy
dipenta@unisannio.it

Abstract. Software maintenance is, together with testing, one of the most critical and effort-prone activities in software development. Surveys conducted in the past [4] have estimated that up to 80% of the overall software development cost is due to maintenance activities. For such a reason, automated techniques aimed at supporting developers in their daunting tasks are highly appealing. Problems developers often face off include understanding undocumented software, improving the software design and source code structure to ease future maintenance tasks, and producing patches to fix bugs.

Finding a solution for the aforementioned problems is intrinsically NP-Hard, and therefore such problems are not tractable with conventional algorithmic techniques; this is particularly true in presence of very complex and large software systems. For this reason, search-based optimization techniques have been—and currently are—successfully applied to deal with all the aforementioned problems. Noticeable examples of successful applications of search-based software engineering (SBSE) to software maintenance include software modularization [2], software refactoring [3], or automated bug fixing [1].

Despite the noticeable achievements, there are still crucial challenges researchers have to face-off. First, software development is still an extremely human-centric activity, in which many decisions concerning design or implementation are really triggered by personal experience, that is unlikely to be encoded in heuristics of automated tools. Automatically-generated solutions to maintenance problems tend very often to be meaningless and difficult to be applied in practice. For this reason, researchers should focus their effort in developing optimization algorithms—for example Interactive Genetic Algorithms [5]—where human evaluations (partially) drive the production of problem solutions. This, however, requires to deal with difficulties occurring when involving humans in the optimization process: human decisions may be inconsistent and, in general, the process tend to be fairly expensive in terms of required effort.

A further challenge where SBSE meets software maintenance concerns systems which necessitate rapid reconfigurations at run-time, e.g., highly dynamic service-oriented architectures, or autonomic systems. If, on the one hand, such reconfigurations imply complex decisions where the space of possible choices is high, on the other hand such decisions have to be taken in a very short time. This implies a careful choice and performance assessment of the search-based optimization techniques adopted, aspect

G. Fraser (Ed.): SSBSE 2012, LNCS 7515, pp. 27–28, 2012.

that is often not particularly considered when search-based optimization techniques are applied offline, e.g. for test data generation or for software modularization.

Last, but not least, many modern software systems—open source and not only—are part of large and complex software ecosystems where the evolution of a component might impact others, e.g., by creating incompatibilities from a technical or even legal point of view. Also in this case, SBSE has a great potential to support the explore such large spaces of software configurations.

References

1. Le Goues, C., Nguyen, T., Forrest, S., Weimer, W.: Genprog: A generic method for automatic software repair. IEEE Trans. Software Eng. 38(1), 54–72 (2012)
2. Mitchell, B.S., Mancoridis, S.: On the automatic modularization of software systems using the bunch tool. IEEE Trans. Software Eng. 32(3), 193–208 (2006)
3. O'Keeffe, M.K., Cinnéide, M.Ó.: Search-based refactoring for software maintenance. Journal of Systems and Software 81(4), 502–516 (2008)
4. Standish: The scope of software development project failures. Tech. rep., The Standish Group, Dennis, MA (1995), http://www.standishgroup.com/chaos.html
5. Takagi, H.: Interactive evolutionary computation: Fusion of the capacities of ec optimization and human evaluation. Proceedings of the IEEE 89(9), 1275–1296 (2001)

Tutorial: High Performance SBSE
Using Commodity Graphics Cards

Simon Poulding

Department of Computer Science, University of York,
Deramore Lane, York, YO10 5GH, UK
smp@cs.york.ac.uk

In contrast to manual software engineering techniques, search-based software engineering (SBSE) is able to exploit high performance computing resources in order to improve scalability and to solve hitherto intractable engineering problems. This is an increasingly viable approach: affordable high performance computing is readily available using cloud services, server clusters, and—perhaps surprisingly—commodity graphics cards that may already be in your laptop or PC.

Modern graphics cards render high resolution, high frame-rate graphics in real-time to support the demands of applications such as gaming. To achieve this performance, the architecture of the graphics processing unit (GPU) is designed to execute a large number of threads in parallel, and to minimise the overhead of memory access. Over the last few years, this high performance parallel computing environment has become available for use by applications other than graphics rendering, a technique known as general-purpose computing on graphics processing units (GPGPU).

This tutorial is an introduction to GPGPU and how it can be used by SBSE techniques. Firstly, I will demonstrate how to develop applications that run on GPUs, highlighting the important architectural differences between CPUs and GPUs, and discussing the tools that support development. Secondly, I will illustrate how the high-performance GPGPU environment may be exploited by SBSE techniques (a) to accelerate fitness evaluation, and, (b) to apply sophisticated search algorithms.

G. Fraser (Ed.): SSBSE 2012, LNCS 7515, p. 29, 2012.

Evolving Robust Networks
for Systems-of-Systems

Jonathan M. Aitken, Rob Alexander,
Tim Kelly, and Simon Poulding

Department of Computer Science, University of York
Heslington, York, UK
{jonathan.aitken,rob.alexander,tim.kelly,simon.poulding}@york.ac.uk

Abstract. Software systems that rely on ad-hoc networks are becoming increasingly complex and increasingly prevalent. Some of these systems provide vital functionality to military operations, emergency services and disaster relief; such systems may have significant impact on the safety of people involved in those operations. It is therefore important that those networks support critical software requirements, including those for latency of packet transfer. If a network ceases to meet the software's requirements (e.g. due to a link failure) then engineers must be able to understand it well enough to reconfigure the network and restore it to a requirement-satisfying state. Given a complex network, it is difficult for a human to do this under time pressure. In this paper we present a search-based tool which takes a network defined using the Network Description Language (NDL), annotated with a set of network-hosted applications and a set of latency requirements between each. We then evolve variants of the network configuration which meet the requirements and are robust to single link failures. We use network calculus tools to get a fast, conservative evaluation of whether a given network meets its requirements. We demonstrate that this approach is viable, designs networks much faster than a human engineer could, and is superior to a random generate-and-test approach.

Keywords: network calculus, systems of systems, network enabled capability, genetic algorithms.

1 Introduction

Many software systems are distributed across a network, and rely on that network having certain properties, such as adequate bandwidth provision and acceptable latency. This is particularly true in a Network Enabled Capability (NEC) System of Systems (SoS) where the networked software is used to coordinate the real-world actions of teams and organisations. For example, in a disaster relief scenario where a network is quickly set up to co-ordinate a response to flooding, conditions on the ground will rapidly change, and it is highly likely that network links will be compromised. Yet operation co-ordinators will rely on the network to provide information under all conditions, for example

G. Fraser (Ed.): SSBSE 2012, LNCS 7515, pp. 30–44, 2012.

to direct rescuers to critical areas. Any loss of this information could lead to misdirecting rescuers to irrelevant or overly dangerous areas. The network has thus taken on an important safety role.

Many network problems (e.g. unacceptable latency between two nodes) can be solved by reconfiguration, particularly if new links can be added. However, as the SoS grows (perhaps organically, while deployed), reconfiguration becomes a more complex task. First, a repair attempt motivated by problems with one service may inadvertently compromise another service. Second, a repair may succeed in the short term, but leave the network in a perilous condition where one more failure could lead to a serious network outage.

A real world example of the danger of operating with a marginal network can be seen in the Überlingen mid-air collision [15], where air-traffic controllers unknowingly relied on a network that provided them with no connection to the outside world. It was thus impossible for other air-traffic stations to warn the controllers that they had set two aircraft on a collision course.

In this paper, we describe our initial work on a search-based method that can design networks that robustly support a set of software applications with strict latency requirements. The main question for our research is "is it possible to produce a software tool that evolves reconfigurations of a given network (built using Commercially available Off-The-Shelf (COTS) technology) to fix an inability to meet a delay requirement, e.g. as a result of link failure, while taking account of the robustness of the resultant network against future failures?"

1.1 Contributions

The main contributions of this paper are to present our method for evolving reconfigurations of robust networks, and to demonstrate that our method can:

- Create viable networks far faster than is conceivable for a human engineer
- Outperform random search
- Take account of factors that matter for safety (e.g. robustness to link failure) and practicality (e.g. network complexity) to produce feasible solutions

2 Background

2.1 Hardening Networks

Elia et al. [9] review various modifications to standard Ethernet that are designed to 'harden' it against various types of failure and hence make it more reliable (specifically, they bring it up to the level required by the safety-critical electronics standard IEC-61508). Unfortunately, those protocols place many restrictions on how networks can be configured, which make them impractical for ordinary deployments using mainstream software.

Given the above, the work described in this paper concentrates on what can be achieved with mainstream equipment and protocols. Advanced protocols such as Avionics Full-Duplex Switched Ethernet (AFDX) have a low failure rate which enables it to be used to satisfy a 1×10^{-9} per hour dangerous failure rate for critical systems [18]. This is a reasonable target for many safety-critical functions (where their failure may cause a serious accident without any other failure in the system). In our work, however, we are not concerned with such *critical* functions — we are concerned with cases where the network is only one part of the safety mitigation in the SoS, where a failure rate several orders of magnitude higher might be acceptable. We refer to these as *safety-related* systems.

2.2 Assessing Network Latency Performance

Skeie et al. [20] suggest two possible methods for assessing the performance of a network — network simulation and network calculus.

Simulation techniques (such those implemented by the ns2, ns3 and OM-NeT++) provide highly detailed representations of the network that follow each packet as it is transmitted through a complete virtual network including routing tables. This produces a wealth of information about the characteristics of the network, but takes considerable computational effort to get statistical confidence in the results. More importantly, these simulation packages do not produce any *guarantees* about the latency-characteristics on the network; each run provides a selection of the possible latencies that may be measured within the network. There is no guarantee that the simulation will exercise all corner cases in a given run (or indeed set of runs), and therefore worst-case scenarios may be overlooked.

In order to develop a network model that can provide hard guarantees about latency, Cruz and others developed Network Calculus [6,7]. This mathematical technique calculates a hard bound on latencies, buffer allocations and throughput. Network Calculus is less demanding on computation time than packet-based simulation packages and therefore "can be suitable for online assessment" [20]. Network traffic is generated from source nodes and flows through to sink nodes via a path passing through intermediate routing points. Whenever one flow crosses another (for example by sharing an output or input port) it creates interference that can increase latency. Network calculus represents these flows using arrival curves which represent the data travel within the network. Traffic at various points in the network is bounded by the "leaky bucket" function, $b(t) = \sigma + \rho t$, where $b(t)$ is an upper bound of the traffic generated based on σ the maximum burst size and ρ the sustainable data rate.

Tool kits now exist that provide good solutions to these equations, allowing worst case delays to be calculated. Boyer [4] presents an overview of the available tools that implement network calculus. The DISCO [17] tool has been selected for use in this paper. DISCO allows the creation of service curves and provides a high degree of flexibility, including different methods for flow analysis. It provides an API written in Java allowing for easy integration with other components.

The model of the network used by network calculus is much simpler than simulators such as OMNeT++. The representation consists of a basic network topology, link capacities, sustainable data rate (in bits per second) and a maximum packet size in bits. This is a significant abstraction of the network, but it is sufficient for basic network modelling including periodic and periodic sources. In addition, the leaky bucket is a simple equation that does not consider data bursts; it assumes the maximum data burst at all times. This is suitable for safety-related work, as it does cover worst-case scenarios.

Here, we have only given the briefest of introductions to network calculus; fuller explanations can be found in [3,6,7]. The technique has been used successfully to find guaranteed delivery times for the AFDX protocol [5,16] running on large and complex networks with interference between different data streams.

2.3 Representing Network Configurations in a Machine-Readable Way

We cannot, of course, apply network calculus to a network until we have a representation of the network in human and machine-readable form. NDL [12] provides such a representation, and as it is defined in Extensible Markup Language (XML) and based on the Resource Description Framework (RDF). Whilst NDL has primarily been targeted at optical networks, the schemas are portable to wired and wireless topologies.

NDL was developed to represent ad-hoc hybrid networks, helping users understand the likely behaviour of the network and thus be able to modify it to get the macroscopic behaviours they desire.

The focus on ad-hoc networks makes NDL naturally applicable to SoS networks, where a key goal is reconfigurability — the aim is to provide a flexible network with many potential configurations that support different operations and different capability configurations. On a practical level, NDL breaks down the network into a bi-directional graph. This representation is of the form used by the DISCO tool, allowing direct coupling of Network Calculus to NDL.

2.4 Evolving Networks

We are not the first to use evolutionary algorithms to generate networks. Leu et al. [13] search for optimal network configurations for large networks, with a trade-off between robustness (where the number of paths between nodes is increased at the expense of a densely connected graph) and cost (where the number of connections is limited at the expense of robustness). They define fitness functions that either produce dense robust networks, or sparse networks, but not hybrids. They do not consider satisfaction of latency requirements.

Georges et al. [10,11] use network calculus as a tool to evaluate the performance of switched networks. In [11] they couple network calculus with a genetic algorithm to create a network configuration that is capable of meeting a requirement under worst case performance. However, this work takes no account of the robustness of the network created (its resilience to failures and disruptions). George et al. are interested in evolving a tree-like network structures that minimises cabling distances but leave one path between end nodes; in contrast, we are interested in developing mesh networks which are more representative of the networks used in SoS where end-to-end connections often need to share links.

Ali et al. [1] evolve networks to meet Quality of Service constraints such as latency, but do not consider the robustness of the network to component failure. Newth and Ash [14] do consider the robustness of the networks they generate, but their focus is on power distribution networks rather than data networks.

Ducatelle et al. [8] investigate techniques to mitigate against failure events in wireless and optical networks. They acknowledge that faults occur regularly, and therefore dependability is a key concern, especially in wireless ad-hoc networks. They develop a two-stage algorithm. The first stage verifies that a solution is possible. The second stage optimises a set of constraints, such as capacity using a simulated model. A routing algorithms is then used on the simulated model (similar to an Ant Colony Optimisation) to discover routes and calculate their latency. This provides some of the information needed for our purposes, but only measures average latency rather than the worst case information offered by network calculus, and does not optimise the network to ensure that the latencies are acceptable.

3   Proposed Approach for Evolving Robust Networks

We have combined several tools and technique in a combination that:

- provides an abstract model of connections and individual nodes so that it is applicable to a wide range of networks and technologies
- performs worst case analysis to provide assurance that requirements can be met in that system configuration
- generates robust networks without single point failures

3.1   Robustness

Ideally, we would like to evolve a network that supports all aspects of dependability, including availability, safety and maintainability, as defined by Avižienis et al. in [2]. For this work we have limited ourselves to a simpler notion of *robustness*, which we define informally to mean 'the ability of the system to meet its requirements under a range of representative failure conditions'. The requirements we are concerned with are (a) the ability of each safety-related service to work at all (to make the connections that it needs) and (b) for such services to get a tolerable level of network latency (each service specifies a maximum end-to-end delay that is acceptable to it). We follow Georges et al. [11] in treating the latency specification as a hard deadline which must not be exceeded.

We take a simple view of failure, considering only total failures of single links within the network. Ducatelle et al. [8] notes that link failures are frequent in deployed networks, being considerably more frequent than node failures. They are therefore sufficient for this initial study.

A network configuration is highly robust (for a given set of services running over that network) if all the services can get the connections they need with the latency they specify under all possible single-link failures. A configuration that does not give the connections and latencies that are needed under some proportion of single-link failures is proportionally less robust.

3.2   Process Overview

The process we use to evolve network configurations has four basic components, shown in slightly more detail in Figure 1(a). We start from a representation of the available network components (nodes and possible connections) and our software system (applications, node assignments and latency requirements) that is understandable to both humans and computers. We have adopted NDL for this purpose; details are in Section 4.1.

Once we have our description of the available components, we need to generate a range of candidate network designs. We therefore convert our NDL description into a genome, where each gene is a pair of integers describing a connection from one network port to another. This representation can be mutated (starting from the base case of a completely unconnected network) to generate new candidates as part of a search process, and as also suitable for crossover. The genome is explained in Section 4.2.

Each candidate design needs to be evaluated to assess whether it meets the robustness criteria defined in Section 3.1, given the latency requirements of the software system. As noted in Section 2.2, we use the DISCO network calculus tool to perform this evaluation. As well as evaluating the candidate in its default state, we also evaluate it for each possible single-link failure, and produce a robustness measure that combines all this information. Section 4.3 explains this in more detail.

Now that we have a way to generate candidates and assess their robustness, it is possible to search for acceptable networks. We use a genetic algorithm as implemented by Java Evolutionary Computation Toolkit (ECJ), with the core of the fitness function being a measure of the robustness of the network. It is easy, however, to create a robust network by using many redundant connections, which will be expensive. We have therefore introduced an additional fitness term that steers the search towards smaller networks. Section 4.4 explains how the search and fitness function are implemented.

4   Implementing the Process

Our implementation of the process uses a number of different software tools; the tools and their interaction is shown in Figure 1(b). The NDL library (supported by the RDF library) imports the model from the human-editable description, ECJ performs the evolutionary search, and the Java Universal Network/Graph and DISCO frameworks perform the network representation and analysis.

4.1   Describing the Network

The configuration of the network is described using NDL. NDL describes devices in terms of: name (ndl:Device), latitude (geo:Lat), longitude (geo:Long) and a list of interface ports (ndl:hasInterface).

An NDL Python parser is used to import the information into a machine readable form. This creates a description of the network devices and internal

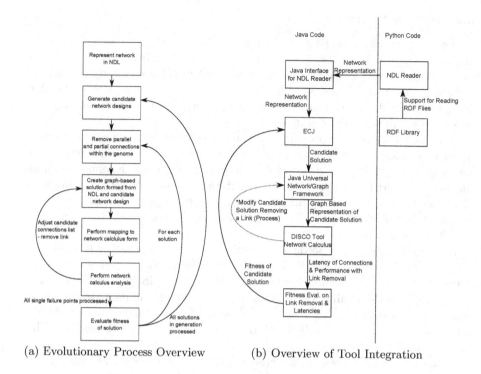

(a) Evolutionary Process Overview   (b) Overview of Tool Integration

Fig. 1. Process and Implementation Overview

connections. This also contains a list of all possible interface ports within the network, extracted from each of the individual device descriptions in the NDL.

Given the geographical information provided by NDL, the latencies of any given connection be calculated using the Haversine formula [19] to determine crow-flies cabling lengths (adjusted using a scaling factor of 1.5, to allow for deviations or extensions), and assuming a propagation velocity of two-thirds the speed of light.

Expanding NDL for Our Chosen Scenario. We have expanded the standard NDL description to represent applications that produce and consume User Datagram Protocol (UDP) video streams. These video streams prove readily analysable using the DISCO tool, as they contain fixed transmission rates and packet sizes. We have added the following fields to a standard NDL description of a network:

- ndl:hasVideoServer - represents a video server present at the node, giving a reference name.
- ndl:hasVideoServerConnection - represents a video server client present at the node, giving a reference name.

Each ndl:hasVideoServer is linked to a ndl:VideoServer instance, by reference name, which describes the stream properties:

- The time between packets - ndl:videoServerWaitInterval.
- The packet data size - ndl:videoServerPacketLength.
- The packet data content - ndl:videoServerContentLength.

Each ndl:hasVideoServerConnection is linked to a single instance of a ndl:VideoServerConnection which describes the nature of the connection at the sink end, containing:

- The name of the server to which the client is connected - ndl:applicationToServerConnection.
- The latency requirement - ndl:maxDelay.

4.2   Generating and Interpreting the Genome

We use a Genetic Algorithm (GA) comprised of individuals whose genotype is a 1250 gene sequence. Each gene is a pair of integers describing a connection from one port to another. The length of the genome limits the possible number of connections possible, it has been chosen to provide space for junk connections which allow the network to expand and contract the number of links. Each integer maps to one of p possible port names specified within the NDL. The integer values run from -1 to $p - 1$, where -1 represents no connection. When the NDL is parsed a table is made which maps these integers to physical port names; the genome is interpreted by performing lookups on this table.

4.3   Generating and Analysing the Network Model

After generating the genome for a candidate configuration, an initial filter is applied — this ensures that the network can be analysed by DISCO. This filter uses the following criteria:

- Double connections from one port - each port can only have one connection.
- Parallel connections between devices - the standard network calculus algorithms do not allow parallel wiring of devices. (Parallel wiring is a valid method to solve robustness problems, although it is not ideal as any environmental condition that causes a link failure may be common mode to both; for example, in a wired network a digger may cut a cable or in a wireless network a source of electromagnetic interference. We will consider ways to support parallel wiring in future work.)
- Disconnected source or sink nodes — DISCO requires that each source and sink node is connected to at least one other node.

Once through this initial filter, the connection list is made bi-directional before the Java Universal Network/Graph (JUNG) Framework is used to create a in-memory graph representation of the separate nodes and connections. Analysis is carried out on the graph based model using the DISCO tool which applies network calculus to derive the worst-case latencies (using Open Shortest Path First (OSPF) to calculate routes) on each flow specified by the NDL. Our system then repeats this analysis for each single-failure variant of the network; here, each failure is a failure of a (bi-directional) link. This provides:

– The worst-case performance of the network under all single point failures. This is expressed as the amount of headroom that is available on each flow when the worst-case delay is compared to the specification — the difference between the latency required and the latency achieved. A negative headroom indicates a failure to meet the requirement.
– The number of single-failure cases that completely fail to provide some required connection

For a network to be considered robust, all of the services must have the connections that they require within their specified latency under any possible single-link failure.

4.4   Search Algorithm

We used the Evolutionary Computation for Java (ECJ) as a harness for a GA to perform the search functionality. The GA has a population size of 2000 and uses a tournament selection of size 10, crossing over the best individual within the population with the best from the tournament. Mutation then occurs on each gene within the child with a probability of 0.1. When mutation occurs it changes the destination of the connection to either be one from the list specified within the NDL (0 to $p - 1$) or to become not connected (-1).

Fitness Function. When evaluating a networks for robustness under our case study situation, four criteria stand out as particularly important:

1. Can represented in, and analysed by, the DISCO tool.
2. Is tolerant to single-point failures.
3. Meets timing constraints for every flow in the network under all single point failure conditions.
4. Contains as few connections as possible.

This can be represented as a multi-objective fitness function that is composed of four separate components. Some of these components are simply attributes to be optimise, but others form hard constraints; for example, in order for a network to be robust under the criteria in Section 3.1 it must meet points 2 and 3 in the list above. A network cannot, therefore, merely not be evolved to contain as few connections as possible, as it may fail timing constraints or be intolerant to single-point failures. Each of the objectives is represented by a simple equation that must be maximised in order to produce a network that meets the robustness criteria. The fitness functions for each attribute are shown in a corresponding order in Equations 1-4.

$$fitness(0) = \sum_{n_{sink}, n_{source}} \frac{1}{c_{sinksource}} \tag{1}$$

$$fitness(1) = 1 - \frac{n_{missedcases}}{n_{totalcases}} \tag{2}$$

$$fitness(2) = \begin{cases} H_{total} & \text{if all specified latencies met} \\ 0 & \text{otherwise} \end{cases} \tag{3}$$

$$fitness(3) = \begin{cases} 1 - \frac{c_{network}}{c_{initial}} & \text{starting point, large, network found} \\ 0 & \text{otherwise} \end{cases} \tag{4}$$

Equation 1 represents connections between the sources and sinks, where $c_{sinksource}$ represents the number of sinks and sources within the network. Maximising this ensures that all sources and sinks are connected. In the case of the sample network in Section 5.1 there are 17 software-connections from 12 individual nodes therefore the maximum value is 12/17.

Equation 2 represents the tolerance of the network to single-point failure. This is a key component in the fitness function, ensuring that the solutions produced do not contain single links that are critical to their operation. The term $n_{totalcases}$ represents the total number of possible single point failures possible within the network, evaluated for each individual source-to-sink connection pairing, $n_{missedcases}$ represents the number of these cases where a single point failure will disrupt the source-to-sink connection pairing. In a network resilient to all single-point failures, $n_{missedcases} = 0$, so $fitness(1)$ has a maximum value of 1.

Equation 3 represents the amount of headroom (the amount of extra delay that could be tolerated before exceeding the specification) that DISCO predicts. It is summed over all source-to-sink connections to provide this value, H_{total} - this value is specific for each simulated individual and will be dependent on the network configuration and inter-stream interference due to routing. Together Equations 2 and 3 represent the robustness of the network.

The final of these four criteria is shown in Equation 4, and is used to represent the size of the network. The term will increase as smaller networks are found. The denominator, $c_{initial}$, represents the size of the first network found which is robust. This is likely to be a large network with many connections, as more fully connected networks are easier to find. Every subsequent individual is then compared with this initial network. This acts as a starting point, as the network size of an individual ($c_{network}$) decreases, $fitness(3)$ increases. By maximising this term the network size will be made as small as possible. This term ensures that small networks are promoted, however, these networks must still meet the specification criteria in order to be considered viable. $fitness(3)$ can be both positive and negative - negative values indicate a larger network than the first found.

Therefore to optimise the size of a network in Equation 4, Equation 1 must be maximised (constraint), Equation 2 must be equal to unity (constraint) and Equation 3 must be greater than zero (constraint, but can be optimised to be as high as possible).

The fittest networks maximised values for $fitness(0)$, an essential value of 1 for $fitness(1)$, and as large values of $fitness(2)$ and $fitness(3)$ as possible.

5   Experimentation

The main aim of our experimentation is to check whether our prototype tool can produce a viable, robust network configuration faster than would be practical for a human engineer. For a sanity check, to show that this is a computationally non-trivial problem, we also compare our tool using a GA with a version of our tool that merely uses random search.

5.1   Choice and Definition of Sample Application

The example used in this paper is a system that transmits video streams, with the assumption that these are then used to take operational decisions under time pressure. As was noted in Section 1, networked software can be used to monitor the environment, streaming videos to keep key people updated on local situations. In order to do this, the network must be tolerant to failure and meet specified latency requirements.

Table 1 describes the software system that is to run over our example network, in terms of the nodes where applications (and users) are located, the nature of the data that is sent, and the latency requirements. Each sink node is connected to two source node streams, with the node streams. These streams represent a video encoded with a fixed bit rate of 512kbps, transmitted using a UDP connection with standard size packets.

Table 1. Node Connections in Sample Application

Source Node	Sink Nodes	Packet Period (ms)	Packet Size (bytes)	Data Size (bytes)	Delay Spec.(s)
0	6 and 7	20.6	1344	1316	0.01
1	7 and 8	20.6	1344	1316	0.01
2	8 and 9	20.6	1344	1316	0.01
3	9 and 10	20.6	1344	1316	0.01
4	10 and 11	20.6	1344	1316	0.01
5	11	20.6	1344	1316	0.01

A fictional network of 21 nodes (12 end-nodes and 9 routing-nodes) has been spaced across the University of York Heslington West Campus. The end nodes are placed in major department buildings, while routing nodes are in key locations such as bridges, walkways and nucleus buildings. We assume that connections can be made as necessary.

5.2   Experimental Protocol

The software was implemented as outlined in Section 4. The NDL file was written to contain a network as described in Section 5.1. The GA was set up using the parameters outlined in Section 4.4.

In order to investigate the performance of the process we ran the tool on our sample network 5 times, each time for 500 generations (this was the only stopping criterion). As a sanity check, to confirm that the example we had chosen was genuinely challenging, we also repeated those experiments using a random search for the same number of generations. As a further evaluation, random search was also applied to the problem in a series of longer runs of 1500 generations.

We will make the code used to conduct the experiments and the NDL used to describe the network available on the lead author's website.

5.3   Evaluation of the Process on Sample Network

Our tool consistently produced multiple robust network designs early on in the search. The first such robust network appeared after an average of 40.5 ± 13.5s; the average size of this was 33.2 ± 1.3 connections, which is not optimal but is practical. It is unlikely that a human engineer could achieve this performance.

Figure 2(a) shows the smallest networks found that meet the latency specification for the GA and (as a comparison) the random search, as time passes averaged across five optimisations. Networks appearing on this graph have been filtered to ensure that they both meet the specification and are tolerant to any single-point failure. Clearly the GA produces smaller networks much more quickly than a random search.

It is clear that both the GA and the random search are capable of producing robust networks that meet the specification, but over the same number of candidates generated the GA found significantly smaller robust networks — the average number of connections in the robust networks found was 25.8 ± 0.4 for the GA compared to 30.4 ± 0.8 for the random search.

In terms of compute effort (as opposed to simple number of generations), the GA produces large numbers of networks that are testable - an average of $25,663 \pm 1,393$ per experiment, compared with 87 ± 9 by the random search (both out of $1,000,000$ candidates generated). Each testable network must be built, tested using network calculus and then examined under single-link failure; in contrast, a non-testable network can be ignored. This leads to a longer average processing time per individual — for the GA these 500 generations took an average of $1,028.2 \pm 406.7$s, compared with 234.7 ± 22.8s for random search.

The longer random runs, however, took an average of 2346.3 ± 95.4s — twice that for 500 generations of the GA. The average size of the network found was 30.6 ± 1.1, no better than for the 500-generation random search and substantially worse than the 25.8 ± 0.4 achieved by the 500-generation GA. It would appear from this that random search performance will not improve rapidly with increased computing power.

The Pareto fronts for the GA networks are shown in Figures 2(b), 2(c) and 2(d). Note that the graphs show actual network size, rather than the than the relative size metric used in the fitness function. The Pareto fronts produced by the GA show clear relationships between the different objectives. However, the filtering process used to identify networks that meet the robustness criteria is a key component in identifying viable solutions. Figures 2(b), 2(c) and 2(d) show combinations of the fronts of optimisation for the GA. As can be seen the only clear trade-off that can be made is for Number of Connections versus proportion of single-point failures (Figure 2(c)) — typically networks with fewer connections will prove to be more susceptible to single-point failures. However, different levels of headroom can be achieved at any level of single-point failure (thee specification only demands a headroom exists). As expected, networks with fewest connections, with headroom and robustness prove most desirable.

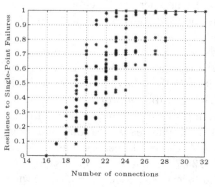

(a) Time Comparison of Network Size for GA and Random Search

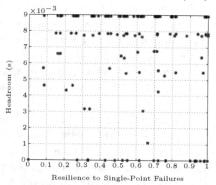

(b) Pareto Front for Number of Connections Versus Headroom Achieved (GA)

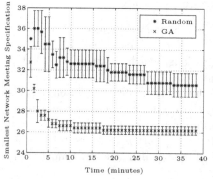

(c) Pareto Front for Number of Connections Versus Resilience to Single-Point Failures (GA)

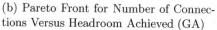

(d) Pareto Front for Resilience to Single-Point Failures Versus Headroom (GA)

Fig. 2. Investigating the GA and Random Search Over 500 Generations

6   Conclusions

This paper has described the implementation of a prototype network design tool. This tool can be used to design SoS networks to meet the latency requirements of a given software system, and make those networks robust to single link failures.

We have a applied the tool to a simple software system running over a small network. Our experimental results show that the tool can produce viable robust solutions more quickly than is feasible for a human engineer, given the the cognitive effort required to consider network robustness under many different failure conditions. They show that both random and GA search are possible approaches, but that the GA consistently produces smaller networks.

It is plausible that our automated approach will scale well to cover other types of failure, and to more complex networks. For a human engineer, by contrast, such changes will make the task even more impractical.

6.1   Future Work

A priority for the future is to apply this tool to larger and more complex networks, including wireless ad-hoc networks. As part of this, we plan to develop a network generator that produces random networks and software specifications with particular attributes (e.g. size, connectivity, range of latency requirements). This will help us explore what factors the performance of the tool is sensitive to — in particular, how it scales with larger networks and higher complexity.

The current work has allowed the search free reign in configuring networks from the components available, without reference to any current configuration of the network. In order to explicitly tackle the problem of reconfiguration, an algorithm should be developed to calculate change cost from one configuration to another, and this should be incorporated into the fitness function.

The model of robustness used here is quite limited — it only considers total failures of individual links. A richer failure model will be needed to accurately represent real network problems. Modelling more sophisticated failures (e.g. intermittent link failure) is not possible DISCO, so future work will need to consider simulation tools such as ns2, ns3 or OMNeTT++. These tools use much more compute time than DISCO, so an optimal solution may involve an initial search using DISCO followed by fine-tuning for subtle failures using simulation.

Acknowledgements. The authors wish to thank Jeroen van der Ham of the University of Amsterdam for assistance with NDL.

References

1. Ali, S., Maciejewski, A.A., Siegel, H.J., Kim, K.J.: Robust resource allocation for sensor-actuator distributed computing systems. In: Proceedings of the International Conference on Parallel Processing (2004)
2. Avižienis, A., Laprie, J.C., Randell, B., Landwehr, C.E.: Basic concepts and taxonomy of dependable and secure computing. IEEE Transactions on Dependable and Secure Computing 1(1), 11–33 (2004)

3. Boudec, J.Y.L., Thiran, P.: A Theory of Deterministic Queuing Systems for the Internet. Springer (2011)
4. Boyer, M.: NC-Maude: A Rewriting Tool to Play with Network Calculus. In: Margaria, T., Steffen, B. (eds.) ISoLA 2010, Part I. LNCS, vol. 6415, pp. 137–151. Springer, Heidelberg (2010)
5. Charara, H., Scharbarg, J.L., Ermont, J., Fraboul, C.: Methods for bounding end-to-end delays on an AFDX network. In: 18th Euromicro Conference on Real-Time Systems (2006)
6. Cruz, R.L.: A calculus for network delay, part I: Network elements in isolation. IEEE Transactions on Information Theory 37(1), 114–131 (1991)
7. Cruz, R.L.: A calculus for network delay, part II: Network analysis. IEEE Transactions on Information Theory 37(1), 132–141 (1991)
8. Ducatelle, F., Gambardella, L.M., Kurant, M., Nguyen, H.X., Thiran, P.: Algorithms for Failure Protection in Large IP-over-fiber and Wireless Ad Hoc Networks. In: Kohlas, J., Meyer, B., Schiper, A. (eds.) Dependable Systems: Software, Computing, Networks. LNCS, vol. 4028, pp. 231–259. Springer, Heidelberg (2006)
9. Elia, A., Ferrarini, L., Veber, C.: Analysis of ethernet-based safe automation networks according to iec-61508. In: IEEE Conference on Emerging Technology in Factory Automation, Prague, Czech Republic (2006)
10. Georges, J.P., Divoux, T., Rondeau, E.: Network calculus: Application to switched real-time networking. In: 5th International ICST Conference on Performance Evaluation Methodologies and Tools, ValueTools (2011)
11. Georges, J.P., Krommenacker, N., Divoux, T., Rondeau, E.: A design process of switched ethernet architectures according to real-time application constraints. Engineering Applications of Artificial Intelligence 19, 335–344 (2006)
12. van der Ham, J., Grosso, P., van der Pol, R., Toonk, A., de Laat, C.T.A.M.: Using the network description language in optical networks. In: Integrated Network Management (2007)
13. Leu, G., Namatame, A.: Evolving Failure Resilience in Scale-Free Networks. In: Gen, M., Green, D., Katai, O., McKay, B., Namatame, A., Sarker, R.A., Zhang, B.-T. (eds.) Intelligent and Evolutionary Systems. SCI, vol. 187, pp. 49–59. Springer, Heidelberg (2009)
14. Newth, D., Ash, J.: Evolving cascading failure resilience in complex networks. In: Proceedings of the 8th Asia Pacific Symposium on Intelligent and Evolutionary Systems (2004)
15. Nunes, A., Laursen, T.: Identifying the factors that contributed to the Überlingen mid-air collision. In: Proceedings of the 48th Annual Chapter Meeting of the Human Factors and Ergonomics Society (2004)
16. Scharbarg, J.L., Ridouard, F., Fraboul, C.: A probabilistic analysis of end-to-end delays on an AFDX avionic network. IEEE Transactions on Industrial Informatics 5(1), 38–48 (2009)
17. Schmitt, J.B., Zdarsky, F.A.: The DISCO network calculator. In: First International Conference on Performance Evaluation Methodologies and Tools (2006)
18. Sghairi, M., Aubert, J.J., Brot, P., de Bonneval, A., Crouzet, Y., Laarouchi, Y.: Distributed and reconfigurable architecture for flight control system. In: Digital Avionics Systems Conference (2009)
19. Sinnott, R.W.: Virtues of the haversine. Sky and Telescope 68(2), 159 (1984)
20. Skeie, T., Johannessen, S., Holmeide, O.: Timeliness of real-time IP communication in switched industial ethernet networks. IEEE Transactions on Industrial Informatics 2(1), 25–39 (2006)

On the Application of SAT Solvers
to the Test Suite Minimization Problem

Franco Arito, Francisco Chicano, and Enrique Alba

Departamento de Lenguajes y Ciencias de la Computación,
University of Málaga, Spain
{franco,chicano,eat}@lcc.uma.es

Abstract. The Test Suite Minimization problem in regression testing is a software engineering problem which consists in selecting a set of test cases from a large test suite that satisfies a given condition, like maximizing the coverage and/or minimizing the oracle cost. In this work we use an approach based on SAT solvers to find optimal solutions for the Test Suite Minimization Problem. The approach comprises two translations: from the original problem instance into Pseudo-Boolean constraints and then to a propositional Boolean formula. In order to solve a problem, we first translate it into a SAT instance. Then the SAT instance is solved using a state-of-the-art SAT solver. Our main contributions are: we create an encoding for single and multi-objective formulations of the Test Suite Minimization Problem as Pseudo-Boolean constraints and we compute optimal solutions for well-known and highly-used instances of this problem for future reference.

Keywords: Test suite minimization, satisfiability problem, multi-objective optimization.

1 Introduction

In the last years the performance of Boolean satisfiability (SAT) solvers has been boosted by the introduction of techniques like clause learning, watched literals, and random restarts [1]. Nowadays it is possible to solve SAT instances up to half million variables[1], covering a search space of roughly 2^{500000}. If we compare this cardinality with the cardinality of several combinatorial optimization problems, the difference is considerable in favor of SAT. Thus, we wonder if we can take advantage of this progress in the SAT community to solve interesting optimization problems. In particular, we wonder if we can use the algorithms and tools developed for the SAT problem to find optimal solutions in $\mathcal{NP}$-hard Software Engineering optimization problems, for which metaheuristic techniques are being used at the moment. The answer is yes, and even the SAT community itself has explored other applications of SAT solvers to problems that arise in model checking, planning, and test-pattern generation, among others [12].

---

[1] The reader can visit http://www.satcompetition.org/ for details.

G. Fraser (Ed.): SSBSE 2012, LNCS 7515, pp. 45–59, 2012.

The main challenge to solve combinatorial optimization problems using SAT solvers is the translation of the target problem to a Boolean Propositional formula. Unfortunately, few problems have an obvious representation as a propositional formula. To overcome this, a common technique is to introduce an intermediate representation for the original problem closer to a Boolean formula. One kind of intermediate representation are Pseudo-Boolean (PB) constraints, which are closely related to SAT. PB constraints provide big expressive power and could be translated to SAT in an automatic way [7].

In this work, we present an approach to solve two variants of the Test Suite Minimization Problem (TSMP) [17] up to optimality using SAT solvers. This is done by modelling TSMP instances as a set of Pseudo-Boolean constraints that are later translated to SAT instances. With the help of a SAT solver the instances are solved and the resulting variable assignment provides an optimal solution for TSMP. Hsu and Orso [10] have tackled this problem in a manner closely related to us, however our contribution over theirs is twofold: we provide the optimal solutions for instances from SIR [11,5], and we apply the approach to a multi-objective formulation of TSMP, obtaining the *Pareto Front* (and a *Pareto optimal set*). A similar approach based on Integer linear programming has been used by Zhang *et al.* for the Time-Aware Test-Case Prioritization [18]. Test-Case prioritisation is a problem related to the TSMP, in which the goal is to find an optimal order in which to execute test cases.

The remainder of this article is structured as follows. In Section 2 we introduce background concepts of SAT solvers and Pseudo-Boolean constraints. In Section 3 we introduce the TSMP and two formulations for this problem: single and multi-objective formulations. In Section 4 we present the application of the proposed approach to the TSMP. Section 5 shows experimental results applying the proposed approach for a set of open well-known instances under the single and multi-objective formulations. Finally with Section 7 we conclude the paper.

2   Background

The Boolean Satisfiability problem (SAT) consists in determining if there exists a Boolean variable assignment that makes *true* a given a propositional Boolean formula. SAT was the first decision problem shown to be $\mathcal{NP}$-Complete [2] and is one of the most important and extensively studied problems, since any other $\mathcal{NP}$ decision problem can be translated into SAT in polynomial time. Thus, if there exists a polynomial time algorithm to solve the SAT problem, then $\mathcal{P} = \mathcal{NP}$. This would answer one of the more important questions in computer science and would bring a great revolution in complexity theory. The algorithms known to solve this problem have complexity $O(2^n)$ in the worst case.

The propositional Boolean formulas are frequently expressed in *Conjunctive Normal Form* (CNF) when they are used as input for the SAT solvers. That is, a Boolean formula is in this case a conjunction of *clauses*, each one consisting in a disjunction of literals (variables negated or not). Let us denote with x_i the Boolean variables for $1 \leq i \leq n$, that is, $x_i \in \{true, false\}$. A *clause* C_j is a

Boolean formula of the form $C_j = x_{j_1} \lor x_{j_2} \lor \ldots \lor x_{j_k} \lor \neg x_{j_{k+1}} \lor \neg x_{j_{k+2}} \lor \ldots \lor \neg x_{j_{k+l}}$. A propositional formula in CNF takes the form $F = \bigwedge_{j=1}^{m} C_j$. When a Boolean formula is expressed in CNF, a solution to a SAT instance consists in an assignment which satisfies all the clauses.

2.1 SAT Solvers

In 1962, Davis, Longemann and Loveland [3] presented a backtracking algorithm based on a systematic search which is the base of current SAT solvers. A backtracking algorithm works by selecting at each step a variable and a Boolean value for branching. In each branching step either *true* or *false* can be assigned to a variable. Then, the logical consequences of each branching are evaluated. Each time a clause becomes unsatisfiable, a backtrack is performed. The backtrack corresponds to undoing branching until a variable is reached for which only one possible Boolean value has been explored. These steps are repeated until the root is reached.

The current state-of-the-art SAT solvers, commonly named *Conflict Driven Clause Learning* (CDCL) solvers, introduce improvements over the described backtracking algorithm. Some of these improvements are:

- **Clause learning:** consists in identifying conflicts between assignments and adding clauses that express these conflicts [15].
- **Non-chronological backtracking:** when a conflict occurs, allows to backtrack to specific decision levels [16].
- **Variable (value) selection heuristic:** establishes rules for determining which variable should be selected and which value the variable should take [14].
- **Random restarts:** allows the restart of the search from scratch. Usually performed as a function of the number of backtracks [9].

These are some of the main techniques included in current CDCL solvers. The interested reader can deepen on these techniques in [1] (chapter 4). In this paper we use the SAT solver MiniSat [6], which is a CDCL solver that includes the techniques described above. MiniSat was designed to be easily extensible, is implemented in C++ (the original source code is under 600 lines) and has been awarded in several categories of the SAT Competition.

2.2 Optimization Problems and Pseudo-boolean Constraints

SAT is a decision problem, that is, it answers a question (the Boolean formula) with a yes/no answer (satisfiable or unsatisfiable). However, we are interested in optimization problems, in which the goal is to minimize or maximize an objective function. Thus, we need to transform the optimization problem into one or several decision problems that can be translated into a Boolean formula. Let us denote with $f : X \to \mathbb{Z}$ the objective function of the optimization problem[2] and

---

[2] We focus on integer functions but this is not a hard constraint in practice since floating point numbers in the computers have a finite representation and could be represented with integer numbers.

let us suppose without loss of generality that we want to find a solution $x^* \in X$ that minimizes[3] f, that is, $f(x^*) \leq f(x)$ for all the solutions $x \in X$. This optimization problem can be transformed in a series of decision problems in which the objective is to find a solution $y \in X$ for which the constraint $f(y) \leq B$ holds, where $B \in \mathbb{Z}$ takes different integer values. This series of decision problems can be used to find the optimal (minimal) solution of the optimization problem. The procedure could be as follows. We start with a value of B low enough for the constraint to be unsatisfiabe. We solve the decision problem to check that it is unsatisfiable. Then, we enter a loop in which the value of B is increased and the constraint is checked again. The loop is repeated until the result is satisfiable. Once the loop finishes, the value of B is the minimal value of f in the search space and the solution to the decision problem is an optimal solution of the optimization problem.

If the optimization problem has several objective functions $f_1, f_2, \ldots, f_m$ to minimize, we need one constraint for each objective function:

$$f_1(y) \leq B_1$$
$$f_2(y) \leq B_2$$
$$\vdots$$
$$f_m(y) \leq B_m$$

In order to use SAT solvers to solve optimization problems, we still need to translate the constraints $f(y) \leq B$ to Boolean formulas. To this aim the concept of *Pseudo-Boolean constraint* plays a main role. A Pseudo-Boolean (PB) constraint is an inequality on a linear combination of Boolean variables:

$$\sum_{i=1}^{n} a_i x_i \odot B \tag{1}$$

where $\odot \in \{<, \leq, =, \neq, >, \geq\}$, $a_i, B \in \mathbb{Z}$, and $x_i \in \{0, 1\}$. A PB constraint is said to be *satisfied* under an assignment if the sum of the coefficients a_i for which $x_i = 1$ satisfies the relational operator $\odot$ with respect to B.

PB constraints can be translated into SAT instances. The simplest approaches translate the PB constraint to an equivalent Boolean formula with the same variables. The main drawback of these approaches is that the number of clauses generated grows exponentially with respect to the variables. In practice, it is common to use one of the following methods for the translation: network of adders, binary decision diagrams and network of sorters [1] (chapter 22). All of these approaches introduce additional variables to generate a formula which is semantically equivalent to the original PB constraint. Although the translation of a non-trivial PB constraint to a set of clauses with some of these methods have also an exponential complexity in the worst case, in practice it is not common to have exponential complexity [7] and the translation can be done in a reasonable time.

---

[3] If the optimization problem consists in maximizing f, we can formulate the problem as the minimization of $-f$.

Boolean formula of the form $C_j = x_{j_1} \vee x_{j_2} \vee \ldots \vee x_{j_k} \vee \neg x_{j_{k+1}} \vee \neg x_{j_{k+2}} \vee \ldots \vee \neg x_{j_{k+l}}$. A propositional formula in CNF takes the form $F = \bigwedge_{j=1}^{m} C_j$. When a Boolean formula is expressed in CNF, a solution to a SAT instance consists in an assignment which satisfies all the clauses.

2.1 SAT Solvers

In 1962, Davis, Longemann and Loveland [3] presented a backtracking algorithm based on a systematic search which is the base of current SAT solvers. A backtracking algorithm works by selecting at each step a variable and a Boolean value for branching. In each branching step either *true* or *false* can be assigned to a variable. Then, the logical consequences of each branching are evaluated. Each time a clause becomes unsatisfiable, a backtrack is performed. The backtrack corresponds to undoing branching until a variable is reached for which only one possible Boolean value has been explored. These steps are repeated until the root is reached.

The current state-of-the-art SAT solvers, commonly named *Conflict Driven Clause Learning* (CDCL) solvers, introduce improvements over the described backtracking algorithm. Some of these improvements are:

- **Clause learning:** consists in identifying conflicts between assignments and adding clauses that express these conflicts [15].
- **Non-chronological backtracking:** when a conflict occurs, allows to backtrack to specific decision levels [16].
- **Variable (value) selection heuristic:** establishes rules for determining which variable should be selected and which value the variable should take [14].
- **Random restarts:** allows the restart of the search from scratch. Usually performed as a function of the number of backtracks [9].

These are some of the main techniques included in current CDCL solvers. The interested reader can deepen on these techniques in [1] (chapter 4). In this paper we use the SAT solver MiniSat [6], which is a CDCL solver that includes the techniques described above. MiniSat was designed to be easily extensible, is implemented in C++ (the original source code is under 600 lines) and has been awarded in several categories of the SAT Competition.

2.2 Optimization Problems and Pseudo-boolean Constraints

SAT is a decision problem, that is, it answers a question (the Boolean formula) with a yes/no answer (satisfiable or unsatisfiable). However, we are interested in optimization problems, in which the goal is to minimize or maximize an objective function. Thus, we need to transform the optimization problem into one or several decision problems that can be translated into a Boolean formula. Let us denote with $f : X \to \mathbb{Z}$ the objective function of the optimization problem[2] and

---

[2] We focus on integer functions but this is not a hard constraint in practice since floating point numbers in the computers have a finite representation and could be represented with integer numbers.

let us suppose without loss of generality that we want to find a solution $x^* \in X$ that minimizes[3] f, that is, $f(x^*) \leq f(x)$ for all the solutions $x \in X$. This optimization problem can be transformed in a series of decision problems in which the objective is to find a solution $y \in X$ for which the constraint $f(y) \leq B$ holds, where $B \in \mathbb{Z}$ takes different integer values. This series of decision problems can be used to find the optimal (minimal) solution of the optimization problem. The procedure could be as follows. We start with a value of B low enough for the constraint to be unsatisfiabe. We solve the decision problem to check that it is unsatisfiable. Then, we enter a loop in which the value of B is increased and the constraint is checked again. The loop is repeated until the result is satisfiable. Once the loop finishes, the value of B is the minimal value of f in the search space and the solution to the decision problem is an optimal solution of the optimization problem.

If the optimization problem has several objective functions $f_1, f_2, \ldots, f_m$ to minimize, we need one constraint for each objective function:

$$f_1(y) \leq B_1$$
$$f_2(y) \leq B_2$$
$$\vdots$$
$$f_m(y) \leq B_m$$

In order to use SAT solvers to solve optimization problems, we still need to translate the constraints $f(y) \leq B$ to Boolean formulas. To this aim the concept of *Pseudo-Boolean constraint* plays a main role. A Pseudo-Boolean (PB) constraint is an inequality on a linear combination of Boolean variables:

$$\sum_{i=1}^{n} a_i x_i \odot B \tag{1}$$

where $\odot \in \{<, \leq, =, \neq, >, \geq\}$, $a_i, B \in \mathbb{Z}$, and $x_i \in \{0, 1\}$. A PB constraint is said to be *satisfied* under an assignment if the sum of the coefficients a_i for which $x_i = 1$ satisfies the relational operator $\odot$ with respect to B.

PB constraints can be translated into SAT instances. The simplest approaches translate the PB constraint to an equivalent Boolean formula with the same variables. The main drawback of these approaches is that the number of clauses generated grows exponentially with respect to the variables. In practice, it is common to use one of the following methods for the translation: network of adders, binary decision diagrams and network of sorters [1] (chapter 22). All of these approaches introduce additional variables to generate a formula which is semantically equivalent to the original PB constraint. Although the translation of a non-trivial PB constraint to a set of clauses with some of these methods have also an exponential complexity in the worst case, in practice it is not common to have exponential complexity [7] and the translation can be done in a reasonable time.

---

[3] If the optimization problem consists in maximizing f, we can formulate the problem as the minimization of $-f$.

The PB constraints make easier the translation of combinatorial optimization problems into a SAT instance, since we can use PB constraints as an intermediate step in the translation. This step from the original problem formulation to the set of PB constraints requires human intervention. It is desirable to model the problem using a low number of Boolean variables and PB constraints in order to avoid an uncontrolled increase of the search space. In Section 4 we detail the translation to PB constraints of the Test Suite Minimization Problem.

2.3   Multi-objective Optimization

A general multi-objective optimization problem (MOP) [4] can be formally defined as follows (we assume minimization without loss of generality).

Definition 1 (MOP). *Find a vector* $\boldsymbol{x}^* = (x_1^*, x_2^*, \ldots, x_n^*)$ *which satisfies the* m *inequality constraints* $g_i(\boldsymbol{x}) \geq 0, i = 1, 2, \ldots, m$, *the* p *equality constraints* $h_i(\boldsymbol{x}) = 0, i = 1, 2, \ldots, p$, *and minimizes the vector function* $\boldsymbol{f}(\boldsymbol{x}) = (f_1(\boldsymbol{x}), f_2(\boldsymbol{x}), \ldots, f_k(\boldsymbol{x}))$, *where* $\boldsymbol{x} = (x_1, x_2, \ldots, x_n)$ *is the vector of decision variables.*

The set of all values satisfying the constraints defines the *feasible region* Ω and any point $\boldsymbol{x} \in \Omega$ is a *feasible solution*. It is common in MOPs that not all the objective functions can be simultaneously minimized, there are some conflicts between them. This means that decreasing the value of one objective function implies increasing the value of another one. For this reason, the goal of multi-objective search algorithms is not to find an optimal solution, but a set of *non-dominated solutions* which form the so-called *Pareto optimal set*. We formally define these concepts in the following.

Definition 2 (Pareto Optimality). *A point* $\boldsymbol{x}^* \in \Omega$ *is Pareto optimal if for every* $\boldsymbol{x} \in \Omega$ *and* $I = \{1, 2, \ldots, k\}$ *either* $\forall_{i \in I} f_i(\boldsymbol{x}) = f_i(\boldsymbol{x}^*)$ *or there is at least one* $i \in I$ *such that* $f_i(\boldsymbol{x}) > f_i(\boldsymbol{x}^*)$.

This definition states that $\boldsymbol{x}^*$ is Pareto optimal if no feasible vector $\boldsymbol{x}$ exists which would improve one objective without causing a simultaneous worsening in at least another objective. Other important definitions associated with Pareto optimality are the following:

Definition 3 (Pareto Dominance). *A vector* $\boldsymbol{u} = (u_1, \ldots, u_k)$ *is said to dominate* $\boldsymbol{v} = (v_1, \ldots, v_k)$ *(denoted by* $\boldsymbol{u} \preccurlyeq \boldsymbol{v}$*) if and only if* $\boldsymbol{u}$ *is partially smaller than* $\boldsymbol{v}$, *i.e.,* $\forall i \in I, u_i \leq v_i \wedge \exists i \in I : u_i < v_i$.

Definition 4 (Pareto optimal set). *For a given MOP* $\boldsymbol{f}(\boldsymbol{x})$, *the Pareto optimal set is defined as* $\mathcal{P}^* = \{\boldsymbol{x} \in \Omega | \neg \exists \boldsymbol{x}' \in \Omega, \boldsymbol{f}(\boldsymbol{x}') \preccurlyeq \boldsymbol{f}(\boldsymbol{x})\}$.

Definition 5 (Pareto Front). *For a given MOP* $\boldsymbol{f}(\boldsymbol{x})$ *and its Pareto optimal set* $\mathcal{P}^*$, *the Pareto front is defined as* $\mathcal{PF}^* = \{\boldsymbol{f}(\boldsymbol{x}) | \boldsymbol{x} \in \mathcal{P}^*\}$.

Obtaining the Pareto optimal set and the Pareto front of a MOP are the main goals of multi-objective optimization.

3   Test Suite Minimization Problem

When a piece of software is modified, the new software is tested using some previous test cases in order to check if new errors were introduced. This check is known as regression testing. One problem related to regression testing is the Test Suite Minimization Problem (TSMP). This problem is equivalent to the Minimal Hitting Set Problem which is $\mathcal{NP}$-hard [8]. Let $\mathcal{T} = \{t_1, t_2, \cdots, t_n\}$ be a set of tests for a program where the cost of running test t_i is c_i and let $\mathcal{E} = \{e_1, e_2, \cdots, e_m\}$ be a set of elements of the program that we want to cover with the tests. After running all the tests $\mathcal{T}$ we find that each test can cover several program elements. This information is stored in a matrix $\mathbf{M} = [m_{ij}]$ of dimension $n \times m$ that is defined as:

$$m_{ij} = \begin{cases} 1 & \text{if element } e_j \text{ is covered by test } t_i \\ 0 & \text{otherwise} \end{cases}$$

The single-objective version of this problem consists in finding a subset of tests $X \subseteq \mathcal{T}$ with minimum cost covering all the program elements. In formal terms:

$$minimize \quad cost(X) = \sum_{\substack{i=1 \\ t_i \in X}}^{n} c_i \tag{2}$$

subject to:

$$\forall e_j \in \mathcal{E}, \exists t_i \in X \qquad \text{such that element } e_j \text{ is covered by test } t_i, \text{ that is, } m_{ij} = 1.$$

The multi-objective version of the TSMP does not impose the constraint of full coverage, but it defines the coverage as the second objective to optimize, leading to a bi-objective problem. In short, the bi-objective TSMP consists in finding a subset of tests $X \subseteq \mathcal{T}$ having minimum cost and maximum coverage. Formally:

$$minimize \quad cost(X) = \sum_{\substack{i=1 \\ t_i \in X}}^{n} c_i \tag{3}$$

$$maximize \quad cov(X) = |\{e_j \in \mathcal{E} | \exists t_i \in X \text{ with } m_{ij} = 1\}| \tag{4}$$

There is no constraint in this bi-objective formulation. We should notice here that solving the bi-objective version (2-obj in short) of TSMP implies solving the single-objective version (1-obj). In effect, let us suppose that we solve an instance of the 2-obj TSMP, then a solution for the related 1-obj TSMP is just the set $X \subseteq \mathcal{T}$ with $cov(X) = |\mathcal{E}|$ in the Pareto optimal set, if such a solution exists. If there is no solution of 2-obj TSMP with $cov(X) = |\mathcal{E}|$, then the related 1-obj TSMP is not solvable.

4 Solving TSMP Instances Using PB Constraints

In this section, we will present the proposed approach for solving the TSMP using SAT solvers. First, we detail how the two versions of TSMP can be translated into a set of PB constraints and then we present the algorithms used to solve both versions of TSMP with the help of the SAT solvers.

4.1 Translating the TSMP

The single-objective formulation of TSMP is a particular case of the bi-objective formulation. Then, we can translate the 2-obj TSMP into a set of PB constraints and then infer the translation of the 1-obj TSMP as a especial case.

Let us introduce n binary variables $t_i \in \{0, 1\}$: one for each test case in $\mathcal{T}$. If $t_i = 1$ then the corresponding test case is included in the solution and if $t_i = 0$ the test case is not included. We also introduce m binary variables $e_j \in \{0, 1\}$: one for each program element to cover. If $e_j = 1$ then the corresponding element is covered by one of the selected test cases and if $e_j = 0$ the element is not covered by a selected test case.

The values of the e_j variables are not independent of the t_i variables. A given variable e_j must be 1 if and only if there exists a t_i variable for which $m_{ij} = 1$ and $t_i = 1$. The dependence between both sets of variables can be written with the following $2m$ PB constraints:

$$e_j \leq \sum_{i=1}^{n} m_{ij} t_i \leq n \cdot e_j \qquad 1 \leq j \leq m. \qquad (5)$$

We can see that if the sum in the middle is zero (no test is covering the element e_j) then the variable $e_j = 0$. However, if the sum is greater than zero $e_j = 1$. Now we need to introduce a constraint related to each objective function in order to transform the optimization problem in a decision problem, as we described in Section 2.2. These constraints are:

$$\sum_{i=1}^{n} c_i t_i \leq B, \qquad (6)$$

$$\sum_{j=1}^{m} e_j \geq P, \qquad (7)$$

where $B \in \mathbb{Z}$ is the maximum allowed cost and $P \in \{0, 1, \dots, m\}$, is the minimum coverage level. We required a total of $n + m$ binary variables and $2m + 2$ PB constraints for the 2-obj TSMP.

For the 1-obj TSMP the formulation is simpler. This is a especial case of the 2-obj formulation in which $P = m$. If we include this new constraint in (7) we have $e_j = 1$ for all $1 \le j \le m$. Then we don't need the e_j variables anymore because they are constants. Including these constants in (5) we have:

$$1 \le \sum_{i=1}^{n} m_{ij} t_i \le n \qquad 1 \le j \le m, \qquad (8)$$

which is equivalent to:

$$\sum_{i=1}^{n} m_{ij} t_i \ge 1 \qquad 1 \le j \le m, \qquad (9)$$

since the sum is always less than or equal to n. Thus, for the 1-obj TSMP the PB constraints are (8) and (9).

4.2   Translation Example

In this section we show through a small example how to model with PB constraints an instance of the TSMP according to the methodology above described. Let $\mathcal{T} = \{t_1, t_2, t_3, t_4, t_5, t_6\}$, $\mathcal{E} = \{e_1, e_2, e_3, e_4\}$ and $\mathbf{M}$:

	e_1	e_2	e_3	e_4
t_1	1	0	1	0
t_2	1	1	0	0
t_3	0	0	1	0
t_4	1	0	0	0
t_5	1	0	0	1
t_6	0	1	1	0

If we want to solve the 2-obj TSMP we need to instantiate Eqs. (5), (6) and (7). The result is:

$$e_1 \le t_1 + t_2 + t_4 + t_5 \le 4e_1 \qquad (10)$$

$$e_2 \le \quad t_2 + t_6 \quad \le 4e_2 \qquad (11)$$

$$e_3 \le \quad t_1 + t_3 + t_6 \quad \le 4e_3 \qquad (12)$$

$$e_4 \le \quad t_5 \quad \le 4e_4 \qquad (13)$$

$$t_1 + t_2 + t_3 + t_4 + t_5 + t_6 \le B \qquad (14)$$

$$e_1 + e_2 + e_3 + e_4 \quad \ge P \qquad (15)$$

where $P, B \in \mathbb{N}$.

If we are otherwise interested in the 1-obj version the formulation is simpler:

$$t_1 + t_2 + t_4 + t_5 \ge 1 \qquad (16)$$

$$t_2 + t_6 \ge 1 \qquad (17)$$

$$t_1 + t_3 + t_6 \geq 1 \tag{18}$$

$$t_5 \geq 1 \tag{19}$$

$$t_1 + t_2 + t_3 + t_4 + t_5 + t_6 \leq B \tag{20}$$

4.3   Algorithms

This section describes the procedures used to find the optimal solutions to the single- and multi-objective formulation of TSMP. Algorithm 1 shows the steps needed to find the optimal solution in the single-objective formulation. We assume, without loss of generality, that full coverage can be reached. If this is not the case we can just remove from $\mathcal{E}$ the program elements that are not covered by any test case.

Algorithm 1. Procedure to compute the optimal solution for 1-obj TSMP

Input: TSMP matrix **M**
Output: optimal solution $\mathbf{S}^*$
 1: $B \leftarrow 1$
 2: $result \leftarrow$ false
 3: **while** not($result$) **do**
 4:      Translate $(\mathbf{M}, B)$ into a set of PB constraints: Eqs. (8) and (9)
 5:      Transform the set of PB constraints into a SAT instance I
 6:      Run the SAT solver with I as input
 7:      **if** SAT solver found solution **then**
 8:          $\mathbf{S}^* \leftarrow$ assignment found
 9:          $result \leftarrow$ true
10:      **else**
11:          $B \leftarrow B + 1$
12:      **end if**
13: **end while**

We can observe that the pseudocode in Algorithm 1 follows the description we introduced in Section 2.2 to solve an optimization problem using SAT solvers. In this case the procedure is adapted to solve the 1-obj TSMP. When the algorithm ends, the value of B is the minimal number of tests required to get full coverage. In the algorithm, the value of B is increased in 1 unit in each iteration (line 11). However, it is possible to use a search strategy based on a binary search in the interval $[1, n]$ for the B value.

In Algorithm 2 we show the procedure used to find a Pareto optimal set for the 2-obj TSMP. In this case the initial value of B will be the one found by the Algorithm 1, which is run in line 2. Each iteration of the loop in line 6 of Algorithm 2 can be seen as a modification of the value P in Eq. (6). This way the algorithm computes the maximum number of elements covered by B test cases. The value of B is decreased in each iteration of the external loop in order to explore the complete Pareto front. Again a binary search could be applied to P or B in order to accelerate the search.

Algorithm 2. Procedure to compute the Pareto optimal set for the 2-obj TSMP

Input: TSMP matrix $\mathbf{M}$
Output: Pareto optimal set
 1: Pareto optimal set $= \emptyset$
 2: Run Algorithm 1
 3: $P = |\mathcal{E}|$
 4: **while** $B \geq 1$ **do**
 5:     *found* $\leftarrow$ false
 6:     **while** $(P \geq 1)$ and (not(*found*)) **do**
 7:         Translate $(\mathbf{M}, B, P)$ into a set of PB constraints: Eqs. (5), (6) and (7)
 8:         Transform the set of PB constraints into a SAT instance I
 9:         Run the SAT solver with I as input
10:         **if** SAT solver found solution **then**
11:             Add the assignment found to the Pareto optimal set
12:             *found* $\leftarrow$ true
13:         **else**
14:             $P \leftarrow P - 1$
15:         **end if**
16:     **end while**
17:     $B \leftarrow B - 1$
18: **end while**

---

5   Experimental Results

We performed an experiment to check our approach using the programs from the Siemens suite [11] available at SIR[4] (Software-artifact Infrastructure Repository). The Siemens programs perform a variety of tasks: `printtokens` and `printtokens2` are lexical analyzers, `tcas` is an aircraft collision avoidance system, `schedule` and `schedule2` are priority schedulers, `totinfo` computes statistics given input data, and `replace` performs pattern matching and substitution. The coverage matrix $\mathbf{M}$ can be obtained from the data in the SIR. For the cost values c_i we considered that all the costs are 1: $c_i = 1$ for $1 \leq i \leq n$. As a consequence, the cost function of a set of test cases is just the number of test cases. We implemented Algorithms 1 and 2 as shell scripts in Linux and we used MiniSat+ [7] as Pseudo-Boolean solver. MiniSat+ translates PB constraints into Boolean formulas (in CNF), and uses MiniSat [6] as SAT solver engine.

In a second experiment we transformed the instances into equivalent ones with fewer test cases. We can do this because we are considering that all test cases have the same cost. Under this assumption we can remove any test case for which there is another test case covering at least all the elements covered by the first one. In formal terms, if test case t_i covers program elements $E_i \subseteq \mathcal{E}$ and test case t_h covers program elements $E_h \subseteq \mathcal{E}$ where $E_i \subseteq E_h$, then we remove t_i from the original test suite and the Pareto front of the instance does not change, because any solution having test case t_i cannot get worse after replacing t_i by t_h.

---

[4] http://sir.unl.edu/portal/index.php

The result is an instance with fewer test cases but having the same Pareto front. These transformed instances were solved using Algorithm 2. Table 1 shows the size of the test suites with and without the reduction for each program. We can observe a really great reduction in the number of test cases when the previous approach is used.

Table 1. Details of the instances used in the experiments

Instance	Original Size	Reduced Size	Elements to cover
printtokens	4130	40	195
printtokens2	4115	28	192
replace	5542	215	208
schedule	2650	4	126
schedule2	2710	13	119
tcas	1608	5	54
totinfo	1052	21	117

In Table 2 we present the Pareto optimal set and the Pareto front for the instances described above. The columns "Tests" and "Elements" correspond to the functions *cost* and *cov* of the 2-obj TSMP. The column "Coverage" is the number of covered elements divided by the total number of elements. The optimal solution for the 1-obj TSMP can be found in the lines with 100% coverage, as explained in Section 3. It is not common to show the Pareto optimal set or the Pareto front in numbers in the multi-objective literature because only approximate Pareto fronts can be obtained for $\mathcal{NP}$-hard problems. However, in this case we obtain the exact Pareto fronts and optimal sets, so we think that this information could be useful for future reference. Figure 1 shows the Pareto front for all the instances of Table 1: they present the same information as Table 2 in a graphical way. The information provided in the tables and the figures is very useful for the tester, knowing beforehand which are the most important test cases and giving the possibility to make a decision taking into account the number of tests necessary to assure a particular coverage level or vice versa.

We show in Table 3 the running time of Algorithm 2, which includes the execution of Algorithm 1. The experiments were performed on a Laptop with an Intel CORE i7 running Ubuntu Linux 11.04. Since the underlying algorithm is deterministic the running time is an (almost) deterministic variable. The only source of randomness for the SAT solver comes from limited random restarts and the application of variable selection heuristics. Additionally, we compared the running time of our approach with the performance of two heuristic algorithms: a local search (LS) algorithm and a genetic algorithm (GA) for the 1-obj formulation of the TSMP. The LS algorithm is based on an iterative best improvement process and the GA is a steady-state GA with 10 individuals in the population, binary tournament selection, bit-flip mutation with probability $p = 0.01$ of flipping a bit, one-point crossover and elitist replacement. The stopping condition is to equal the running time of the SAT-based method for each reduced instance.

Table 2. Pareto optimal set and Front for the instances of SIR

Instance	Elements	Tests	Coverage	Solution
printtokens	195	5	100%	$(t_{2222}, t_{2375}, t_{3438}, t_{4100}, t_{4101})$
	194	4	99.48%	$(t_{1908}, t_{2375}, t_{4099}, t_{4101})$
	192	3	98.46%	$(t_{1658}, t_{2363}, t_{4072})$
	190	2	97.43%	(t_{1658}, t_{3669})
	186	1	95.38%	(t_{2597})
printtokens2	192	4	100%	$(t_{2521}, t_{2526}, t_{4085}, t_{4088})$
	190	3	98.95%	$(t_{457}, t_{3717}, t_{4098})$
	188	2	97.91%	(t_{2190}, t_{3282})
	184	1	95.83%	(t_{3717})
replace	208	8	100%	$(t_{306}, t_{410}, t_{653}, t_{1279}, t_{1301}, t_{3134}, t_{4057}, t_{4328})$
	207	7	99.51%	$(t_{309}, t_{358}, t_{653}, t_{776}, t_{1279}, t_{1795}, t_{3248})$
	206	6	99.03%	$(t_{275}, t_{290}, t_{1279}, t_{1938}, t_{2723}, t_{2785})$
	205	5	98.55%	$(t_{426}, t_{1279}, t_{1898}, t_{2875}, t_{3324})$
	203	4	97.59%	$(t_{298}, t_{653}, t_{3324}, t_{5054})$
	200	3	96.15%	$(t_{2723}, t_{2901}, t_{3324})$
	195	2	93.75%	(t_{358}, t_{5387})
	187	1	89.90%	(t_{358})
schedule	126	3	100%	$(t_{1403}, t_{1559}, t_{1564})$
	124	2	98.41%	(t_{1570}, t_{1595})
	122	1	96.82%	(t_{1572})
schedule2	119	4	100%	$(t_{2226}, t_{2458}, t_{2462}, t_{2681})$
	118	3	99.15%	$(t_{101}, t_{1406}, t_{2516})$
	117	2	98.31%	(t_{2461}, t_{2710})
	116	1	97.47%	(t_{1584})
tcas	54	4	100%	$(t_5, t_{1191}, t_{1229}, t_{1608})$
	53	3	98.14%	$(t_{13}, t_{25}, t_{1581})$
	50	2	92.59%	(t_{72}, t_{1584})
	44	1	81.48%	(t_{217})
totinfo	117	5	100%	$(t_{62}, t_{118}, t_{218}, t_{1000}, t_{1038})$
	115	4	98.29%	$(t_{62}, t_{118}, t_{913}, t_{1016})$
	113	3	96.58%	$(t_{65}, t_{216}, t_{913})$
	111	2	94.87%	(t_{65}, t_{919})
	110	1	94.01%	(t_{179})

For the two heuristic algorithms we show the average coverage and number of test cases over 30 independent runs.

Regarding the computational time, we observe that all the instances can be solved in much less time using the reduction. The speed up for the SAT-based approach ranges from more than 200 for tcas to more than 2000 for printtokens2. All the instances can be solved in around 2 seconds with the exception of replace, which requires almost 6 minutes. In the case of the heuristic algorithms, we observe that LS reaches full coverage in all the instances and independent runs. However, the required number of test cases is non-optimal in printtokens, printtokens2 and replace. LS obtains optimal solutions in the

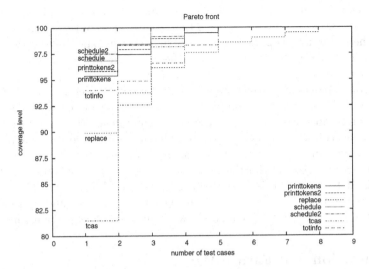

Fig. 1. Pareto front for the SIR instances

Table 3. Information about clauses-to-variables ratio, computation time of Algorithm 2, average coverage and number of test cases for the two heuristic algorithms for the instances from SIR

Instance	Ratio	Algorithm 2		Local Search		Genetic Algorithm	
		Original (s)	Reduced (s)	Avg. Cov.	Avg. Tests	Avg. Cov.	Avg. Tests
printtokens	4.61	3400.74	2.17	100.00%	6.00	99.06%	5.16
printtokens2	4.61	3370.44	1.43	100.00%	4.60	99.23%	3.56
replace	4.62	1469272.00	345.62	100.00%	10.16	99.15%	15.46
schedule	2.19	492.38	0.24	100.00%	3.00	99.84%	2.90
schedule2	4.61	195.55	0.27	100.00%	4.00	99.58%	3.70
tcas	4.61	73.44	0.33	100.00%	4.00	95.80%	3.23
totinfo	4.53	181823.50	0.96	100.00%	5.00	98.89%	5.13

rest of the programs. However, we should recall here that LS cannot ensure that the result is an optimal solution, as the SAT-based approach does. In the case of GA, it is not able to reach full coverage in any program.

It is interesting to remark that almost all the resulting SAT instances obtained from the translation are in the phase transition of SAT problems except the one for schedule. It has been shown experimentally that most of the instances where the ratio of clauses-to-variables is approximately equal to 4.3 are the hardest to be solved [13].

6   Threats to Validity

We identified two major threats to validity in our approach. The first one is related to the inherent difficulty of the SAT problem for which there is no known efficient algorithm to solve it. This implies that there is a limit in the number of test cases that can be addressed with our method. For example, in the experiments we observed a high execution time to solve two of the seven instances without the reduction: `replace` and `totinfo`. Further experiments with more and larger instances could show what are the actual limitation of our approach in real-world software.

The second one concerns the complexity of the translation from the PB constraints to SAT clauses. This complexity is exponential in the worst-case but not in practice. In our experiments the time required for this translation was very small. We think this is a consequence of having unitary cost for the tests. If we consider non-unitary cost for the tests, the translation could require a non-negligible amount of time. Furthermore, the resulting SAT instance could be more difficult to solve in this case, making the approach impractical. To study these issues we need more experiments using larger instances and non-unitary costs for the tests. These costs could be based on the tests execution time.

7   Conclusion and Future Work

In this work we show an approach to optimally solve the TSMP. This approach comprises two translations to obtain a SAT instance which is solved by a state-of-the-art SAT solver. The power of current SAT solvers give us the possibility of solving to optimality TSMP instances that were previously solved in an approximate way using metaheuristic algorithms. With the help of MiniSat+ we solved well-known and highly-used instances of a single- and a bi-objective formulation of the TSMP problem. Most of the instances were solved in less than 2 seconds and all of them required less than 6 minutes.

The approach presented here to solve the TSMP problem can be easily extended to other hard problems in Search-Based Software Engineering and other domains. As future work we plan to consider different cost coefficients for the different test cases in the TSMP.

Acknowledgements. This research has been partially funded by the Spanish Ministry of Science and Innovation and FEDER under contract TIN2011-28194 (RoadME project) and the Andalusian Government under contract P07-TIC-03044 (DIRICOM project).

References

1. Biere, A., Heule, M., van Maaren, H., Walsh, T.: Handbook of Satisfiability. Frontiers in Artificial Intelligence and Applications, vol. 185. IOS Press, Amsterdam (2009)
2. Cook, S.A.: The complexity of theorem-proving procedures. In: Proceedings of the Third Annual ACM Symposium on Theory of Computing, STOC 1971, pp. 151–158. ACM, New York (1971)
3. Davis, M., Logemann, G., Loveland, D.: A machine program for theorem-proving. Commun. ACM 5(7), 394–397 (1962)

4. Deb, K.: Multi-objective optimization using evolutionary algorithms. John Wiley & Sons (2001)
5. Do, H., Elbaum, S., Rothermel, G.: Supporting controlled experimentation with testing techniques: An infrastructure and its potential impact. Empirical Software Engineering 10, 405–435 (2005)
6. Eén, N., Sörensson, N.: An Extensible SAT-solver. In: Giunchiglia, E., Tacchella, A. (eds.) SAT 2003. LNCS, vol. 2919, pp. 502–518. Springer, Heidelberg (2004)
7. Eén, N., Sörensson, N.: Translating Pseudo-Boolean Constraints into SAT. Journal on Satisfiability, Boolean Modeling and Computation 2(1-4), 1–26 (2006)
8. Garey, M.R., Johnson, D.S.: Computers and Intractability: A Guide to the Theory of NP-Completeness. W. H. Freeman & Co., New York (1979)
9. Gomes, C., Selman, B., Crato, N.: Heavy-tailed Distributions in Combinatorial Search. In: Smolka, G. (ed.) CP 1997. LNCS, vol. 1330, pp. 121–135. Springer, Heidelberg (1997)
10. Hsu, H.Y., Orso, A.: MINTS: A general framework and tool for supporting test-suite minimization. In: IEEE 31st International Conference on Software Engineering, ICSE 2009, pp. 419–429 (May 2009)
11. Hutchins, M., Foster, H., Goradia, T., Ostrand, T.: Experiments of the effectiveness of dataflow- and controlflow-based test adequacy criteria. In: Proceedings of the 16th International Conference on Software Engineering, ICSE 1994, pp. 191–200. IEEE Computer Society Press, Los Alamitos (1994)
12. Marques-Silva, J.: Practical applications of boolean satisfiability. In: 9th International Workshop on Discrete Event Systems, WODES 2008, pp. 74 –80 (May 2008)
13. Mitchell, D., Selman, B., Levesque, H.: Hard and easy distributions of SAT problems. In: Proceedings of the Tenth National Conference on Artificial Intelligence, AAAI 1992, pp. 459–465. AAAI Press (1992)
14. Moskewicz, M.W., Madigan, C.F., Zhao, Y., Zhang, L., Malik, S.: Chaff: engineering an efficient SAT solver. In: Proceedings of the 38th Annual Design Automation Conference, DAC 2001, pp. 530–535. ACM, New York (2001)
15. Silva, J.P.M., Sakallah, K.A.: Grasp: A new search algorithm for satisfiability. In: Proceedings of the 1996 IEEE/ACM International Conference on Computer-aided Design, pp. 220–227. IEEE Computer Society, Washington, DC (1996)
16. Stallman, R.M., Sussman, G.J.: Forward reasoning and dependency-directed backtracking in a system for computer-aided circuit analysis. Artificial Intelligence 9(2), 135–196 (1977)
17. Yoo, S., Harman, M.: Regression testing minimization, selection and prioritization: a survey. Software Testing, Verification and Reliability (2010), http://dx.doi.org/10.1002/stvr.430, doi: 10.1002/stvr.430
18. Zhang, L., Hou, S.S., Guo, C., Xie, T., Mei, H.: Time-aware test-case prioritization using integer linear programming. In: Proceedings of the Eighteenth International Symposium on Software Testing and Analysis, ISSTA 2009, pp. 213–224. ACM, New York (2009)

Evaluating the Importance of Randomness in Search-Based Software Engineering

Márcio de Oliveira Barros

Post-graduate Information Systems Program – PPGI/UNIRIO
Av. Pasteur 458, Urca – Rio de Janeiro, RJ – Brazil
marcio.barros@uniriotec.br

Abstract. Random number generators are a core component of heuristic search algorithms. They are used to build candidate solutions and to reduce bias while transforming these solutions during the search. Despite of their usefulness, random numbers also have drawbacks, as one cannot guarantee that all portions of the search space are covered and must run an algorithm many times to statistically evaluate its behavior. In this paper we present a study in which a Hill Climbing search with random restart was applied to the software clustering problem under two configurations. First, the algorithm used pseudo-random numbers to create the initial and restart solutions. Then, the algorithm was executed again but the initial and restart solutions were built according to a quasi-random sequence. Contrary to previous findings with other heuristic algorithms, we observed that the quasi-random search could not outperform the pseudo-random search for two distinct fitness functions and fourteen instances.

Keywords: pseudo-random number generation, quasi-random sequences, hill climbing, software module clustering.

1 Introduction

Heuristic search relies on random numbers to find suitable solutions for optimization problems without analyzing the whole spectrum of their search space. For instance, consider the situation in which one has to select a subset of N test cases to maximize the coverage of a program's source-code given a constraint that the subset must run up to a certain length of time. Since the number of test case combinations is exponential to the number of cases (2^N), the analysis of all alternatives becomes unfeasible as N grows. Randomization allows unbiased selection of as many as possible alternatives and can be combined with transformation heuristics to guide the search.

In heuristic algorithms, the dependence on random numbers occurs while drawing candidate solutions (as in Random Search), generating a starting point for the search (as in Hill Climbing), determining whether a solution might replace a better one in the search process (as in Simulated Annealing), creating a population of initial solutions (as in Genetic Algorithms), selecting solutions for reproduction and mutation, among other solution generation and solution composition operations.

G. Fraser (Ed.): SSBSE 2012, LNCS 7515, pp. 60–74, 2012.

Despite the advantages of unbiased selection and manipulation, randomness makes heuristic algorithms strongly susceptible to chance [1]. Distinct runs of an algorithm usually produce different results. The time required to find solutions of acceptable quality frequently varies in distinct runs. Therefore, multiple runs are required to allow drawing probability distributions of an algorithm's outputs and to statistically analyze its efficiency and effectiveness.

Instead of using random numbers one might think of traversing the search space through points positioned in a regular-spaced grid. The grid covers the search space uniformly, but the number of divisions of that space must be established *a priori*, rendering the algorithm unable to evaluate more points if resources are available or fewer points if resources shorten. An alternative to random numbers or regular-spaced grids would be using quasi-random sequences. These are sequences of numbers which are spread uniformly in an N-dimensional space, though the order in which they appear in the sequence is not straightforward as in a grid. They have been successfully used in Monte-Carlo integration and have outperformed pseudo-random numbers when used in heuristic optimization [18][19].

In this paper we present an experimental study addressing the efficiency and effectiveness of using a Sobol sequence (a quasi-random sequence) instead of pseudo-random numbers to guide a Hill Climbing search with random restarts in the context of software module clustering. We use two fitness functions and compare the solution quality and execution time under different sequences over 14 problem instances of different sizes. Contrary to previous reports on the usage of quasi-random sequences to guide algorithmic optimization, experimental results show that the Sobol sequence does neither run on par nor outperforms pseudo-random numbers.

Besides this introduction, this paper is organized in six sections. Next, we present background information about quasi-random sequences and the software clustering problem. Then, the design of the proposed experiment is presented in section 3. Sections 4 and 5 discuss the results obtained for each fitness function used in the experiment. Section 6 comprises considerations on potential validity threats, while future works and conclusions are drawn in section 7.

2   Background

2.1   Quasi-random Sequences

A *quasi-random sequence* of dimension N is a sequence of points defined in a subset of the R^N hyperspace (usually a unitary hypercube) that covers the selected subspace more uniformly than uncorrelated random points [22]. These sequences frequently replace pseudo-random numbers or even-spaced grids in calculations. For instance, it has been shown that the estimation of integrals for high-dimension, complex functions through Monte Carlo simulation converges faster if quasi-random sequences are used instead of pseudo-random numbers [19].

Quasi-random sequences seek to bridge the gap between the flexibility of pseudo-random numbers and the advantages of a regular grid. They are designed to be uniformly distributed in multidimensional space, but unlike pseudo-random numbers they are not statistically independent [13]. On the other hand, points can be added to fill the sample space in regular intervals as the calculation advances and, unlike grids, the granularity of the search does not have to be stated beforehand.

There are many alternative quasi-random sequences, such as the van der Corput numbers, the Halton sequences, the Faure sequences, and Sobol sequences [18]. The Sobol sequence is possibly the most tested sequence and has shown good results when used in simulation and optimization [7][24]. It is an ordered-set of numbers in the [0, 1] interval, represented as binary fractions of w bits and created by bitwise *XOR-ing* a set of w direction numbers [22]. The direction numbers are calculated from polynomials and are used as parameters for the Sobol sequence generator. Figure 1 (a) shows 2000 points from a 2D Sobol sequence in the [0, 100] interval, while (b) shows 2000 points created by a pseudo-random generator. It can be observed that the Sobol sequence fills the semi-plane much regularly than the pseudo-random points.

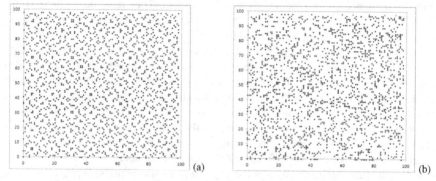

Fig. 1. 2D sequences of (a) Sobol numbers and (b) pseudo-random numbers

In the realm of heuristic optimization, Maaranen et al. [14] used a quasi-random sequence to build the initial population for a GA and found that it outperformed a pseudo-random based GA. This result was endorsed by Kimura and Matsumura [11], who used a quasi-random sequence to create the initial population and affect crossover off-springs in a GA. The algorithm was compared to a traditional GA over a set of complex functions and the authors obtained better solutions when the quasi-random sequence was used. Quasi-random numbers were also used to seed the initial population of a particle swarm search [20], again producing better results than a pseudo-random swarm. Similar results were found by Thangaraj et al. [24] while using quasi-random sequences to seed a Differential Evolution search.

2.2   Software Module Clustering

Software module clustering addresses the distribution of modules representing domain concepts and computational constructs composing a software system into larger, container-like structures. A proper module distribution aids in identifying modules responsible for a given functionality [1], provides easier navigation among software parts [8] and enhances comprehension of the source-code [12] [17]. Therefore, it supports the development and maintenance of a software system. Besides, experiments have shown strong correlation between bad module distributions and the presence of faults in software systems [4].

The quality of a module distribution is frequently addressed by two metrics: coupling and cohesion [27]. Coupling measures the dependency between clusters, while cohesion measures the internal strength of a cluster. A module distribution is considered a good decomposition for a software system if it presents low coupling and high cohesion. These metrics are calculated from the observed dependencies among modules. A module A is said to depend on a module B if it requires some function, procedure or definition declared in module B to perform its duties.

To evaluate its module distribution, a software system is usually represented as a Module Dependency Graph, or MDG [16]. The MDG is a directed graph in which modules are depicted as nodes, dependencies are presented as edges, and clusters are partitions. Weights may be assigned to edges in order to represent the strength of the dependency between modules. The coupling of a cluster is calculated by summing the weights of edges leaving or entering the partition (inter-edges), while its cohesion is calculated by summing the weights of edges whose source and target modules pertain to the partition (intra-edges) [21]. In unweighted MDG, the weight of every edge is set to 1 and coupling and cohesion are calculated by counting intra- and inter-edges.

The software module clustering problem can be modeled as a graph partition problem which minimizes coupling (inter-edges) and maximizes cohesion (intra-edges). This problem is known to be NP-hard and bears two aspects that make it interesting for SBSE: a large search space and quickly calculated fitness functions. Mancoridis et al. [16] have been the first to present a search-based assessment of the problem. They propose a single-objective Hill Climbing search to find the best module distribution for a system. The search is guided by a fitness function called modularization quality (MQ), calculated by the equations below, where N represents the number of clusters, C_k represents a cluster, i and j respectively represent the number of intra-edges and inter-edges in C_k. MQ looks for a balance between coupling and cohesion, rewarding clusters with many intra-edges and penalizing them for dependencies on other clusters. The proposed search aims to find partitions with the highest MQ possible. Doval et al. [5] presents a genetic algorithm to address the software clustering problem using MQ as a single-objective fitness function. The genetic algorithm, however, has been found less effective and less efficient than the Hill Climbing search.

$$MQ = \sum_{k=1}^{N} MF(C_k) \qquad MF(C_k) = \begin{cases} 0, & i = 0 \\ \dfrac{i}{i + j/2}, & i > 0 \end{cases}$$

Harman et al. [9] compare MQ with a second clustering fitness function called EVM [25] on regard of their robustness to the presence of noise in the MDG. EVM is calculated as a sum of cluster scores. The score of a given cluster starts as zero and is calculated as follows: for each pair of modules in the cluster, the score is incremented if there is dependency between the modules; otherwise, it is decremented. The fitness function rewards modules which depend on other modules pertaining to the same cluster and penalizes those depending on modules from other clusters. Distinctly from

MQ, EVM does not take dependency weights into account. Therefore, it does not make any distinction between weighted and non-weighted MDGs. When compared to a baseline module distribution, clusterings based on EVM were found to degrade more slowly in the presence of noise (represented by adding random dependencies on the original module dependency structure) than clusterings based on MQ.

Mahdavi et al. [15] address the software clustering problem by using a procedure comprised of two sets of Hill Climbing searches. First, 23 parallel and independent Hill Climbing searches are executed upon the MDG representation of the software. Next, a partial solution is constructed by fixing the distribution of modules that share the same cluster in at least $\alpha\%$ of the solutions yielded by the initial hill climbers, where α is a parameter defined is the [10%, 100%] interval. Next, a second round of Hill Climbing searches is performed to distribute the modules that were not assigned to clusters during the partial solution construction phase. Finally, the best solution found by the second round of hill climbers is selected as the best solution for the problem. The authors observed improvements on solutions found by the second run of hill climbers when compared to those produced by the first round, in particular for large instances or small instances combined with small values for the α parameter.

Other approaches where applied to software clustering, such as different measures to evaluate module distribution quality (information-theoretic complexity measures, Distance from Main Sequence, among others), clustering analysis and design documents, and multi-objective optimization. These issues lie out of the scope of this paper, which concentrates on Hill Climbing and the fitness functions discussed above. Other SBSE approaches to software clustering can be found in [10], while a broader survey on SBSE application to software design is provided in [23].

3   Experimental Setup

To compare the results produced by Hill Climbing searches powered by pseudo-random numbers and quasi-random sequences while addressing the software clustering problem, problem instances will be proposed and optimized according to the MQ and EVM fitness functions presented in Section 2.2. Next, results collected from both searches will be compared in terms of fitness and execution time. This section conveys the design of the experiment, presenting the algorithm, the configurations under which it was executed, selected instances, and research questions.

3.1   Parameterization and Instrumentation

The experiment used a Hill Climbing search to find solutions for the software clustering problem. Given a software system with N modules, the search started by creating N clusters and assigning each module randomly to a single cluster. By "randomly" it is meant that a number was sampled according to the generator under analysis (pseudo-random or quasi-random) and a decision on the cluster to which a given module should be assigned was based on this number. Numbers were sampled in the [0, 1) interval, multiplied by N, and truncated to their integer radix to identify a cluster. Thus, clusters were numbered in the integer interval [0, N-1].

After the initial distribution of modules to clusters was complete, the fitness of the solution (known as *seed solution*) was calculated according to the fitness function under interest (MQ or EVM). The solution was stored as the best solution so far and the main loop of the search followed. The loop attempted to find solutions with better fitness by moving a single module to a distinct cluster in each iteration. The first module was selected and a move to any other cluster than the currently occupied was evaluated. After all clusters were evaluated for the first module, the search followed a similar procedure to the second module, the third, and so on. Whenever a solution with higher fitness than the best known solution was found, the main loop repeated its trials from the first module. If no movement could improve fitness, the search triggered a *random restart* in which a new, random solution (*restart solution*) was created by using the same procedure that built the seed solution. The restart solution was then used as starting point to the main loop. The search stopped when a predefined budget of fitness evaluations was consumed. In our experiments, we have used a budget of 2000 times N^2 evaluations, where N is the number of modules. Except for using random restarts, the search approach followed Mancoridis et al. [16] and the size of the evaluation budget was compatible with further works on software clustering [16][19][21]. The source code and instances used in the experiment are available at http://www.uniriotec.br/~marcio.barros/qrsobol.

The generation of the seed and restart solutions are the only nondeterministic parts of the Hill Climbing search. A single sample from an N-dimensional random number generator (one that produces N random numbers at a request) is required for a Hill Climbing search without random restarts. Thus, random restarts were used to allow a fair comparison of solutions found by driving the search with a quasi-random sequence and those produced by using pseudo-random numbers.

For each fitness function and problem instance under evaluation, the search was executed under two configurations, hereafter referred to as C_S and C_P. Configuration C_S used a Sobol sequence to create the seed and restart solutions. Configuration C_P relied on pseudo-random numbers to create these solutions. We have used JMetal's [6] implementation of a pseudo-random number generator, which is based on Knuth's portable subtractive generator [22]. Configuration C_P represented the most commonly used procedure to setup a Hill Climbing search, serving as baseline for the comparison. Configuration C_S represented potential improvements achieved by using a quasi-random sequence instead of pseudo-random numbers. Our primary objective was to evaluate whether C_S could produce competitive solutions when compared to C_P both in terms of fitness and processing time.

3.2   Problem Instance Selection

The experiment was executed upon 14 (fourteen) real-world instances, all represented as unweighted MDG. We have selected open-source or free-software projects of distinct size, all developed using the Java programming language. We have also used a small IS developed for a Brazilian company (the SEEMP instance). Module dependency data was collected using the PF-CDA open-source static analysis tool. Table 1 presents the characteristics of the instances.

These instances were selected to cover a wide range of software products of different sizes, smoothly covering the space from small systems with less than 30 modules to mid-size systems with about 200 modules. For all analysis further on, a module is a source-code file, possibly conveying more than a single class. Since larger applications are usually divided into smaller deployment units, which can be independently subjected to the clustering process, we believe that the selected instances are representative of real systems. Finally, the concentration on the Java language was due to the availability of an open-source static analysis tool.

Table 1. Characteristics of the instances used in the experiment

Problem Instance	Modules	Dependencies
JODA Money: money management library	26	102
JXLS Reader: library for reading Excel files	27	73
SEEMP: small information system	31	61
Java OCR: written text recognition library	59	155
Java Servlets API	63	131
JXLS Core: library to represent Excel files	83	330
JPassword: password management program	96	361
Java XML DOM Classes	100	276
JUnit: unit testing library	119	209
Tiny TIM: Topic Maps Engine	134	564
Java CC: Yacc for Java	154	722
Java XML API	184	413
JMetal: heuristic search algorithms	190	1137
DOM4J: alternative XML API for Java	195	930

3.3   Data Collection

The configuration using pseudo-random numbers (C_P) was executed 30 times for each instance. Each running cycle resulted in a single best solution whose fitness value was collected. Execution time, represented by the wall-clock time required to run each cycle, was also collected. Thus, a series of 30 observations of fitness value and execution time for solutions found under C_P was recorded for each instance.

The configuration using a Sobol sequence (C_S) was executed a single time for each instance. Since Sobol sequences are deterministic, repeated searches based on them produce exactly the same results and, thus, there is no need to run multiple cycles. For each instance, the search produced a single best solution whose fitness value was collected along with the time required to run the algorithm.

All executions were performed in a computer with an eight-core i7 Intel processor running at 2,600 MHz and 4 gigabytes of RAM memory. The optimization ran over Windows 7 Professional and minimal platform software required to run the Java implementation of the Hill Climbing search.

Configurations were compared on a per instance and per fitness function basis, that is, results produced under C_S for the *dom4j* instance were compared to those found configuration by C_P for the same instance. Smaller execution times for a given configuration indicate that it is more efficient than the other. On the other hand, larger fitness values denote that the configuration produces more effective results than its competing one. These values were subjected to statistical inference tests to ascertain whether there was significant difference between the configurations. Conclusions for the MQ fitness function are limited to unweighted MDG, since no weighted graph was used during the experimental study.

Two research questions (RQ1 and RQ2, stated below) were addressed by our experiment for each fitness function. Section 4 presents the data analysis for the MQ function, evaluating these questions based on data collected after executing the experiment. Section 5 presents a similar analysis for the EVM function.

RQ1: Can a Hill Climbing search using a Sobol sequence produce solutions with higher fitness than the same algorithm powered by pseudo-random numbers?

RQ2: Can a Hill Climbing search using a Sobol sequence run in less wall-clock time than the same algorithm powered by pseudo-random numbers?

4 Results Based on MQ and Analysis

Table 2 summarizes the results of running Hill Climbing searches for the selected instances under configurations C_S and C_P and using Modularization Quality (MQ) as fitness function. The C_S column shows the value of the fitness function for the best solution found using a Sobol sequence (configuration C_S). The next column, *MIN C_P*, shows the MQ value for the best solution with lower fitness found over the 30 cycles which were executed under configuration C_P. Next, the C_P column presents means and standard deviations for fitness value of the best solution found over the cycles executed under C_P. The following column, *MAX C_P*, shows MQ for the fittest solution found over these execution cycles.

Table 2. Fitness values found by the search using Sobol and pseudo-random numbers (for MQ)

	C_S	MIN C_P	C_P	MAX C_P	PV	ES
JODAMONEY	2.73	2.72	2.74 ± 0.01	2.75	0.44	72%
JXLSREADER	3.60	3.59	3.60 ± 0.00	3.60	0.77	43%
SEEMP	4.65	4.62	4.65 ± 0.01	4.65	0.77	43%
JAVAOCR	8.93	8.95	9.00 ± 0.02	9.02	0.09	100%
SERVLETS	9.47	9.48	9.52 ± 0.02	9.55	0.10	100%
JXLSCORE	8.67	8.93	9.15 ± 0.10	9.34	0.06	100%
JPASSWORD	9.49	10.19	10.29 ± 0.04	10.36	0.10	100%
JUNIT	10.33	11.09	11.09 ± 0.00	11.09	< 0.01	100%
XMLDOM	10.26	10.85	10.92 ± 0.02	10.92	0.06	100%
TINYTIM	11.67	12.31	12.38 ± 0.04	12.44	0.06	100%
JAVACC	8.82	10.42	10.49 ± 0.04	10.59	0.06	100%
XMLAPI	16.09	18.77	18.93 ± 0.07	19.02	0.10	100%
JMETAL	8.82	12.33	12.40 ± 0.04	12.50	0.06	100%
DOM4J	14.07	18.28	18.44 ± 0.08	18.62	0.06	100%

Regarding RQ1, modularization quality values under C_S are on average 9% lower than under C_P. This difference substantially decreases for smaller instances: it is on average 3% for instances with up to 100 modules. Moreover, the minimum MQ value under configuration C_P is higher than the respective MQ value under C_S for every

instance except the three smallest ones. Therefore, Hill Climbing searches guided by
MQ and based on a Sobol sequence can neither behave on par nor outperform a similar search powered by pseudo-random numbers, except for small instances.

Besides fitness value ranges, we calculated the p-value for the non-parametric Wilcoxon-Mann-Whitney statistical test (*PV* column) and the effect-size for each instance
(*ES* column). These values were calculated using the R Statistical Computing system
v2.13.0. P-values closer to zero indicate strong confidence that the results being compared are statistically different. Regarding Table 2, p-value is significantly small (under 0.05) for a single instance (*junit*), denoting that the average MQ under C_P is
significantly higher than under C_S with at least 99% confidence for that instance. The
difference is also significant for ten other instances if 90% confidence is acceptable.
These findings enforce the results discussed above: the Sobol-powered search is unable to attain the same effectiveness of the pseudo-random search.

Effect-size measures, such as the non-parametric Vargha and Delaney's $\hat{A}_{12}$ statistics [1] used in Table 2, assess the magnitude of improvement in a pair-wise comparison. Given a measure M for observations collected after applying treatments A and B,
$\hat{A}_{12}$ measures the probability that treatment A yields higher M values than B. If both
treatments are equivalent, then $\hat{A}_{12} = 0.5$; otherwise, it indicates the frequency of improvement. For instance, $\hat{A}_{12} = 0.8$ denotes that higher results would be obtained 80%
of the time with A. Regarding Table 2, effect-sizes show that except for the three
smallest instances 100% of the runs under C_P will produce a solution with higher MQ
than under configuration C_S, adding evidence to the findings discussed above.

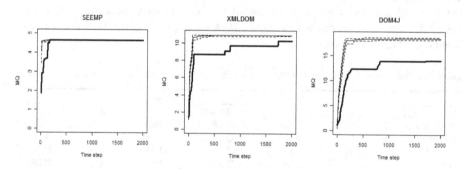

Fig. 2. Evolution profile for three software systems (for MQ)

Figure 2 shows the convergence of the search processes overtime. The x-axis
shows time, represented as a fraction of the fitness evaluation budget consumed up to
each position in the axis. As each instance was given $2000.N^2$ evaluations (N being
the number of modules in the instance), the charts in Figure 2 are represented in terms
of time steps of size N^2. The vertical axis represents fitness value (MQ for Figure 2).
The thick continuous line represents the Sobol search, while the thin continuous line
represents the pseudo-random search, the latter depicting the average value over the
30 execution cycles for each time step. Dotted lines represent both the maximum and
minimum fitness values found for the pseudo-random search over its execution
cycles. Due to space requirements, only three systems are presented in the figure: a

small instance (*SEEMP*, with 31 modules), a medium-sized instance (*xmldom*, with 119 modules), and the largest one (*dom4j*, with 195 modules). The pseudo-random search converges faster to the best observed value for the fitness function, while the Sobol search takes more time, especially for large instances. We ran the Sobol search with a larger fitness evaluation budget and observed that it would require about 5,400 time steps to produce solutions within 1% distance of the fitness value yielded by the pseudo-random search for the *xmldom* instance. For *dom4j*, the maximum fitness value attained by the Sobol search after 8,000 time steps was still 15.25, well below the value produced by the pseudo-random search.

Regarding RQ2, execution time under C_P is on average 1.5% higher than under C_S. Recent advancements have allowed fast generation of Sobol sequences and they can now be sampled as fast as pseudo-random numbers. Sobol sequence generators, on the other hand, tend to consume more memory than pseudo-random generators, given that direction numbers must be read from a file before creating the sequence. Detailed execution times are presented in Table 3, which shows execution time under configuration C_S, means and standard deviations under C_P, p-values and effect-sizes for comparing C_S and C_P, as well as the number of random restarts for each configuration (RR_S for C_S and RR_P for C_P). P-values for the Wilcoxon-Mann-Whitney test are under 10% and effect-sizes are 100% (C_P consumes more time than C_S) for all except the three smallest instances, showing that there is significant difference in execution time in favor of the Sobol search for all but the smallest instances. The latter, on the other hand, are more susceptible to external factors (such as disturbances due to programs and services competing with the optimization process for computing resources) since a complete cycle runs in about a second.

Table 3. Execution time for searches using Sobol and pseudo-random numbers (for MQ)

	C_S	C_P	PV	ES	RR_S	RR_P
JODAMONEY	0.8	0.8 ± 0.01	0.23	13%	132	133.5 ± 2.0
JXLSREADER	0.8	0.9 ± 0.01	0.11	98%	107	103.3 ± 1.6
SEEMP	1.3	1.3 ± 0.01	0.23	13%	130	134.9 ± 1.7
JAVAOCR	13.7	14.1 ± 0.03	0.10	100%	35	42.5 ± 1.0
SERVLETS	17.6	18.0 ± 0.03	0.10	100%	33	45.1 ± 0.8
JXLSCORE	54.3	55.7 ± 0.05	0.10	100%	23	33.4 ± 0.9
JPASSWORD	97.7	99.8 ± 0.13	0.10	100%	17	22.5 ± 0.6
JUNIT	110.6	113.0 ± 0.23	0.10	100%	20	28.8 ± 0.6
XMLDOM	210.2	211.5 ± 0.52	0.10	100%	19	26.4 ± 0.5
TINYTIM	356.7	364.1 ± 0.48	0.10	100%	12	17.1 ± 0.3
JAVACC	625.5	640.3 ± 1.30	0.10	100%	12	15.0 ± 0.4
XMLAPI	1172.9	1184.7 ± 1.6	0.10	100%	11	14.0 ± 0.4
JMETAL	1462.6	1494.9 ± 3.2	0.06	100%	8	9.9 ± 0.4
DOM4J	1614.1	1636.5 ± 1.7	0.06	100%	8	11.6 ± 0.5

5   Results Based on EVM and Analysis

Table 4 summarizes the results of running Hill Climbing searches for the selected instances under configurations C_S and C_P and using EVM as fitness function. The table follows the same structure as Table 2: column C_S shows the EVM value of the best solution found using the Sobol sequence; *MIN C_P* shows the minimum EVM for best solutions found over the 30 cycles which were executed under C_P; column C_P presents means and standard deviations for the best EVM over these cycles; *MAX C_P* shows the maximum EVM value found over the cycles; *PV* shows the p-value for the Wilcoxon-Mann-Whitney test comparing EVM values produced under C_P and C_S; and, finally, ES shows $\hat{A}_{12}$ effect-size on the frequency under which the pseudo-random search results in higher EVM than the Sobol search.

Table 4. Fitness values found by the search using Sobol and pseudo-random numbers (for EVM)

	C_S	MIN C_P	C_P	MAX C_P	PV	ES
JODAMONEY	28	28	28.0 ± 0.0	28	N/A	50%
JXLSREADER	26	26	26.0 ± 0.0	26	N/A	50%
SEEMP	19	19	19.0 ± 0.0	19	N/A	50%
JAVAOCR	46	46	46.0 ± 0.0	46	N/A	50%
SERVLETS	50	50	50.0 ± 0.0	50	N/A	50%
JXLSCORE	87	87	87.8 ± 0.6	89	0.15	88%
JPASSWORD	91	90	91.2 ± 0.7	92	0.71	62%
JUNIT	51	53	53.8 ± 0.6	55	0.06	100%
XMLDOM	42	44	44.9 ± 0.7	46	0.08	100%
TINYTIM	133	133	134.3 ± 0.9	136	0.15	92%
JAVACC	155	152	154.4 ± 1.3	157	0.64	35%
XMLAPI	105	103	105.5 ± 1.3	108	0.73	62%
JMETAL	176	173	176.0 ± 1.5	179	0.95	53%
DOM4J	183	180	182.5 ± 1.2	185	0.60	33%

Regarding RQ1, EVM values under C_S are on average 1% lower than their counterparts under C_P. This difference decreases to zero if only small instances are taken into account: all runs, either Sobol or pseudo-random, produced the same results for the five smallest instances. This is confirmed by data in Table 4 which shows zero standard deviation under C_P and $\hat{A}_{12}$ equal to 50%. However, the analysis becomes less straightforward as we move to larger instances: differences on EVM remain as small as 1.5% for instances with more than 100 modules and results produced by both sequences are very close. The Sobol search slightly outperforms the pseudo-random one for the *javacc* and *dom4j* instances. On the other hand, the pseudo-random search slightly beats Sobol for the *jxlscore, jpassword, xmldom, junit, tinytim, xmlapi,* and *jmetal* instances. Neither of these improvements is statistically significant with 95% or more confidence and only the *jxlscore, xmldom, junit,* and *tinytim* instances present large effect-sizes favoring the pseudo-random search. So, in average pseudo-random search shows higher performance than Sobol search, but evidence is not as strong as for the MQ fitness function. This may be a sign that the fitness landscape for EVM is easier to transverse than MQ's landscape, maybe due to the former's resilience to noise [9], and adds to its usefulness in software module clustering.

Figure 3 shows the evolution profile for the EVM fitness function using the same instances selected in Section 4. By comparing Figure 3 to Figure 2, we observe that the Sobol search performs better with EVM than with MQ, showing rapid convergence even for the largest systems. The first points of the lines representing the Sobol search were hidden in the charts because they are related to very unfit solutions with large, negative EVM. If these values were presented in the charts, particularly for *dom4j*, there would not be enough resolution to observe differences between the Sobol search and the lines representing the pseudo-random search.

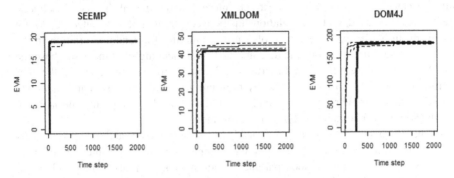

Fig. 3. Evolution profile for three software systems (for EVM)

Table 5. Execution time for searches using Sobol and pseudo-random numbers (for EVM)

	C_S	C_P	PV	ES	RR_S	RR_P
JODAMONEY	0.48	0.5 ± 0.0	0.81	41.6%	176	182.0 ± 1.7
JXLSREADER	0.55	0.5 ± 0.0	0.39	25%	180	185.2 ± 2.1
SEEMP	0.91	0.9 ± 0.0	0.25	15%	232	259.3 ± 3.2
JAVAOCR	11.4	11.3 ± 0.0	0.10	0%	52	74.4 ± 0.8
SERVLETS	14.9	14.6 ± 0.0	0.10	0%	42	79.1 ± 0.8
JXLSCORE	43.0	43.7 ± 0.0	0.10	100%	22	53.7 ± 0.6
JPASSWORD	75.4	77.7 ± 0.1	0.10	100%	16	50.2 ± 0.7
JUNIT	86.1	90.4 ± 0.1	0.10	100%	16	63.6 ± 0.9
XMLDOM	167.8	179.4 ± 1.4	0.10	100%	14	67.9 ± 0.7
TINYTIM	286.0	295.3 ± 0.2	0.10	100%	9	34.1 ± 0.5
JAVACC	478.7	499.8 ± 0.4	0.10	100%	8	26.8 ± 0.4
XMLAPI	970.7	997.4 ± 1.5	0.10	100%	6	30.3 ± 0.5
JMETAL	1133.3	1161.7 ± 1.7	0.10	100%	7	25.9 ± 0.4
DOM4J	1226.7	1284.6 ± 1.7	0.06	100%	5	27.0 ± 0.2

Regarding RQ2, configuration C_S is on average 2% faster than C_P. This result is compatible with the small differences observed under MQ. Table 5 shows information about execution time and random restarts, using the same structure as Table 3. The percentile difference in execution time grows accordingly to instance size (Pearson's linear correlation of 84.6%). Also noticeable, the number of restarts decreases significantly for larger instances, especially for the Sobol search. Linear correlation between the percentile differences in number of restarts between the two searches is inversely proportional to instance size (Pearson 86.6%), showing that the Sobol search restarts much less than the pseudo-random one for large instances. Except for the smallest instances, p-values are equal or below 0.10, showing that there is significant difference in execution time between C_P and C_S for all instances, always favoring C_S. Also, effect-sizes are always 100% for all but the smallest instances, showing that all pseudo-random searches for medium or large instances take more time than the quasi-random search.

6   Threats to Validity

Wohlin et al [26] classify the threats to validity that may affect an experimental study into four categories: conclusion, construct, internal, and external threats. Conclusion threats are concerned with the relationship between treatment and outcome. Major conclusion in SBSE experiments involve not accounting for random variation in the search, poor summarization of the data and lack of a meaningful comparison baseline [18]. These issues were addressed by running 30 cycles for each instance for the configuration depending on randomness, by presenting central and dispersion measures as well as non-parametric p-values and effect sizes, and by comparing the new approach (Sobol) with a traditional one (pseudo-random).

Internal threats evaluate if the relationship between treatment and outcome is causal or results from factors which the researcher cannot control. Major internal threats in SBSE experiments involve poor parameterization, lack of real-world problem instances, and not discussing code instrumentation and data collection procedures. In this paper we have reused parameter values from former experiments addressing the same problem, have used 14 real-world instances of distinct sizes, have described the data collection procedure, and made the source code available for research purposes. Also, results presented for the MQ function are applicable only for unweighted MDG, given that no weighted MDG was evaluated.

Construct threats are concerned with the relation between theory and observation, ensuring that the treatment reflects the construct of the cause and that the outcome reflects the construct of the effect. Major construct threats in SBSE experiments involve using unreliable efficiency and effectiveness measures and not discussing the underlying model subjected to optimization. On regard of these issues, we have worked with a well-known problem presented and discussed in Section 2, and have used two measures (MQ and EVM) that were previously tested and repeatedly used either in SBSE research works or in the broad heuristic clustering literature. One may question whether these metrics represent the perceptions of real software engineering practitioners. Though this is an important question for automated clustering, such analysis if out of the scope of the present research. Finally, using wall-clock time as a measure for efficiency is questionable, but all experiments were executed in similar computers with compatible load.

Finally, external threats are concerned with the generalization of the observed results to a larger population, outside the sample instances used in the experiment. Major external threats to SBSE experiments include the lack of a clear definition of target instances, unclear instance selection strategy, and not having enough diversity in instance size and complexity. We have used 14 instances of different sizes (in terms of number of modules) and complexity (in terms of number of dependencies). These instances were described in section 3.2 and are available for download and further evaluation along with the source code. Finally, one may argue that all instances are restricted to the Java language and so conclusions are restricted to this language. Although we cannot dismiss completely this possibility, the proposed approach to clustering does not dependence on a particular programming language, relying on it only to identify static dependencies between source code modules.

7   Conclusions

This paper evaluated the efficiency and effectiveness of using two distinct number sequences (*pseudo-random* and *quasi-random*) to build the seed and restart solutions used by a Hill Climbing search with random restarts that addressed the software clustering problem. An experiment using two different fitness functions and fourteen real-world instances was designed to evaluate whether one sequence would produce better solutions and require less processing time than the other. Contrary to previous findings related to other kinds of heuristic search algorithms, we found that the quasi-random search (using a Sobol sequence) could not attain the same effectiveness as the pseudo-random search, particularly for the MQ fitness function. Though it was slightly more efficient than the pseudo-random search, the Sobol search produced solutions with in average 10% (for MQ) or 1% (for EVM) lower fitness than the pseudo-random search. Therefore, at least in the context of Hill Climbing searches with random restart applied to the software module clustering problem, quasi-random numbers cannot successfully replace pseudo-random numbers. Future extensions of this work might consider addressing the usage of other types of quasi-random sequences (such as the alternatives to Sobol presented in Section 2.1) and other heuristic algorithms (such as genetic algorithms or simulated annealing).

Acknowledgments. The author would like to express his gratitude for FAPERJ and CNPq, the research agencies that financially supported this project.

References

1. Arcuri, A., Briand, L.: A Practical Guide for Using Statistical Tests to Assess Randomized Algorithms in Software Engineering. In: Proc. of the 33th International Conference on Software Engineering, ICSE 2011, Hawaii, EUA (2011)
2. Barros, M.O.: An Analysis of the Effects of Composite Objectives in Multiobjective Software Module Clustering. In: Proceedings of the Genetic and Evolutionary Computing Conference (GECCO 2012), Philadelphia, USA (2012)
3. Barros, M.O., Dias-Neto, A.C.: Threats to Validity in Search-based Software Engineering Empirical Studies, Technical Report DIA/UNIRIO, No. 6, Rio de Janeiro, Brazil (2011), http://www.seer.unirio.br/index.php/monografiasppgi/article/viewFile/1479/1307
4. Briand, L.C., Morasca, S., Basili, V.R.: Defining and Validating Measures for Object-based High-Level Design. IEEE Trans. on Software Engineering 25(5) (1999)
5. Doval, D., Mancoridis, S., Mitchell, B.: Automatic Clustering of Software Systems using a Genetic Algorithm. In: Proc. of the International Conference on Software Tools and Engineering Practice, STEP 1999 (1999)
6. Durillo, J.J., Nebro, A.J., Luna, F., Doronsoro, B., Alba, E.: JMetal: A Java Framework for Developing Multi-objective Optimization Metaheuristics. TR ITI-2006-10, Dept. de Lenguajes y Ciencias de Computacion, Univ. Málaga (2006)
7. Georgieva, A., Jordanov, I.: Global optimization based on novel heuristics, low-discrepancy sequences and genetic algorithms. European Journal of Operational Research (196), 413–422 (2009)

8. Gibbs, S., Tsichritzis, D., et al.: Class Management for Software Communities. Communications of the ACM 33(9), 90–103 (1990)
9. Harman, M., Swift, S., Mahdavi, K.: An Empirical Study of the Robustness of two Module Clustering Fitness Functions. In: Proceedings of the Genetic and Evolutionary Computing Conference (GECCO 2005), Washington DC, USA (2005)
10. Harman, M., Masouri, S.A., Zhang, Y.: Search Based Software Engineering: A Comprehensive Analysis and Review of Trends Techniques and Applications, Dept. of Computer Science, King's College London, Technical Report TR-09-03 (April 2009)
11. Kimura, S., Matsumura, K.: Genetic Algorithms using Low-Discrepancy Sequences. In: Proceedings of GECCO 2005, Washington DC, USA (2005)
12. Larman, C.: Applying UML and Patterns: An Introduction to Object-Oriented Analysis and the Unified Process. Prentice Hall, Upper Saddle River (2002)
13. Levy, G.: An introduction to quasi-random numbers. Numerical Algorithms Group Ltd., http://www.nag.co.uk/IndustryArticles/introduction_to_quasi_random_numbers.pdf (last accessed in April 10, 2012)
14. Maaranen, H., Miettinen, K., Makela, M.M.: Quasi-Random Initial Population for Genetic Algorithms. Computers and Mathematics w/ Applications (47), 1885–1895 (2004)
15. Mahdavi, K., Harman, M., Hierons, R.M.: A Multiple Hill Climbing Approach to Software Module Clustering. In: Proceedings of the International Conference on Software Maintenance, Amsterdan, pp. 315–324 (2003)
16. Mancoridis, S., Mitchell, B.S., Chen, Y., Gansner, E.R.: Bunch: A Clustering Tool for the Recovery and Maintenance of Software System Structures. In: Proceedings of the IEEE International Conference on Software Maintenance, pp. 50–59 (1999)
17. McConnell, S.: Code Complete, 2nd edn. Microsoft Press (2004)
18. Morokoff, W.J., Caflish, R.E.: Quasi-random Sequences and their Discrepancies. SIAM Journal of Scientific Computing 15(6), 1251–1279 (1994)
19. Niederreiter, H.G.: Quasi-Monte Carlo methods and pseudo-random numbers. Bulletin of the American Mathematical Society 84(6), 957–1041 (1978)
20. Pant, M., Thangaraj, R., Grosan, C., Abraham, A.: Improved Particle Swarm Optimization with Low-Discrepancy Sequences. In: Proceedings of the IEEE Congress on Evolutionary Computation (CEC 2008), Hong Kong, pp. 3011–3018 (2008)
21. Praditwong, K., Harman, M., Yao, X.: Software Module Clustering as a Multiobjective Search Problem. IEEE Transactions on Software Engineering 37(2), 262–284 (201)
22. Press, W.H., Teukolsky, S.A., Vetterling, W.T., Flannery, B.P.: Numerical Recipes: The Art of Scientific Computing, 2nd edn. Cambridge University Press, NY (1992)
23. Räihä, O.: A Survey on Search-Based Software Design. Technical Report D-2009-1, Department of Computer Sciences University Of Tampere (March 2007)
24. Thangaraj, R., Pant, M., Abraham, A., Badr, Y.: Hybrid Evolutionary Algorithm for Solving Global Optimization Problems. In: Corchado, E., Wu, X., Oja, E., Herrero, Á., Baruque, B. (eds.) HAIS 2009. LNCS, vol. 5572, pp. 310–318. Springer, Heidelberg (2009)
25. Tucker, A., Swift, S., Liu, X.: Grouping Multivariate Time Series via Correlation. IEEE Transactions on Systems, Man, & Cybernetics, Part B: Cybernetics 31(2), 235–245 (2001)
26. Wohlin, C., Runeson, P., Höst, M., Ohlsson, M., Regnell, B., Wesslén, A.: Experimentation in Software Engineering. Kluwer Academic Publishers, Norwell (2000)
27. Yourdon, E., Constantine, L.L.: Structured Design: Fundamentals of a Discipline of Computer Program and Systems Design. Yourdon Press (1979)

Putting the Developer in-the-Loop: An Interactive GA for Software Re-modularization

Gabriele Bavota[1], Filomena Carnevale[1], Andrea De Lucia[1],
Massimiliano Di Penta[2], and Rocco Oliveto[3]

[1] University of Salerno, Via Ponte don Melillo, 84084 Fisciano (SA), Italy
[2] University of Sannio, Palazzo ex Poste, Via Traiano, 82100 Benevento, Italy
[3] University of Molise, Contrada Fonte Lappone, 86090 Pesche (IS), Italy
{gbavota,adelucia}@unisa.it, flmn.carnevale@gmail.com,
dipenta@unisannio.it, rocco.oliveto@unimol.it

Abstract. This paper proposes the use of Interactive Genetic Algorithms (IGAs) to integrate developer's knowledge in a re-modularization
task. Specifically, the proposed algorithm uses a fitness composed of
automatically-evaluated factors—accounting for the modularization
quality achieved by the solution—and a human-evaluated factor, penalizing cases where the way re-modularization places components into
modules is considered meaningless by the developer.

The proposed approach has been evaluated to re-modularize two software systems, SMOS and GESA. The obtained results indicate that IGA
is able to produce solutions that, from a developer's perspective, are more
meaningful than those generated using the full-automated GA. While
keeping feedback into account, the approach does not sacrifice the modularization quality, and may work requiring a very limited set of feedback
only, thus allowing its application also for large systems without requiring a substantial human effort.

1 Introduction

Software is naturally subject to change activities aiming at fixing bugs or introducing new features. Very often, such activities are conducted within a very
limited time frame, and with a limited availability of software design documentation. Change activities tend to "erode" the original design of the system. Such
a design erosion mirrors a reduction of the cohesiveness of a module, the increment of the coupling between various modules and, therefore, makes the system
harder to be maintained or, possibly, more fault-prone [8]. For this reason, various automatic approaches, aimed at supporting source code re-modularization,
have been proposed in literature (see e.g., [11,14,20]). The underlying idea of
such approaches is to (i) group together in a module highly cohesive source code
components, where the cohesiveness is measured in terms of intra-module links;
and (ii) reduce the coupling between modules, where the coupling is measured
in terms of inter-module dependencies. Such approaches use various techniques,

G. Fraser (Ed.): SSBSE 2012, LNCS 7515, pp. 75–89, 2012.
© Springer-Verlag Berlin Heidelberg 2012

such as clustering [1,11,20], formal concept analysis [15] or search-based optimization techniques [13,14] to find (near) optimal solutions for such objectives, e.g., cohesion and coupling.

While automatic re-modularization approaches proved to be very effective to increase cohesiveness and reduce coupling of software modules, they do not take into account developers' knowledge when deciding to group together (or not) certain components. For example, a developer may decide to place a function in a given module even if, in its current implementation, the function does not communicate a lot with other functions in the same module. This is because the developer is aware that, in future releases, such a function will strongly interact with the rest of the module. Similarly, a developer may decide that two functions must be placed in two different modules even if they communicate. This is because the two functions have different responsibilities and are used to manage semantically different parts of the system. In the past, some authors proposed approaches to account for developers' knowledge in software re-modularization [6]. However, such approaches assume the availability of a whole set of constraints before the re-modularization starts. This is often difficult to be achieved, especially for very large systems.

This paper proposes the use of Interactive Genetic Algorithms (IGAs) [17] to integrate, into a re-modularization approach, a mechanism allowing developers to feed-back automatically produced re-modularizations. IGAs are a variant of Genetic Algorithms (GAs) in which the fitness function is partially or entirely evaluated by a human while the GA evolves. Recently, IGAs have been applied to software engineering problems such as requirement prioritization [18] or upstream software design [16]. In our approach, part of the fitness (capturing aspects such as intra-module, extra-module dependencies, or modularization quality) is automatically evaluated, while the human adds penalties for artifacts that are not where they should be. Summarizing, the specific contributions of the paper are:

1. Different variants of IGAs, allowing the integration of feedback provided by developers upon solutions produced during the GA evolution. Specifically, the paper presents the integration of feedback in both a single-objective GA, using the Modularization Quality (MQ) measure [13], and a multi-objective GA proposed by Praditwong et al. [14].

2. The empirical evaluation of the proposed IGAs over two software systems. Although IGAs are conceived to allow a "live" feedback seeding, in this paper we simulated such a mechanism using constraints randomly identified from the actual system design. Results indicate that the IGAs are able to produce re-modularizations that better reflect the developer intents, without however sacrificing the modularization quality.

The paper is organized as follows. Section 2 describes the related work, while Section 3 describes the proposed IGA-based re-modularization. Section 4 reports the empirical study conducted to evaluate the proposed approach, while Section 5 concludes the paper and outlines directions for future work.

2   Background and Related Work

Several approaches have been proposed in the literature to support software re-modularization. Promising results have been achieved using clustering algorithms [1,11,20] and formal concept analysis [15]. In the following we focus only on search-based approaches.

Mancoridis *et al.* [10] introduce a search-based approach using hill-climbing based clustering to identify the modularization of a software system. This technique is implemented in Bunch [13], a tool supporting automatic system decomposition. To formulate software re-modularization as a search problem, Mancoridis *et al.* define (i) a representation of the problem to be solved (i.e., software module clustering) and (ii) a way to evaluate the modularizations generated by the hill-climbing algorithm. Specifically, the system is represented by the Module Dependency Graph (MDG), a language independent representation of the structure of the code components and relations [10]. The MDG can be seen as a graph where nodes represent the system entities to be clustered and edges represent the relationships among these entities. An MDG can be weighted (i.e., a weight on an edge measures the strength of the relationship between two entities) or unweighted (i.e., all the relationships have the same weight).

Starting from the MDG (weighted or unweighted), the output of a software module clustering algorithm is represented by a partition of this graph. A good partition of an MDG should be composed by clusters of nodes having (i) high dependencies among nodes belonging to the same cluster (i.e., high cohesion), and (ii) few dependencies among nodes belonging to different clusters (i.e., low coupling). To capture these two desirable properties of the system decompositions (and thus, to evaluate the modularizations generated by Bunch) Mancoridis *et al.* [10] define the Modularization Quality (MQ) metric as:

$$MQ = \begin{cases} (\frac{1}{k}\sum_{i=1}^{k} A_i) - (\frac{1}{\frac{k(k-1)}{2}}\sum_{i,j=1}^{k} E_{i,j}) & \text{if } k > 1 \\ A_1 & \text{if } k = 1 \end{cases}$$

where A_i is the Intra-Connectivity (i.e., cohesion) of the ith cluster and $E_{i,j}$ is the Inter-Connectivity (i.e., coupling) between the ith and the jth clusters. The Intra-Connectivity is based on the number of intra-edges, that is the relationships (i.e., edges) existing between entities (i.e., nodes) belonging to the same cluster, while the Inter-Connectivity is captured by the number of inter-edges, i.e., relationships existing between entities belonging to different clusters.

Single-objective genetic algorithms have been used to improve the subsystem decomposition of a software system by Doval *et al.* [7]. The objective function is defined using a combination of quality metrics, e.g., coupling, cohesion, and complexity. However, hill-climbing have been demonstrated to ensure higher quality and more stable solutions than a single objective genetic algorithm [12]. Praditwong *et al.* [14] introduce two multi-objective formulations of the software re-modularization problem, in which several different objectives are represented separately. The two formulations slightly differ for the objectives embedded in the multi-objective function. The first formulation–named Maximizing Cluster

Approach (MCA)—has the following objectives: (i) maximizing the sum of intra-edges of all clusters, (ii) minimizing the sum of inter-edges of all clusters, (iii) maximizing MQ, (iv) maximizing the number of clusters, and (v) minimizing the number of isolated clusters (i.e., clusters composed by only one class). The second formulation—named Equal-Size Cluster Approach (ECA)—attempts at producing a modularization containing clusters of roughly equal size. Its objectives are exactly the same as MCA, except for the fifth one (i.e., minimizing the number of isolated clusters) that is replaced with (v) minimizing the difference between the maximum and minimum number of entities in a cluster. The authors compared their algorithms with Bunch. The conducted experimentation provides evidence that the multi-objective approach produces significantly better solutions than the existing single-objective approach though with a higher processing cost.

Based on the results achieved by Praditwong *et al.* [14], this paper defines an interactive version of the single-objective GA and of the multi-objective GA (based on the MCA algorithm). This allows to analyze the benefits provided by developers' feedback to solve a re-modularization problem.

3   The Proposed Interactive Genetic Algorithms

This section describes the IGA we use to integrate software engineers' feedback into the single-objective [10] and multi-objective [14] re-modularization process.

3.1   Solution Representation, Operators, and Fitness Function

The solution representation (chromosome) and GA operators are the same for both single- and multi-objective GAs. Given a software system composed of n software components (e.g., classes) the chromosome is represented as a n-sized integer array, where the value $0 < v \leq n$ of the ith element indicates the cluster which the ith component is assigned. A solution with the same value (whatever it is) for all elements means that all software components are placed in the same cluster, while a solution with all possible values (from 1 to n) means that each cluster is composed of one component only.

The crossover operator is a one-point crossover, while the mutation operator randomly identifies a gene (i.e., a position in the array), and modifies it by assigning to it a random value $0 < v \leq n$. This means moving a component to cluster v. The selection operator is the roulette-wheel selection.

The single-objective GA uses as fitness function (to be maximized) the MQ metric, while as said in Section 2 the multi-objective GA—implemented as Non-Dominating Sorting Genetic Algorithm (NSGA-II) [5]—considers five different objectives, related to maximizing MQ, intra-cluster connectivity and number of clusters, and minimizing the inter-cluster connectivity and the number of isolated clusters.

Algorithm 1. R-IGA: IGA for providing feedback about pairs of components.

1: **for** $i = 1 \ldots nInteractions$ **do**
2:    Evolve GA for $nGens$ generations
3:    Select the solution having the highest MQ
4:    **for** $j = 1 \ldots nFeedback$ **do**
5:        Randomly select two components c_i and c_j
6:        Ask the developer whether c_i and c_j must go together or kept separate
7:    **end for**
8:    Repair the solution to meet the feedback
9:    Create a new GA population using the repaired solution as starting point
10: **end for**
11: Continue (non-interactive) GA evolution until it converges or it reaches $maxGens$

3.2   Single-Objective Interactive GAs

The basic idea of the IGA is to periodically add a constraint to the GA such that some specific components shall be put in a given cluster among those created so far. Thus, the IGA evolves exactly as the non-interactive GA. Then, every $nGens$ generations, the best individual is selected and shown to the developer. Then, the developer analyzes the proposed solutions and provides feedback (which can be seen as constraints to the re-modularization problem), indicating that certain components shall be moved from a cluster to another. After enacting the developer's indications, a new GA population is created from such a best solution, and then the GA evolves for further $nGens$ generations, keeping into account the provided constraints. One crucial point is choosing how to guide the developer to provide feedback. In principle, one could ask developers any possible kind of feedback. However, this would make the developer's task quite difficult. For this reason, we propose to guide developers in providing feedback, by means of two different kinds of IGAs. The first one—referred to as R-IGA and described by Algorithm 1—takes the best solution produced by the GA, randomly selects two components (from the same cluster or from different clusters), and then asks the developer whether, in the new solutions to be generated, such components must be placed in the same cluster (i.e., stay together) or whether they should be kept separated.

As the algorithm indicates, every $nGens$ generations the developer is asked to provide feedback about a number $nFeedback$ of component pairs from the best solution (in terms of MQ) contained in the current population. The feedback can either be (i) "c_i and c_j shall stay together" or (ii) "c_i and c_j shall be kept separate". After feedback is provided, the solution is repaired by enforcing the constraints, e.g., by randomly moving one of c_i and c_j away if the constraint tells that they shall be kept separate. After all $nFeedback$ have been provided, a new population is created by randomly mutating such a repaired solution. Then, the GA starts again. When creating the new population and when evolving it, the GA shall ensure that the new produced solutions meet the feedback collected so far. Hence, we add a penalty factor to the fitness function (as proposed by Coello Coello [3]), aiming at penalizing solutions violating the constraints imposed by the developers. Given $CS \equiv cs_1, \ldots cs_m$ the set of feedback collected by the users, the fitness $F(s)$ for a solution s is computed as follows:

$$F(s) = \frac{MQ(s)}{1 + k \cdot \sum_{i=1}^{m} vcs_{i,s}}$$

where $k > 0$ is an integer constant weighting the importance of the feedback penalty, and $vcs_{i,s}$ is equal to one if solution s violates cs_i, zero otherwise. After *nInteractions* have been performed, the GA continues its evolution in a non-interactive way until it reaches stability or the maximum number of generations. One consideration needs to be made about the selection of the pairs for which asking feedback. While in our experiments the selection is random (see Section 4), in a realistic scenario the developer could pick component pairs based on her knowledge, or else further heuristics could be used for such purposes.

The second IGA we propose—called IC-IGA and described by Algorithm 2—focuses on specific parts of the re-modularization produced by the GA. Among others, very small clusters should be subject to manual changes by the developer. In fact, automatic re-modularization approaches often tend to create a large number of many small clusters, that seldom reflect the actual or desired system decomposition. For this reason, the second variant of our IGA asks feedback on the *nClusters* smallest clusters in the best solution (in terms of MQ). Then, for each of these clusters, if it is an isolated cluster (i.e., composed of one component only), the developer is asked to specify a different cluster where the isolated component must be placed while for not isolated clusters the developer is asked to specify for each pair of components whether they must stay together or not. It is worth noting that the developer does not specify the cluster, she rather indicates whether, when moving such components to a different (randomly selected) cluster, they should be moved together or it must be made sure they are kept separate. Besides the nature of the collected feedback, it works similarly to R-IGA (the fitness function does not change).

Clearly, several other kinds of heuristics could be used to ask feedback to the developer (e.g., a combination of the two approaches presented in this paper, with feedback required on both random couples of elements and on elements belonging to small clusters). However, this is out of the scope of this paper.

3.3   Multi-objective Interactive GAs

The multi-objective variants of our IGA are quite similar to the single-objective ones. Also in this case, we propose one—referred as R-IMGA—where feedback is provided on randomly selected pairs of components, and one—referred as IC-IMGA—where feedback is provided on components belonging to isolated (or smallest) clusters.

A crucial point in the multi-objective variant is the selection of the best individual for which the developer shall provide feedback and, after applying the feedback, to be used for generating the new GA population. The single-objective GA selects individual having the highest MQ, i.e., the highest fitness value. As for

---

Algorithm 2. IC-IGA: IGA for handling small and isolated clusters

---

1: **for** $i = 1 \ldots nInteractions$ **do**
2:     Evolve the GA for $nGens$ generations
3:     Select the solution having the highest MQ
4:     Find the $nClusters$ smallest clusters and store them in C
5:     **for all** $c_i \in C$ **do**
6:         **if** c_i is an isolated cluster **then**
7:             Specify the cluster where the component must be placed
8:         **else**
9:             Specify for each component pair whether the components must stay together or not
10:         **end if**
11:         Repair the solution to meet the feedback
12:         Create a new GA population using the repaired solution as starting point
13:     **end for**
14: **end for**
15: Continue (non-interactive) GA evolution until it converges or it reaches $maxGens$

---

the multi-objective GA, we again select solutions with the highest MQ, although they might not be the ones with the highest values for the other objectives. The motivation is similar to the one of Praditwong *et al.* [14], which used MQ to select the best solution in the NSGA-II Pareto fronts. While other objectives are useful to drive the population evolution, MQ is a measure that characterize the "overall" quality of a re-modularization solution, thus can be used to select—among a set of Pareto-optimal solutions—the one that would likely better pursue the re-modularization objectives. Finally, to ensure that the new produced solutions meet the feedback provided by the developer, also for the multi-objective GA we add a penalty factor to each fitness function following the same approach adopted for the single-objective GA.

4   Empirical Evaluation

This section reports the design and the results of the study we conducted to compare the different variants of IGAs with their non-interactive counterparts in the context of software re-modularization. The experimentation has been carried out on an industrial project, namely GESA, and on a software system, SMOS, developed by a team of Master students at the University of Salerno (Italy) during their industrial traineeship. GESA automates the most important activities in the management of University courses, like timetable creation and classroom allocation. It is operational since 2007 at the University of Molise (Italy)[1]. SMOS is a software developed for high schools, and offers a set of features aimed at simplifying the communications between the school and the student's parents. Table 1 reports the size, in terms of KLOC, number of classes, and number of packages, and the versions of the object systems. The table also reports the

---

[1] http://www.distat.unimol.it/gesa/

Table 1. Characteristics of the software systems used in the case study

System	KLOC	#Classes	#Packages	Quality of modularization			
				Isolated clusters	Intra-edges	Inter-edges	MQ
GESA 2.0	46	297	22	1	330	4,472	2.78
SMOS 1.0	23	121	12	2	155	1,158	2.18

values of some metrics (e.g., MQ) to measure the modularization quality of the systems. Such metrics are also used as fitness functions in the implementation of the different variants of GAs (interactive and not).

4.1   How Is Feedback Provided?

Since we are not performing a user study, we simulated the developer by automatically generating feedback extracted from the original design of the object systems (similarly to [18]). This means that every time the IGA asks whether two classes must go together or be kept separate (or it asks to specify the cluster where the class must be placed) a tool simulated the developer response by finding the answer in the original design. We believe this feedback would be representative of an expert's behavior, since the two object systems have a good modularization quality and have been previously used as gold standard to test other re-modularization approaches [2].

4.2   Study Planning and Analysis Method

Our study aims at answering the following research question:

> **RQ**: *How do IGAs perform—compared to non-interactive GAs—in terms of quality and meaningfulness of the produced re-modularizations?*

To answer this research question, we compare the modularizations achieved with the different variants of IGAs (R-IGA, IC-IGA, R-IMGA, and IC-IMGA) with those achieved applying canonical GA and MGA. For each algorithm, an initial population is randomly generated. All the algorithms have been executed 30 times on each object system to account the inherent randomness of GAs. For all the algorithms we used the same configuration. We calibrated the various parameters of the GA similarly to what done by Praditwong *et al.* [14], by properly adapting some of them to the characteristics of our object systems. Other calibration—mainly related to the number of times the GA interacts with the software engineer, the number of feedback provided, and the number of generations between one interaction and the subsequent one—were calibrated by trial-and-error for our object systems. Therefore, a proper calibration might be needed for larger systems. Specifically:

- We use a population size of n individuals for systems having $n > 150$ software components, a population of $2 \cdot n$ for smaller systems. Praditwong *et al.* [14] used a population size of $10 \cdot n$ instead, because the bigger system used in their study was considerably smaller than the one used in our study (i.e., GESA with 297 classes). Similarly, we consider a maximum number of generations $maxGen$ equal to $20 \cdot n$ for systems having $n > 150$ components, and equal to $50 \cdot n$ for smaller systems. Also for the maximum number of generations, Praditwong $et\ al.$ used in their experimentation a higher number (i.e., $200 \cdot n$).

- The crossover probability is set to 0.8, while the mutation probability to $0.004 \cdot log_2(n)$ (as also done by Praditwong $et\ al.$ [14]).
- The number of times ($nInteractions$) the GA stops the evolution and asks for an interaction is set to 5;
- The number of generations ($nGens$) between one interaction and the subsequent one is set to 10;
- The number of class pairs ($nFeedback$) for which Algorithm 1 asks the developer for a feedback every time is set to 3;
- The number of small/isolated clusters ($nClusters$) for which Algorithm 2 asks the developer for a feedback every time is set to 3;
- The weight k of the penalty factor in the fitness functions is set to 1.

As it can be noticed, the maximum number of feedback provided is equal to $5 \cdot 3 = 15$ class pairs for Algorithm 1 and $c \cdot 3$ (where c is the number of analyzed classes) for Algorithm 2. This might appears as a small amount of feedback, compared to the total number of classes of the object systems. However, in this context we are interested to evaluate how the IGA would have worked with a limited—i.e., cheap and feasible for the developer—interaction.

One parameter used by all the algorithms is the maximum number of clusters to extract. To the best of our knowledge this parameter has not been described in previous works. However, it is crucial for the setting of a GA. Defining the maximum number of clusters $a\ priori$ is not a trivial task. Such a number highly depends by the system under analysis. In this paper we experimented two different heuristics. Specifically, we set the maximum number of clusters to n and $n/2$, respectively, where n is the number of classes in the system to be re-modularized.

When analyzing results, we compare the ability of GAs and IGAs to reach a fair trade-off between the optimization of some quality metrics (that is the main objective of GAs applied to software re-modularization) and the closeness of the proposed partitions to an authoritative one (and thus, their meaningfulness). Note that we use the original design of the object systems as authoritative partition. This choice is justified by the good modularization quality of the object systems, that have been previously used as gold standard to assess other re-modularization approaches [2]. On the one hand, to analyze the impact of provided feedback from the quality metrics point-of-view, we use four quality metrics previously adopted by Praditwong $et\ al.$ [14], namely MQ, intra-edges, inter-edges, and number of isolated clusters. On the other hand, to measure the meaningfulness of the modularizations proposed by the experimented algorithms, we compute the MoJo eFfectiveness Measure (MoJoFM) [19] between the original modularization of the object systems (authoritative partitions) and that proposed by the algorithms. The MoJoFM is a normalized variant of the MoJo distance and it is computed as follows:

$$MoJoFM(A, B) = 100 - (\frac{mno(A, B)}{max(mno(\forall A, B))} \times 100)$$

where $mno(A, B)$ is the minimum number of *Move* or *Join* operations one needs to perform in order to transform the partition A into B, and $max(mno(\forall A, B)$ is the maximum possible distance of any partition A from the gold standard partition B. Thus, $MoJoFM$ returns 0 if a clustering algorithm produces the farthest partition away from the gold standard; it returns 100 if a clustering algorithm produces exactly the gold standard.

We also statistically analyze whether the results achieved by different algorithms (interactive and not) significantly differ in terms of quality metrics or authoritativeness of the modularizations. In particular, the values of all the employed metrics (e.g., MQ) achieved in the 30 runs by two algorithms are statistically compared using the *Mann-Whitney* test [4]. In all our statistical tests we reject the null hypotheses for p-values < 0.05 (i.e., we accept a 5% chance of rejecting a null hypothesis when it is true [4]). We also estimate the magnitude of the difference between the employed metrics. We use the Cliff's Delta (or d), a non-parametric effect size measure [9] for ordinal data. The effect size is small for $d < 0.33$ (positive as well as negative values), medium for $0.33 \leq d < 0.474$ and large for $d \geq 0.474$ [9].

4.3   Analysis of Results

This section discusses the results achieved in our study aiming at responding to our research question. Working data sets are available for replication purposes[2].

Tables 2 and 3 report the descriptive statistics of the measured quality metrics. The analysis of Table 2 highlights that on GESA the interactive GAs, in most cases, achieve better results than their non-interactive counterparts. Note that this is true (i) for both single- and multi-objective algorithms, (ii) using both n or $n/2$ as maximum number of clusters, and (iii) using both the random (R-IGA and R-IMGA) or the isolated cluster (IC-IGA and IC-IMGA) heuristic to provide feedback to the interactive algorithm; only R-IGA and R-IMGA for GESA go slightly worse than their non-interactive counterparts. These better performances hold for all the exploited quality metrics. Moreover, the number of clusters generated by the interactive algorithms is much more close to the effective number of clusters of the original system decomposition (i.e., 22). The non-interactive GA and MGA always propose modularizations having a very high number of clusters, mostly composed of few classes. For example, GA[n] organizes the 297 classes of GESA into (on average) 149 clusters, 72 of which (on average) are isolated clusters, i.e., clusters containing only one class. The best configuration is MGA[n/2] that, however, still produces a quite fragmented system decomposition, clustering the 297 classes into an average of 108 clusters, 32 of which isolated. Even if this kind of systems decomposition could (near)

---

[2] http://www.distat.unimol.it/reports/IGA-remodularization/

Table 2. GESA: descriptive statistics of the measured quality metrics

Algorithm	#Clusters			#Isolated Clusters			intra-edges			inter-edges			MQ		
	Mean	Med.	St. Dev.	Mean	Med.	St. Dev.	Mean	Med.	St. Dev.	Mean	Med.	St. Dev.	Mean	Med.	St. Dev.
GA[n]	149	159	5	72	72	7	62	63	5	5,009	5,007	10	3.94	3.95	0.22
IC-IGA[n]	79	72	26	22	12	19	421	439	177	4,290	4,254	353	4.86	4.97	0.36
R-IGA[n]	113	110	11	57	56	6	282	316	100	4,568	4,500	201	4.64	4.60	0.36
MGA[n]	155	156	7	83	84	9	36	34	9	5,060	5,064	17	2.04	2.00	0.34
IC-IMGA[n]	90	69	31	25	9	27	296	346	131	4,539	4,440	262	4.19	4.44	0.55
R-IMGA[n]	155	154	10	91	89	13	58	57	14	5,017	5,018	27	2.14	2.06	0.31
GA[n/2]	107	107	3	28	28	5	79	80	4	4,975	4,973	8	3.99	4.04	0.23
IC-IGA[n/2]	63	59	15	11	10	6	298	299	84	4,536	4,534	168	4.81	4.92	0.39
R-IGA[n/2]	88	89	8	29	29	5	359	315	193	4,415	4,502	386	4.65	4.71	0.29
MGA[n/2]	108	108	5	32	32	6	58	57	13	5,016	5,018	26	2.23	2.20	0.32
IC-IMGA[n/2]	59	51	26	10	6	10	324	329	183	4,485	4,475	366	3.97	4.09	0.61
R-IMGA[n/2]	111	112	4	41	40	6	111	104	35	4,911	4,924	71	2.12	2.08	0.22

Table 3. SMOS: descriptive statistics of the measured quality metrics

Algorithm	#Clusters			#Isolated Clusters			intra-edges			inter-edges			MQ		
	Mean	Med.	St. Dev.	Mean	Med.	St. Dev.	Mean	Med.	St. Dev.	Mean	Med.	St. Dev.	Mean	Med.	St. Dev.
GA[n]	54	54	3	22	21	4	52	52	4	1,364	1,365	7	3.40	3.45	0.21
IC-IGA[n]	35	32	9	9	7	7	86	94	20	1,297	1,281	39	3.67	3.77	0.62
R-IGA[n]	42	41	4	15	15	4	76	77	10	1,316	1,315	19	3.78	3.77	0.22
MGA[n]	63	64	4	32	31	6	30	29	7	1,408	1,411	13	1.91	1.88	0.31
IC-IMGA[n]	37	35	7	10	8	6	82	86	20	1,305	1,296	41	2.66	2.68	0.39
R-IMGA[n]	63	64	4	37	37	5	45	45	10	1,379	1,378	20	1.97	1.95	0.30
GA[n/2]	42	42	3	11	11	3	62	63	4	1,344	1,342	8	3.68	3.68	0.21
IC-IGA[n/2]	20	18	7	1	1	1	114	118	29	1,240	1,233	58	3.00	2.84	0.36
R-IGA[n/2]	29	29	4	8	7	3	160	154	44	1,148	1,160	87	3.68	3.69	0.21
MGA[n/2]	46	47	4	15	15	3	50	47	14	1,368	1,375	28	2.10	2.04	0.31
IC-IMGA[n/2]	17	15	8	2	2	2	244	257	115	980	954	229	2.64	2.57	0.30
R-IMGA[n/2]	47	46	3	19	18	5	74	67	27	1,320	1,335	53	2.07	2.08	0.24

optimize some quality metrics, it represents a poor support for the developer during a software re-modularization task, requiring a substantial effort to manually refine the proposed modularization.

Results obtained for SMOS (Table 3), confirm what is seen for GESA. In addition, also on SMOS, the non-interactive GAs provide quite poor performances, especially in terms of number of clusters produced. Also in this case, the best configuration is obtained with MGA[n/2], that organizes the 121 classes of SMOS in 46 clusters (on average). Among them, 15 are isolated. Its interactive version (IC-IMGA[n/2]) is able to produce a more reasonable partition, composed of 17 clusters, of which only 2 are isolated.

All these considerations are also supported by statistical analyses (see Tables 4 and 5). Therefore, we can conclude that IGAs achieve better quality metrics value as compared to the non-interactive counterparts. This result is quite surprising, since the feedback provided to the IGAs in our experimentation are not targeted to improve some quality metrics (as the fitness function is), but only to integrate in the GA the developers' knowledge. It is worth noting that the feedback provided by the developers help the GA to sensibly reduce the number of produced clusters.

Concerning the authoritativeness of the experimented algorithms, the achieved values of MoJoFM on GESA and SMOS are reported in Table 6. Given the results obtained, it is not surprising that the modularizations proposed by the non-interactive GAs (both single- and multi- objective) are very far from the original

Table 4. GESA: Results of the Mann-Whitney tests

	MQ		# clusters		Isolated clusters		MoJoFM	
	p-value	d	p-value	d	p-value	d	p-value	d
GA[n] vs. IC-IGA[n]	0.00	0.92	0.00	-Inf	0.00	-0.96	0.00	0.91
GA[n] vs. R-IGA[n]	0.00	0.89	0.00	-0.97	0.00	-0.89	0.00	0.96
MGA[n] vs. IC-IMGA[n]	0.00	0.99	0.00	-0.99	0.00	-0.98	0.00	Inf
MGA[n] vs. R-IMGA[n]	0.34	0.21	0.88	-0.02	0.04	0.38	0.01	0.44
IC-IGA[n] vs. R-IGA[n]	0.02	-0.38	0.00	0.69	0.00	0.84	0.02	-0.39
IC-IMGA[n] vs. R-IMGA[n]	0.00	-0.99	0.00	0.99	0.00	0.99	0.00	-0.97
GA[n/2] vs. IC-IGA[n/2]	0.00	0.87	0.00	-Inf	0.00	-0.94	0.00	0.99
GA[n/2] vs. R-IGA[n/2]	0.00	0.92	0.00	-0.98	0.24	0.18	0.00	0.96
MGA[n/2] vs. IC-IMGA[n/2]	0.00	0.94	0.00	-0.93	0.00	-0.88	0.00	0.94
MGA[n/2] vs. R-IMGA[n/2]	0.14	-0.27	0.04	0.37	0.00	0.76	0.14	0.24
IC-IGA[n/2] vs. R-IGA[n/2]	0.04	-0.35	0.00	0.84	0.00	0.96	0.04	-0.33
IC-IMGA[n/2] vs. R-IMGA[n/2]	0.00	-0.96	0.00	0.97	0.00	0.98	0.00	-0.93

Table 5. SMOS: Results of the Mann-Whitney tests

	MQ		# clusters		Isolated clusters		MoJoFM	
	p-value	d	p-value	d	p-value	d	p-value	d
GA[n] vs. IC-IGA[n]	0.00	0.43	0.00	-0.85	0.00	-0.83	0.00	0.94
GA[n] vs. R-IGA[n]	0.00	0.80	0.00	-0.96	0.00	-0.81	0.00	Inf
MGA[n] vs. IC-IMGA[n]	0.00	0.84	0.00	-1.00	0.00	-0.99	0.00	0.99
MGA[n] vs. R-IMGA[n]	1.00	0.10	1.00	-0.01	0.01	0.44	0.00	0.90
IC-IGA[n] vs. R-IGA[n]	0.90	-0.02	0.00	0.55	0.00	0.60	0.00	-0.70
IC-IMGA[n] vs. R-IMGA[n]	0.00	-0.83	0.00	1.00	0.00	1.00	0.00	-0.91
GA[n/2] vs. IC-IGA[n/2]	0.00	-0.84	0.00	-0.97	0.00	-1.00	0.00	0.97
GA[n/2] vs. R-IGA[n/2]	0.99	0.00	0.00	-0.99	0.00	-0.54	0.00	0.94
MGA[n/2] vs. IC-IMGA[n/2]	0.00	0.80	0.00	-0.99	0.00	-Inf	0.00	0.98
MGA[n/2] vs. R-IMGA[n/2]	1.00	0.00	0.69	0.14	0.00	0.61	0.01	0.45
IC-IGA[n/2] vs. R-IGA[n/2]	0.00	0.84	0.00	0.74	0.00	0.99	0.00	-0.72
IC-IMGA[n/2] vs. R-IMGA[n/2]	0.00	-0.87	0.00	1.00	0.00	Inf	0.00	-0.96

design. The best results among the non-interactive GAs are achieved by GA[n/2] with MoJoFM equals to 21 on GESA and 37 on SMOS. Instead, thanks to the few feedback provided, the IGAs achieved much better results (as expected). The best performances are achieved using IC-MGA[n] with a maximum MoJoFM of 53 on GESA and 74 on SMOS. Also in this case, statistical analyses support our findings (see Tables 4 and 5).

To better understand what this difference between the performances of interactive and non-interactive GAs means from a practical point of view, Fig. 1 shows an example extracted from the re-modularization of the SMOS software system. The figure is organized in three parts. The first part (left side) shows how the subsystem *RegisterManagement* appears in the original package decomposition (i.e., which classes it contains) made by the SMOS's developers. This subsystem groups together all the classes in charge to manage information related to the scholar register (e.g., the students' delay, justifications for their absences and so on). The second part (middle) reports the decomposition of the

Table 6. Descriptive statistics of the MoJoFM achieved by the different algorithms

Algorithm	MoJoFM				
	Mean	Med.	St. Dev.	Max	Min
GA[n]	14	14	2	19	11
IC-IGA[n]	29	33	8	38	13
R-IGA[n]	26	27	5	34	14
MGA[n]	10	11	2	13	7
IC-IMGA[n]	38	39	11	53	13
R-IMGA[n]	12	12	3	22	8
GA[n/2]	18	17	1	21	15
IC-IGA[n/2]	29	29	6	38	20
R-IGA[n/2]	25	25	4	32	18
MGA[n/2]	14	14	2	18	10
IC-IMGA[n/2]	38	39	11	53	13
R-IMGA[n/2]	15	15	2	20	11

(a) GESA

Algorithm	MoJoFM				
	Mean	Med.	St. Dev.	Max	Min
GA[n]	26	27	2.8	30	21
IC-IGA[n]	53	54	6.4	62	36
R-IGA[n]	39	39	4.7	48	31
MGA[n]	17	16	3.1	23	12
IC-IMGA[n]	59	62	12	74	22
R-IMGA[n]	26	27	4.6	38	15
GA[n/2]	31	31	2.6	37	25
IC-IGA[n/2]	52	53	8.2	62	30
R-IGA[n/2]	42	41	6.2	57	32
MGA[n/2]	24	23	3.5	32	20
IC-IMGA[n/2]	43	43	7.8	62	22
R-IMGA[n/2]	28	28	4.5	36	20

(b) SMOS

classes contained in *RegisterManagement* proposed by the MGA. Note that some classes not belonging to the *RegisterManagement* were mixed to the original set of classes. These classes are reported in light gray. Finally, the third part (right side) shows the decomposition of the classes contained in *RegisterManagement* proposed by the IC-IMGA. Also in this case, classes not belonging to the original *RegisterManagement* package are reported in light gray. As we can see, the original package decomposition groups 31 classes in the *RegisterManagement* package. When applying MGA, these 31 classes are spread into 27 packages, 13 of which are singleton packages. As for the remaining 14 they usually contain some classes of the *RegisterManagement* package mixed with other classes coming from different packages (light gray in Fig. 1). The solution provided by IC-IMGA is very different. In fact, IC-IMGA spreads the 31 classes in only 5 packages. Moreover, it groups together in one package 26 out of the 31 classes originally belonging to the *RegisterManagement* package. It is striking how much the partition proposed by IC-IMGA is closer to the original one resulting in a higher MoJoFM achieved by IC-IMGA with respect to MGA and thus, a more meaningful partitioning from a developer's point of view.

On summary, results of our study showed as the non-interactive GAs, in both their single- and multi- objective formulations, might produce modularizations that, albeit being good in terms of cohesion and coupling, strongly deviate from the developers' intent. This highlights the need for augmenting the GA with developers's knowledge trough IGAs. That is, IGAs would allow to obtain more meaningful solutions—that for our case studies are even better in terms of modularization quality—representing an acceptable starting point for a developer when performing a software re-modularization.

4.4   Threats to Validity

Threats to *construct validity* may essentially depend on the way we simulated the feedback. As said, we believe that simulating feedback from the original system design would be representative of a developer's behavior (similarly to [18]). Nevertheless, controlled experiments are needed to understand to what extent real developers are able to introduce their knowledge in the genetic algorithm through the feedback mechanism. Threats to *internal validity* can be related to

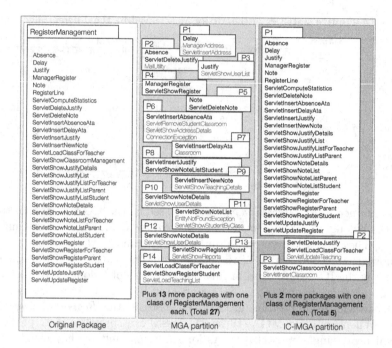

Fig. 1. MGA *vs* IC-IMGA in reconstructing the *RegisterManagement* package of SMOS

the GAs parameter settings. For some of them we used parameters similar to previous studies [14], while for others we used a trial-and-error calibration procedure. Threats to *external validity* concerns the generalization of our results. Although we applied the approach on two realistic systems (one industrial system and one developed by students but actually used), further experimentation is needed, also for example on procedural (e.g., C) systems. Also, we evaluated the advantages of interactive feedback on GAs only, while it would be worthwhile to investigate it also for other heuristics such as hill-climbing. However, it is important to point out that in this paper we were interested to assess the *improvement* in the quality and meaningfulness of the produced solutions when using interactions (with respect to a similar algorithm without interaction), rather than in the absolute quality of the solutions.

5   Conclusion and Future Work

In this paper we proposed the use of IGAs to integrate developers' knowledge during software re-modularization activities. We implemented and experimented both single- and multi-objective IGAs comparing their performances with those achieved by their non-interactive counterparts. The achieved results show that the IGAs are able to propose re-modularizations (i) more meaningful from a developer's point-of-view, and (ii) not worse, and often even better in terms of modularization quality, with respect to those proposed by the non-interactive

GAs. Also, IGAs is applicable with a limited number of feedback, and thus, require a relatively limited effort to the developer.

Future work will be devoted to replicate the empirical evaluation on further software systems and different heuristics such as hill-climbing. Also, we plan to evaluate the usefulness of IGAs during software re-modularization activities.

References

1. Anquetil, N., Lethbridge, T.: Experiments with clustering as a software remodularization method. In: WCRE, pp. 235–255 (1999)
2. Bavota, G., De Lucia, A., Marcus, A., Oliveto, R.: Software re-modularization based on structural and semantic metrics. In: WCRE, pp. 195–204 (2010)
3. Coello Coello, C.A.: Theoretical and numerical constraint-handling techniques used with evolutionary algorithms: A survey of the state of the art. Computer Methods in Applied Mechanics and Engineering 191(11-12) (2002)
4. Conover, W.J.: Practical Nonparametric Statistics, 3rd edn. Wiley (1998)
5. Deb, K.: Multi-Objective Optimization Using Evolutionary Algorithms. Wiley (2001)
6. Di Penta, M., Neteler, M., Antoniol, G., Merlo, E.: A language-independent software renovation framework. JSS 77(3), 225–240 (2005)
7. Doval, D., Mancoridis, S., Mitchell, B.S.: Automatic clustering of software systems using a genetic algorithm. In: STEP, pp. 73–82 (1999)
8. Fowler, M., Beck, K., Brant, J., Opdyke, W., Roberts, D.: Refactoring: Improving the Design of Existing Code. Addison-Wesley Professional (1999)
9. Grissom, R.J., Kim, J.J.: Effect sizes for research: A broad practical approach, 2nd edn. Lawrence Earlbaum Associates (2005)
10. Mancoridis, S., Mitchell, B.S., Rorres, C., Chen, Y.F., Gansner, E.R.: Using automatic clustering to produce high-level system organizations of source code. In: IWPC, pp. 45–53 (1998)
11. Maqbool, O., Babri, H.A.: Hierarchical clustering for software architecture recovery. IEEE TSE 33(11), 759–780 (2007)
12. Mitchell, B.S.: A Heuristic Search Approach to Solving the Software Clustering Problem. Ph.D. thesis, Drexel University, Philadelphia (2002)
13. Mitchell, B.S., Mancoridis, S.: On the automatic modularization of software systems using the bunch tool. IEEE TSE 32(3), 193–208 (2006)
14. Praditwong, K., Harman, M., Yao, X.: Software module clustering as a multiobjective search problem. IEEE TSE 37(2), 264–282 (2011)
15. Siff, M., Reps, T.W.: Identifying modules via concept analysis. IEEE TSE 25(6), 749–768 (1999)
16. Simons, C.L., Parmee, I.C., Gwynllyw, R.: Interactive, Evolutionary Search in Upstream Object-Oriented Class Design. IEEE TSE 36(6), 798–816 (2010)
17. Takagi, H.: Interactive evolutionary computation: Fusion of the capacities of EC optimization and human evaluation. Proceedings of the IEEE 89(9), 1275–1296 (2001)
18. Tonella, P., Susi, A., Palma, F.: Using interactive GA for requirements prioritization. In: SSBSE, pp. 57–66 (2010)
19. Wen, Z., Tzerpos, V.: An effectiveness measure for software clustering algorithms. In: IWPC, pp. 194–203 (2004)
20. Wiggerts, T.A.: Using clustering algorithms in legacy systems re-modularization. In: WCRE, p. 33. IEEE Computer Society (1997)

Optimizing Threads Schedule Alignments to Expose the Interference Bug Pattern

Neelesh Bhattacharya[1], Olfat El-Mahi[1], Etienne Duclos[1], Giovanni Beltrame[1], Giuliano Antoniol[1], Sébastien Le Digabel[2], and Yann-Gaël Guéhéneuc[1]

[1] Department of Computer and Software Engineering
[2] GERAD and Department of Mathematics and Industrial Engineering,
École Polytechnique de Montréal, Québec, Canada
{neelesh.bhattacharya,olfat.ibrahim,etienne.duclos,
giovani.beltrame,giuliano.antoniol,
yann-gael.gueheneuc}@polymtl.ca, sebastien.le.digabel@gerad.ca

Abstract. Managing and controlling interference conditions in multi-threaded programs has been an issue of worry for application developers for a long time. Typically, when write events from two concurrent threads to the same shared variable are not properly protected, an occurrence of the interference bug pattern could be exposed. We propose a mathematical formulation and its resolution to maximize the possibility of exposing occurrences of the interference bug pattern. We formulate and solve the issue as an optimization problem that gives us (1) the optimal position to inject a delay in the execution flow of a thread and (2) the optimal duration for this delay to align at least two different write events in a multi-threaded program. To run the injected threads and calculate the thread execution times for validating the results, we use a virtual platform modelling a perfectly parallel system. All the effects due to the operating system's scheduler or the latencies of hardware components are reduced to zero, exposing only the interactions between threads. To the best of our knowledge, no previous work has formalized the alignment of memory access events to expose occurrences of the interference bug pattern. We use three different algorithms (random, stochastic hill climbing, and simulated annealing) to solve the optimization problem and compare their performance. We carry out experiments on four small synthetic programs and three real-world applications with varying numbers of threads and read/write executions. Our results show that the possibility of exposing interference bug pattern can be significantly enhanced, and that metaheuristics (hill climbing and simulated annealing) provide much better results than a random algorithm.

Keywords: Multi-Threaded Programs Testing, Optimization Techniques, Interference Bug Pattern.

1 Introduction

The advent of multi-core systems has greatly increased the use of concurrent programming. To control concurrent programs and their correct execution, the use of locks, semaphores, and barriers has become a common practice. However, despite of the use

G. Fraser (Ed.): SSBSE 2012, LNCS 7515, pp. 90–104, 2012.

of locks and barriers, it is extremely difficult to detect and remove bugs, specifically data-race conditions and deadlocks. Most bugs are detected by testing, i.e. by multiple runs of a program under various environmental conditions. In fact, the same test case might or might not detect a bug because of a system's non-deterministic components (interrupts, scheduler, cache, etc.), over which the testers have no direct control.

The interference bug pattern is one the most common bugs in concurrent programs. For example, when two or more write events happen in very close proximity on unprotected shared data, the chances of incurring in an interference bug are high. This bug is also one of the hardest to eradicate [21,23].

In this paper, we propose a theoretical formulation that helps maximizing the possibility of exposing interference bug pattern in multi-threaded programs, if it exists. An interference bug might occur when (1) two or more concurrent threads access a shared variable, (2) at least one access is a write, and (3) the threads use no explicit mechanism to enforce strict access ordering. To expose this bug, we want to inject a delay in the execution flow of each thread with the purpose of aligning in time different shared memory access events. Specifically, we want to align two write events occurring in two or more threads that share the same variable to maximize the probability of exposing an interference bug. We use unprotected variables (without locks or semaphores), so that bugs can occur and we can maximize the possibilities of finding them. Our formulation allows the identification of the optimal delays to be injected (positions and durations) using search or optimization algorithms. In particular, we use: random exploration, stochastic hill climbing, and simulated annealing.

We apply our approach to a set of multi-threaded data-sharing programs, called Programs Under Test (PUTs) that comprises of four synthetic programs and three real-world applications, CFFT (Continuous Fast Fourier Transform), CFFT6 (Continuous Fast Fourier Transform 6) and FFMPEG. CFFT computes the Fast Fourier Transform on an input signal, while CFFT6 performs a number of iterations of the Bailey's 6-step FFT to computes 1D FFT of an input signal. FFMPEG is a complete, cross-platform solution to record, convert and stream audio and video. To avoid non-determinism we use a simulation platform (ReSP [2]) which gives us full control over all the components of the system for building a fully-parallel execution environment for multi-threaded programs. We model an environment in which all the common limitations of a physical hardware platform (*i.e.*, the number of processors, memory bandwidth, and so on) are not present and all the operating system's latencies are set to zero. The PUTs are executed in this environment, exposing only the threads' inherent interactions. We collect the exact times of each memory access event and then run, inject delays, and verify whether any interference bugs are exposed.

The rest of this paper is organized as follows: Section 2 presents the relevant previous work; Section 3 recalls some useful notions on interference bug patterns, metaheuristics and optimization; Section 4 presents our formulation of the interference bug pattern and the detection approach; Section 5 describes the context and research questions of the experimental study; Section 6 reports our results while Section 7 discusses the trend observed in the results and specifies the threats to the validity of our results; finally, Section 8 draws some concluding remarks and outlines the scope for future work.

2   Related Work

Our work seeks to maximize the possibility of exposing data-race and interference conditions. There exists significant works on the impact of data-race conditions in concurrent programs. Artho et al. [1] provided a higher abstraction level for data races to detect inconsistent uses of shared variables and moved a step ahead with a new notion of high-level data races that dealt with accesses to set of fields that are related, introducing concepts like *view* and *view consistency* to provide a notation for the new properties. However they focused on programs containing locks and other protections, while our work concerns unprotected programs.

Moving on to research carried out on bug patterns, Hovemeyer et al. [14] used bug pattern detectors to find correctness and perfomance-related bugs in several Java programs and found that there exists a significant class of easily detectable bugs. They were able to identify large number of bugs in real applications. But they faced the problem of knowing the actual population of real bugs in large programs. Farchi et al. [11] proposed a taxonomy for creating timing heuristics for the ConTest tool [9], showing it could be used to enhance the bug finding ability of the tool. They introduced the sleep, losing notify, and dead thread bug patterns. Eytani et al. [10] proposed a benchmark of programs containing multi-threaded bugs for developing testing tools. They asked undergraduates to create some buggy Java programs, and found that a number of these bugs cannot be uncovered by tools like ConTest [9] and raceFinder [3]. Bradbury et al. [4] proposed a set of mutation operators for concurrency programs used to mutate the portions of code responsible for concurrency to expose a large set of bugs, along with a list of fifteen common bug patterns in multi-threaded programs. Long et al. [18] proposed a method for verifying concurrent Java programs with static and dynamic tools and techniques using Petri nets for Java concurrency. They found proper verification tools for each failure. However all these approaches mainly focused on Java programs, so are language specific, while our approach is generic for every programming languages.

In the field of the behavior of concurrent programs, Carver et al. [7] proposed repeatable deterministic testing, while the idea of systematic generation of all thread schedules for concurrent program testing came with works on reachability testing [15,17]. The VeriSoft model checker [13] applied state exploration directly to executable programs, enumerating states rather than schedules. ConTest [9] is a lightweight testing tool that uses various heuristics to create scheduling variance by inserting random delays in a multi-threaded program. CalFuzzer [16] and CTrigger [22] use analysis techniques to guide schedules toward potential concurrency errors, such as data races, deadlocks, and atomicity violations.

One of the most influential tools developed for testing concurrent programs is CHESS [20]. It overcomes most of the limitations of the tools developed before. What set CHESS apart from its predecessors is its focus on detecting both safety and liveness violations in large multi-threaded programs. It relies on effective safety and liveness testing of such programs, which requires novel techniques for preemption bounding and fair scheduling. It allows a greater control over thread scheduling than the other tools and, thus, provides higher-coverage and guarantees better reproducibility. CHESS tries all the possible schedules to find a bug, whereas we create the schedules that maximizes the likelihood of exposing an interference bug, if it is present.

In this paper, we do not intend to compare our approach with CHESS or other mentioned tools. We want to help developers by providing them with the locations and durations of delays to inject in their multi-threaded programs so that they can, subsequently, run their programs to enhance the likelihood of exposing interference bugs, possibly using CHESS. It is to be noted that our work clearly differs from the previous ones, in the sense that none of them played with inserted delays to align the write events ; they explore the various possible schedules to expose a bug.

3   Background Notions

Concurrency is built around the notion of multi-threaded programs. A thread is defined as an execution context or a lightweight process having a single sequential flow of control within a program [8]. Inserting delays in the execution of a thread is an efficient way of disrupting its normal behavior: the inserted delay shifts the execution of the subsequent statements. By doing so, an event in one thread can be positioned in close proximity with another event in another thread, increasing the probability of exposing an interference bug.

Bradbury et al. [4] mentioned fifteen possible bug patterns that could affect the normal behavior of threads and cause severe problems to concurrency. Out of them, we considered the interference bug pattern because it is one of the most commonly encountered and one of the hardest to eradicate [21,23].

3.1   Interference Bug Sequence Diagrams

In Figures 1 and 2, a master thread creates child threads and shares some data with them. The expected behavior (without the bug), illustrated in Figure 1, is that each child thread accesses the shared data sequentially, so that every thread has the last version of the data when it reads it. When we inject a delay just before a child writes its modification to the shared data, as shown in Figure 2, another thread reads a wrong datum and may produce incorrect results.

3.2   Search Algorithms

Given the number of possible thread events in any multi-threaded program, we apply search algorithms to maximize the number of "alignments" between events, i.e., the

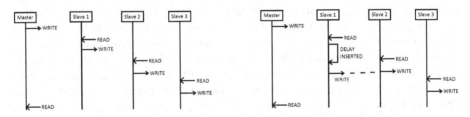

Fig. 1. Behavior of a PUT   **Fig. 2.** PUT with injected delay

number of events in close proximity. To experiment with different optimization techniques we chose the two most commoly used optimization algorithms: stochastic hill climbing (SHC) [25] and simulated annealing (SA) [19], and we validated them against random search (RND).

4 Formulation and Approach

Given a multi-threaded program, the interleaving of threads depends on the hardware (*e.g.*, number of processors, memory architecture, etc.) and the operating system. There are as many schedules as there are environmental conditions, schedule strategies, and policies of the operating system when handling threads. Among these schedules, there could be a subset leading to the expression of one or more interferences. In general, the exhaustive enumeration of all possible schedules is infeasible, and in an ideal situation, all threads would run in parallel.

4.1 Parallel Execution Environment

To provide a deterministic parallel execution environment without external influences we use a virtual platform, namely ReSP [2]. ReSP is a virtual environment for modeling an ideal multi-processor system with as many processors as there are threads. ReSP is also a platform based on an event-driven simulator that can model any multi-processor architecture and that implements an emulation layer that allows the interception of any OS call. ReSP allows access to all execution details, including time of operations in milliseconds, accessed memory locations, thread identifiers, and so on.

To create our parallel execution environment, we model a system as a collection of Processing Elements (PEs), ARM cores in our specific case but any other processor architecture could be used, directly connected to a single shared memory, as shown in Figure 3. Therefore, our environment makes as many PEs available as there are execution threads in a PUT, each thread being mapped to a single PE. PEs have no cache memory, their interconnection is implemented by a 0-latency crossbar, and the memory responds instantaneously, thus each thread can run unimpeded by the sharing of hardware resources. This environment corresponds to an ideal situation in which the fastest possible execution is obtained.

We use ReSP to run our unmodified test cases without (except when injecting delays) and to calculate the threads' execution time for validating our results. We use ReSP's ability to trap any function call being executed on the PEs to route all OS-related activities outside the virtual environment. Thread-management calls and other OS functions (sbrk, file access, and so on) are handled by the host environment, without affecting the timing behaviour of the multi-threaded program. Thus, the PUT perceives that all OS functions are executed instantaneously without access to shared resources. Because all OS functions are trapped, there is no need for a real OS implementation to run the PUT in the virtual environment. This is to say the PUTs are run without any external interference. The component performing the routing of OS functions, referred to as the OS Emulator, takes care of processor initialization (registers, memory allocation, and so on) and implements a FIFO scheduler that assigns each thread to a free PE as soon as it is created.

The main difference between our virtual environment and a real computer system are cache effects, the limited number of cores, other applications running concurrently on the same hardware and other interactions that make scheduling non-deterministic and introduce extra times between events. In our environment, the full parallelism of the PUT is exposed and the only interactions left are inherent to the PUT itself.

4.2 Problem Formalization

Any real hardware/software environment will deviate from the parallel execution environment described above. From the threads' point of view, any deviation will result in one or more delays inserted in their fully-parallel delay-free execution. Figure 4 summarizes the execution of four threads, where each thread performs four read and–or write accesses. For the sake of simplicity, let us assume that just one delay is inserted in a given thread. This delay will shift the thread's computation forward in time, and possibly cause some memory write access(es) to happen in close proximity, leading to the possibility of exposing an interference condition.

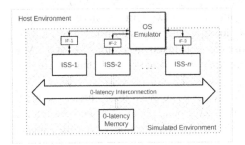

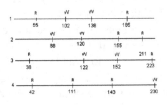

Fig. 3. The virtual platform on which the PUT is mapped: each component except for the processors has zero latency, and OS calls are trapped and executed externally

Fig. 4. Concurrent Threads Example

Concretely, at 102 ms the first thread writes into a memory location a value and, given the schedule, no interference happens. In other words, two threads do not attempt a write at the same time in the same memory location. However, if, for any reason, the thread schedule is different and, for example, thread 2 is delayed by 2 ms after the first writing then a possible interference happens at 122 ms between threads 2 and 3.

The event schedule depends on the operating system, the computer workload, other concurrent programs being run, and the scheduler policy. Therefore, enhancing the possibility of exposing an interference bug via testing for any foreseeable schedule is a challenging problem that has been addressed in several ways, from search-based approaches [5,6] to formal methods [12,24]. The higher the number of available CPUs, the higher the number of scheduled threads and events, and the more difficult it is to manually verify that any two threads will not create an interference under any possible schedule. The example in figure 4 shows a larger system, *i.e.*, with a higher number of threads and a longer execution time. Also in this case, the delays would be inserted in a similar manner between events (taking into account the longer execution time). In general, we believe that our approach would be able to increase the chances of exposing the interference conditions for systems of any size.

Let the PUT be composed on N threads. Let us assume that the ith thread contains M_i events; let $t_{i,j}$ be, for the thread i, the time at which an event (e.g., a memory access, a function call, and so on) happens and let i^* be the thread subject to perturbation, i.e., the thread in which a delay Δ will be injected before an event p. Finally, let $a_{i,j}$ stands for the action performed at time $t_{i,j}$ by the thread i. Because ReSP allows to precisely track memory accesses as well as times, to simplify the formalization, let us further assume that $a_{i,j}$ equals to 1 to show that it is a "write" action to a given memory cell or 0 to show some other action. Then, our objective is to maximize the number of possible interferences $N_{Interference}$:

$$N_{Interference} = \max_{\Delta,p,i*} \left\{ \sum_{i=1,i\neq i*}^{N} \sum_{j=1}^{M_i} \sum_{k=p}^{M_{i*}} \delta(a_{i*,k}, a_{i,j})\delta(t_{i,j}, t_{i*,k} + \Delta) \right\} \quad (1)$$

under the constraint $t_{i,j} \geq t_{i*,k} + \Delta$, and where $\delta(x, y)$ is the Kronecker operator[1].

We want to maximize the numbers of alignments, i.e., two write events coinciding at the same time occurring in the same memory location, using Equation 1. Unfortunately, this equation leads to a staircase-like landscape as it result in a sum of 0 or 1. Any search strategy will have poor guidance with this fitness function and chances are that it will behave akin to a random search.

If we assume that a delay is inserted before each write event in all threads, then all threads events will be shifted. More precisely, if $\Delta_{i,j}$ is the delay inserted between the events $j - 1$ and j of the thread i, all times after $t_{i,j}$ will be shifted. This shift leads to new time τ for the event $a_{i,j}$:

$$\tau_{i,j}(a_{i,j}) = t_{i,j} + \sum_{k=1}^{j} \Delta_{i,k} \quad (2)$$

Considering the difference between $\tau_{i_q,j_q}(a_{i_q,j_q})$ and $\tau_{i_r,j_r}(a_{i_r,j_r})$, when both a_{i_q,j_q} and a_{i_r,j_r} are write events to the same memory location, we rewrite Equation 1 as:

$$N_{Interference}(write) = \max_{\Delta_{1,1},\ldots,\Delta_{N,N}} \left\{ \sum_{i_r}^{N} \sum_{j_r}^{M_{j_r}} \right.$$
$$\left. \sum_{i_q\neq i_r}^{N} \sum_{j_q}^{M_{j_q}} \frac{1}{1 + |\tau_{i_q,j_q}(a_{i_q,j_q}) - \tau_{i_r,j_r}(a_{i_r,j_r})|} \right\} \quad (3)$$

under the constraints $\tau_{i_q,j_q} \geq \tau_{i_r,j_r}$ and $a_{i_q,j_q} = a_{i_r,j_r} = write$ to the same memory location.

---

[1] The Kronecker operator is a function of two variables, usually integers, which is 1 if they are equal and 0 otherwise ($\delta(x, y) = 1$, if $x = y$; or 0 otherwise).

Equation 3 leads to a minimization problem:

$$N_{Interference}(write) = \min_{\Delta_{1,1},\dots,\Delta_{N,N_N}} \left\{ \sum_{i_r}^{N} \sum_{j_r}^{M_{j_r}} \right.$$

$$\left. \sum_{i_q \neq i_r}^{N} \sum_{j_q}^{M_{j_q}} (1 - \frac{1}{1 + |\tau_{i_q,j_q}(a_{i_q,j_q}) - \tau_{i_r,j_r}(a_{i_r,j_r})|}) \right\} \tag{4}$$

For both Equations 3 and 4, given a $\tau_{i_q,j_q}(a_{i_q,j_q})$, we may restrict the search to the closest event in the other threads, typically: $\tau_{i_r,j_r}(a_{i_r,j_r}) \geq \tau_{i_q,j_q}(a_{i_q,j_q})$ & $\tau_{i_q,j_q}(a_{i_q,j_q}) \leq \tau_{i_q,j_s}(a_{i_q,j_s})$. Under this restriction, **Equation 4 is the fitness function used in the search algorithms to inject appropriate delays in threads to maximize the probability of exposing interference bugs (if any).** This equation also solves the staircase-like landscape problem because the fitness function is sum of real numbers, providing a smoother landscape.

4.3 Problem Modeling and Settings

As described above, the key concepts in our thread interference model are the times and types of thread events. To assess the feasibility of modeling thread interferences by mimicking an ideal execution environment, we are considering simple problem configurations. More complex cases will be considered in future works; for example, modeling different communication mechanisms, such as pipes, queues, or sockets. We do not explicitly model resource-locking mechanisms (*e.g.*, semaphores, barriers, etc.) as they are used to protect data and would simply enforce a particular event ordering. Therefore, if data is properly protected, we would simply fail to align two particular events. At this stage, we are also not interested in exposing deadlock or starvation bugs.

From Equations 2 and 4, we can model the problem using the times and types of thread events. Once the occurrence write events has been timestamped using ReSP, we can model different schedules by shifting events forward in time. In practice, for a given initial schedule, all possible thread schedules are obtained by adding delays between thread events (using Equation 2).

Our fitness function 4 (*i.e.*,what an expression of how fit is our schedule to help expose interference bugs) can be computed iterating over all thread write events. The fitness computation has therefore a quadratic cost over the total number of write events. However, for a given write event, only events occurring in times greater or equal to the current write are of interest, thus making the fitness evaluation faster (though still quadratic in theory).

In general, it may be difficult or impossible to know the maximum delay that a given thread will experience. Once a thread is suspended, other threads of the same program (or other programs) will be executed. As our PUTs are executed in an ideal environment (no time sharing, preemption, priority, and so on), there is no need to model the scheduling policy and we can freely insert any delay at any location in the system to increase the chances of exposing the interference bug.

5 Empirical Study

The *goal* of our empirical study is to obtain the conceptual proof of the feasibility and the effectiveness of our search-based interference detection approach and validate our fitness function (see Equation 4).

As a metric for the quality of our model, we take the number of times we succeed in aligning two different write events to the same memory location. Such alignment increases the chances of exposing an interference bug, and can be used by developers to identify where data is not properly protected in the code.

To perform our conceptual proof, we have investigated the following two research questions:

RQ1: Can our approach be effectively and efficiently used on simple as well as real-world programs to maximize the probability of interferences between threads?

RQ2: How does the dimension of the search space impact the performance of three search algorithms: RND, SHC and SA?

The first research question aims at verifying that our fitness function guides the search appropriately, leading to convergence in an acceptable amount of time. The second research question concerns the choice of the search algorithm to maximize the probability of exposing interference bugs. The need for search algorithms is verified by comparing SHC and SA with a simple random search (RND): better performance from SHC and SA increases our confidence in the appropriateness of our fitness function.

One advantage of this approach is that it has no false positives, because it doesn't introduce any functional modification in the code: if a bug is exposed, the data are effectively unprotected. Nevertheless, we do not guarantee that a bug will be exposed even if present, as the approach simply increases the likelihood of showing interferences. The chances of exposure are increased because the manifestation of interference bug depends on the thread schedule and we, unlike other approaches, manufacture the schedules that maximize interference.

It might be argued that a data race detector does not need the timings of write alignments to be so accurate. But as we are dealing with a fully parallelized environment, we target specific instances and align the events with much more precision than what required by a data race detector [1]. Basically, we are pin-pointing the event times with accuracy.

Our approach can also be used in cases where it is enough to have two change the order of two events to verify the correctness of some code. Once the events are aligned, an arbitrarily small additional delay would change the order of two events, possibly exposing data protection issues. The correct use of locks or other data protection measures would prevent a change in the order of the events.

6 Experimental Results

To answer our two research questions, we implemented four small synthetic and three real-world multi-threaded programs with different numbers of threads and read/write events. Table 1 provides the details of these applications: their names, sizes in numbers of lines of code, number of threads, and sequences of read and write access into

Table 1. Application Details

Table 2. Execution Times for Real-World Applications, in milliseconds

PUTs	LOCs	Nbr. of Threads	Events
Matrix Multiplication (MM)	215	4	RWWR, WWWW, RRWW, WWRR
Count Shared Data (CSD)	160	4	WWW, RRW, RWRW, RW
Average of Numbers (AvN)	136	3	W, RW, RW
Area of Circle (AC)	237	5	RW, RWW, RRW, RWRW, RWWRW
CFFT	260	3	WR, WR, WRWRWR
CFFT6	535	3	WRWRWRWRWRWR, WR, WRWRWRWR
FFMPEG	2.9×10^5	4	WRWRRR, WR, WR, WR

	CFFT		CFFT6		FFMPEG	
	1×10^6	1×10^7	1×10^6	1×10^7	1×10^6	1×10^7
SA	3118	5224	27443	20416	1562	4672
HC	3578	4976	27328	21943	1378	5100
RND	113521	107586	342523	339951	59599	133345

memory. The "Events" column shows the various events with a comma separating each thread. For example, the thread events in column 4 for row 3 (*Average of Numbers*) should be read as follows: Thread 1 has just one write event, thread 2 has a read, followed by a write event, and thread 3 has a read followed by a write event.

We believe that our results are independent of the execution system or architecture: our approach aligns write events among threads on a virtual system, and exposes data protection issues regardless of the final execution platform. Similarly, when applied to applications of any size, the interference conditions are exposed irrespective of the number of lines of code that the application may contain.

6.1 RQ1: Approach

RQ1 aims at verifying if our approach can effectively identify time events configurations, and thus schedules, leading to possible thread interferences. We experimented with RND, SHC, and SA. RND is used as a sanity check and to show that a search algorithm is effectively needed.

For the three search algorithms (RND, SHC and SA), we perform no more than 10^6 fitness evaluations. For the three algorithms, we draw the added delays $\Delta_{i,k}$ from a uniform-random distribution, between zero and a maximum admissible delay fixed to 10^7 ms (10^4 seconds).

We configured RND in such a way that we generated at most 10^7 random delays $\Delta_{i,k}$, uniformly distributed in the search space, and applied Equation 2 to compute the actual thread time events. We then evaluated each generated schedule, to check for interferences, using Equation 4.

We set the SHC restart value at 150 trials and we implemented a simple neighbor variable step visiting strategy. Our SHC uses RND to initialize its starting point, then it first attempts to move with a large step, two/three orders of magnitude smaller than the search space dimension, for 1000 times, finally it reduces the local search span by reducing the distance from its current solution to the next visited one. The step is drawn

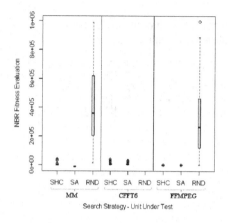
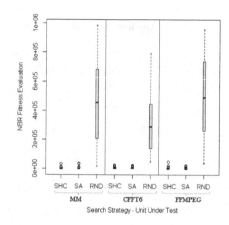

Fig. 5. (RQ1) Algorithm comparison for a search space up to 10 Million sec delay

Fig. 6. (RQ2) Algorithm comparison for a search space up to 1 Million sec delay

from a uniformly-distributed random variable. Thus, for a search space of 10^7, we first attempt to move with a maximum step of 10^4 for 20 times. If we fail to find a better solution, perhaps the optimum is close, and then we reduce the step to 10^3 for another 20 trials, and then to 500 and then to 50. Finally, if we do not improve in 150 move attempts (for the large search space), the search is discarded, a next starting solution is generated and the process restarted from the new initial solution. We selected the values and the heuristic encoded in the SHC via a process of trial and error. Intuitively, if the search space is large, we would like to sample distant regions but the step to reach a distant region is a function of the search space size, so we arbitrarily set the maximum step of two or three orders of magnitude smaller than the size of the search space.

We configured the SA algorithm similarly to the SHC algorithm, except for the cooling factor and maximum temperature. In our experiments, we set $r = 0.95$ and T_{max} depending on the search space, *i.e.*, 0.0001 for a search space of size 110^7 and 0.01 for smaller search spaces.

Regarding the times to compute the solutions with different algorithms, it is a well known fact that SHC and SA scale less than linearly with the design space. This fact can be proven by having a look at the results provided in table 2. Table 2 shows the execution times of various algorithms for computing the alignments 100 times only for the real-world applications with large search space (10^6 and 10^7 ms). It can be seen that despite the increasing size of the design space, SHC and SA converge to solutions in reasonable amounts of times, as compared to RND.

Figure 5 reports the relative performance over 100 experiments of RND, SHC, and SA for *Matrix Multiplication*, CFFT6 and FFMPEG for a search space of 10^7 ms. Each time that our approach has exposed a possible interference, we recorded the number of fitness evaluations and stopped the search. We obtained similar box-plots for the other applications, but do not show them here because they do not bring additional/contrasting insight in the behavior of our approach. The box-plots show that SHC outperforms RND and that SA and SHC perform similarly. However, SA performs marginally better than SHC The missing plots for CFFT6 in Figure 5 are due to the fact that RND did not find any solution in 10^6 iterations, even after 100 runs.

Overall, we can positively answer our first research question: **our approach can effectively and efficiently be used on simple program models to maximize the possibility of interferences between threads.**

6.2   RQ2: Search Strategies

RQ2 aims at investigating the performance of the different strategies for various search-space dimensions. Our intuition is that, if the search space is not large, then even a random algorithm could be used. We set the maximum expected delay to 1 sec (10^3 ms), 10 sec (10^4 ms) and 1000 sec (10^6 ms). As the search space is smaller than that in RQ1, we also reduced the number of attempts to improve a solution before a restart for both SHC and SA as a compromise between exploring a local area and moving to a different search region. We set this number to 50.

Figures 6, Figure 7 and Figure 8 report the box-plots comparing the relative performance of RND, SHC, and SA for the first three synthetic and two real-world applications. Figure 7 has been made more readable by removing a single outlier value of 28,697 for the number of fitness evaluations of *Count Shared Data* when using SHC.

As expected, when the search space is small, RND performs comparably to both SHC and SA, as shown in Figure 7. However, when the search space size increases, SHC and SA perform better than RND, as shown in Figure 8 and 6.

One might argue that in Figure 7 *Average of Numbers* shows instances where some outliers for which SHC reaches almost the maximum number of iterations to align the events, which does not seem to be the case with RND. The explanation is that in small search spaces RND can perform as well as any other optimization algorithm, sometimes even better. It is worth noting that even in a search space of 10^3 ms, there were instances where RND could not find a solution even within the maximum number of iterations (*i.e.*, 10^6 random solutions). SHC and SA were successful in exposing a possible bug each and every time (in some cases with higher number of iterations, which resulted in the outliers).

We also observe differences between the box-plots for the *Count Shared Data* and *Average of Numbers*. We explain this observation by the fact that *Count Shared Data* contains more write actions (see Table 1) than *Average of Numbers*. In other words, it is relatively easier to align two events in *Count Shared Data* than in *Average of Numbers*. Indeed, there are only three possible ways to create an interference in *Average of Numbers* while *Count Shared Data* has 17 different possibilities, a six-fold increase which is reflected into the results of Figure 7.

Once the search space size increases as in Figure 8, RND is outperformed by SHC and SA. In general, SA tends to perform better across all PUTs.

Overall, we can answer our second research question as follows: **the dimension of the search space impacts the performance of the three search algorithms. SA performs the best for the all PUTs and different delays.**

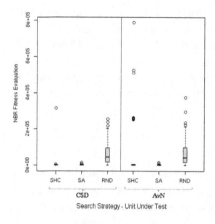
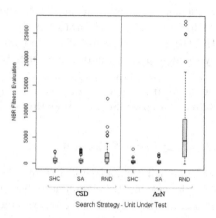

Fig. 7. (RQ2) Algorithm comparison for a search space up to 1 sec delay

Fig. 8. (RQ2) Algorithm comparison for a search space up to 10 sec delay

7   Discussion and Threats

Our results support the conceptual proof of the feasibility and the effectiveness of our search-based interference detection approach. They also show that our fitness function (Equation 4) is appropriate, as well as the usefulness of a virtual environment to enhance the probability of exposing interference bugs.

Exposed interferences are somehow artificial in nature as they are computed with respect to an ideal parallel execution environment. In fact, the identified schedules may not be feasible at all. This is not an issue, as we are trying to discover unprotected data accesses, and even if a bug is found with an unrealistic schedule, nothing prevents from being triggered by a different, feasible schedule. Making sure that shared data is properly protected makes code safer and more robust.

Although encouraging, our results are limited in their validity as our sample size includes only four small artificial and three real-world programs. This is a threat to *construct validity* concerning the relationship between theory and observations. To overcome this threat, we plan to apply our approach on more number of real-world programs in future work.

A threat to *internal validity* concerns the fact that, among the four artificial programs used, we developed three of them. However, they were developed long before we started this work by one of the authors to test the ReSP environment. Thus, they cannot be biased towards exposing interference bugs.

A threat to *external validity* involves the generalization of our results. The number of evaluated programs is small (a total of seven programs). Some of them are artificial, meant to be used for a proof of concept. Future work includes applying our approach to other large, real-world programs.

8   Conclusion

Detecting thread interference in multi-threaded programs is a difficult task as interference depends not only on the source code of the programs but also on the scheduler strategy, the workload of the CPUs, and the presence of other programs.

In this work, we proposed a novel approach based on running the programs under test on an ideal virtual platform, maximizing concurrency and inserting delays to maximize the likelihood of exposing interference between threads.

We used our fitness function and the three search algorithms to find the optimal delays on four small artificial and three real-world small/large multi-threaded programs. Our results show that our approach is viable and requires appropriate search strategies, as a simple random search won't find a solution in a reasonable amount of time.

Future work will be devoted to extending and enriching our interference model with more complex data structures such as pipes, shared memories or sockets.

References

1. Artho, C., Havelund, K., Biere, A., Biere, A.: High-level data races. Journal on Software Testing, Verification & Reliability, STVR (2003)
2. Beltrame, G., Fossati, L., Sciuto, D.: ReSP: a nonintrusive transaction-level reflective MPSoC simulation platform for design space exploration. In: Computer-Aided Design of Integrated Circuits and Systems, pp. 28–40 (2009)
3. Ben-Asher, Y., Farchi, E., Eytani, Y.: Heuristics for finding concurrent bugs. In: Proceedings of the 17th International Symposium on Parallel and Distributed Processing, p. 288.1(2003)
4. Bradbury, J.S., Cordy, J.R., Dingel, J.: Mutation operators for concurrent java (j2se 5.0) 1
5. Briand, L.C., Labiche, Y., Shousha, M.: Stress Testing Real-Time Systems with Genetic Algorithms. In: Proceedings of the 2005 Conference on Genetic and Evolutionary Computation (GECCO 2005), pp. 1021–1028 (2005)
6. Briand, L.C., Labiche, Y., Shousha, M.: Using Genetic Algorithms for Early Schedulability Analysis and Stress Testing in Real-Time Systems. Genetic Programming and Evolvable Machines 7, 145–170 (2006)
7. Carver, R.H., Tai, K.C.: Replay and testing for concurrent programs. IEEE Softw. 8, 66–74 (1991)
8. Drake, D.G.: JavaWorld.com: A quick tutorial on how to implement threads in java (1996), http://www.javaworld.com/javaworld/jw-04-1996/jw-04-threads.html
9. Edelstein, O., Farchi, E., Goldin, E., Nir, Y., Ratsaby, G., Ur, S.: Framework for testing multithreaded java programs. Concurrency and Computation: Practice and Experience 15(3-5), 485–499 (2003)
10. Eytani, Y., Ur, S.: Compiling a benchmark of documented multi-threaded bugs. In: Parallel and Distributed Processing Symposium, International, vol. 17, p. 266a (2004)
11. Farchi, E., Nir, Y., Ur, S.: Concurrent bug patterns and how to test them. In: Proceedings of the 17th International Symposium on Parallel and Distributed Processing. p. 286.2 (2003)
12. Flanagan, C., Freund, S.N.: Atomizer: A dynamic atomicity checker for multithreaded programs. Scientific Computer Program 71(2), 89–109 (2008)
13. Godefroid, P.: Model checking for programming languages using verisoft. In: Proceedings of the 24th ACM Symposium on Principles of Programming Languages, pp. 174–186 (1997)

14. Hovemeyer, D., Pugh, W.: Finding bugs is easy. In: ACM SIGPLAN Notices, pp. 132–136 (2004)
15. Hwang, G.H., Chung Tai, K., Lu Huang, T.: Reachability testing: An approach to testing concurrent software. International Journal of Software Engineering and Knowledge Engineering 5, 493–510 (1995)
16. Joshi, P., Naik, M., Seo Park, C., Sen, K.: Calfuzzer: An extensible active testing framework for concurrent programs
17. Lei, Y., Carver, R.H.: Reachability testing of concurrent programs. IEEE Trans. Softw. Eng. 32, 382–403 (2006)
18. Long, B., Strooper, P., Wildman, L.: A method for verifying concurrent java components based on an analysis of concurrency failures: Research articles. Concurr. Comput.: Pract. Exper. 19, 281–294 (2007)
19. Metropolis, N., Rosenbluth, A., Rosenbluth, M., Teller, A., Teller, E.: Equation of state calculations by fast computing machines. Journal of Chemical Physics 21, 1087–1092 (1953)
20. Musuvathi, M., Qadeer, S., Ball, T.: Chess: A systematic testing tool for concurrent software (2007)
21. Park, A.: Multithreaded programming (pthreads tutorial) (1999),
    http://randu.org/tutorials/threads/
22. Park, S., Lu, S., Zhou, Y.: Ctrigger: exposing atomicity violation bugs from their hiding places. SIGPLAN Not. 44, 25–36 (2009)
23. Software Quality Research Group, Ontario Institute of Technology: Concurrency anti-pattern catalog for java (2010),
    http://faculty.uoit.ca/bradbury/concurr-catalog/
24. Tripakis, S., Stergiou, C., Lublinerman, R.: Checking non-interference in spmd programs. In: 2nd USENIX Workshop on Hot Topics in Parallelism (HotPar 2010), pp. 1–6 (June 2010)
25. Wattenberg, M., Juels, A.: Stochastic hillclimbing as a baseline method for evaluating genetic algorithms. In: Proceedings of the 1995 Conference, vol. 8, p. 430. Kaufmann (1996)

Optimised Realistic Test Input Generation Using Web Services

Mustafa Bozkurt and Mark Harman

CREST Centre, Department of Computer Science,
University College London
Malet Place, London WC1E 6BT, UK
{m.bozkurt,m.harman}@cs.ucl.ac.uk

Abstract. We introduce a multi-objective formulation of service-oriented testing, focusing on the balance between service price and reliability. We experimented with NSGA-II for this problem, investigating the effect on performance and quality of composition size, topology and the number of services discovered. For topologies small enough for exhaustive search we found that NSGA-II finds a pareto front very near (the fronts are a Euclidean distance of ~ 0.00024 price-reliability points apart) the true pareto front. Regarding performance, we find that composition size has the strongest effect, with smaller topologies consuming more machine time; a curious effect we believe is due to the influence of crowding distance. Regarding result quality, our results reveal that size and topology have more effect on the front found than the number of service choices discovered. As expected the price-reliability relationship (logarithmic, linear, exponential) is replicated in the front discovered when correlation is high, but as the price-reliability correlation decreases, we find fewer solutions on the front and the front becomes less smooth.

1 Introduction

Testing is one of the most widely used and important ways in which software engineers gain confidence in systems' behaviour and with which they find faults. Testing is important for all kinds of software systems, but this paper is concerned with web service based systems. Trust is considered as one of the technical barriers' to enterprises transition to such service-centric systems (ScS) [6]. One of the potential solutions for establishing trust among different stakeholders is testing, which potentially provides the necessary assurance in correct functioning of ScS. Not surprisingly, this pressing need has led to a dramatic recent increase in the number of publications on ScS testing [5].

Service-centric systems present many challenges to the tester. The services available for use can change repeatedly, frequently and unpredictably, while many may offer substantially similar services, but with different quality and performance attributes and at different costs. For the tester, this presents the challenge of a rapidly changing, non-deterministic system with important choices to be made about the costs involved in testing; not merely execution costs, but monetary costs accrued from charges for third party service use.

G. Fraser (Ed.): SSBSE 2012, LNCS 7515, pp. 105–120, 2012.

In additional to the inherent complexities of testing of ScSs, there is also a need to construct realistic test cases [1,7,10,13,14,15]; test data that achieves fault revelation and which does so with test cases that are achievable in practice and understandable to the human tester and user alike. For instance, it has been argued that user session data is valuable precisely because it is real data [1,10,13]. Other authors have also developed testing approaches to harness realism in testing, drawing realistic data from the Graphical User Interfaces of the systems under test [7] and from 'web scavenging' for suitable test data [14,15].

Our approach to the challenge of ScS testing is to seek to exploit the flexibility and, more specifically, the composability of services as a key mechanism for finding suitable solutions. Essentially, we seek, not only to ameliorate problems in ScS testing, but to turn these problems around using these very same demanding characteristics as the basis for potential solutions. Our approach recommends composition topologies that deploy existing services to generate realistic tests.

In previous work [3] we introduced the tool ATAM, which implements this approach. ATAM finds services that combine to form a composition topology (an acyclic graph in which nodes are services and edges denote input-output connectivity between each service). We demonstrated that this approach has the potential to help automate the process of finding realistic test data.

ATAM recommends a set of candidate topologies from which a user can select a suitable choice, resulting in an approach to generate realistic test data for the service under test. However, this does not completely solve the tester's problem; there may be many services available from which to choose for each node of the topology. Since services may come at a cost and may offer differing degrees of reliability, there is an inherent multi-objective optimisation problem underpinning the final choice of services to select in order to concretise the topology. This choice is a balance the cost-benefit trade off.

In this paper we study this problem as a bi-objective typed selection problem, in which services of the right types for each node in the composition topology must be selected to balance testing cost against the reliability of the overall composition selected. We used NSGA-II to investigate the effect of three factors (composition size, composition topology and the number of services discovered) on performance (computation time) and quality (approximation to the pareto front). We used real world data on price to inform the cost choices and the results from our previous ATAM study [3] to determine the topologies to consider. However, we have no real-world reliability data available. Therefore, we study the effects of various models of the relationship between cost and reliability (logarithmic, linear and exponential) and various degrees of correlation strength (stronger, medium, weaker) between cost and reliability.

The primary contribution of this paper as follows:

1. We introduce a multi-objective solution to service-oriented test data generation[1]. This approach is the first to apply multi-objective optimisation to service-based testing.

---

[1] This work develops the 'fast abstract' formulation presented at SSBSE 2011 [4].

2. We present an investigation of the behaviour of our approach using different price-reliability models. We confirm that NSGA-II performs almost optimally on those problems for which problem size is sufficiently small to support exhaustive search for the globally optimal pareto front. On larger problems we find that size and topology have more effect on quality and performance than the number of service choices discovered. This is encouraging because it provides evidence that the approach may cope with (widely anticipated) significant increases in the number of services that will become available in the coming years.

The rest of this paper is organised as follows. Section 2 discusses existing test data generation approaches that use optimisation techniques and the concept of QoS in SOA. Section 3 explains the details of the proposed multi-objective optimisation for our approach. Sections 4 present the case study we used in our experiments. Section 5 describes our experimental methodology and presents discussion on the results from the experiments. Section 6 presents future work and concludes the paper.

2   Background

In this section, we discuss the approaches to test data generation that are most closely related to our approach and also provide a brief information on the concept of QoS in SOA. Due to space constraints we did not include how ATAM build topologies and group discovered services. The concept of service-oriented test data generation and details of the ATAM tool are discussed in an earlier paper [3].

2.1   Test Data Generation and Optimisation

Our approach formulates test data generation as application for multi-objective SBSE [11]. We seek service compositions that generate realistic tests of a given type using services not are necessarily designed for testing.

Multi-objective optimisation is not new to test data generation domain. Though our work is the first to use a multi-objective approach to service-based testing. Lakhotia et al. [12], Oster and Saglietti [17] and Pinto and Vergilio [18] already proposed multi-objective test data generation. These approaches focus on structural coverage as the main objective and use additional objectives such as execution time, memory consumption and size of test set. There are other approaches such as Sagarna and Yao [20] where branch coverage is formulated as constrained optimisation problem. The use of multi-objective optimisations in test case selection has also been proposed. For example, Yoo and Harman [22] used objectives such as code coverage and execution cost in test case selection.

Search-based techniques are gaining popularity in testing and earlier approaches applied some of these techniques to SOA. For example, Blanco et al. [2] used scatter search to generating test data for service compositions. Di Penta et al. [8] applied search-based techniques to reveal service level agreement violations.

2.2   Quality of Service Models

Our approach leverages the ability to compose and orchestrate web services to achieve goals for which individual services may not have been defined. One of the advantages SOA is the ability to discover and compare services with similar business logic. The ability to compare services is enabled by a well-known concept called QoS in SOA.

QoS requirements generally refer to several functional and non-functional qualities of services in SOA [5,21]. According to Wan Ab. Rahman [21] the most common QoS characteristics are service response time, performance, reliability, execution price (price from here on), availability and security. The explanation given to the four of these characteristics that are covered in this paper are:

1. **Price**: The monetary cost of invoking a service. There might exist multiple price plans for services but ATAM only considers the pay per use option.
2. **Availability**: The presence of a service for invocation. We consider availability as the percentage of time that the service available for invocation.
3. **Response time**: The amount of time that a service takes to respond various requests. Response time might be provided in different ways such as average or maximum time.
4. **Reliability**: The capability of maintaining service operations and the expected quality. Reliability is regarded as the summary or the estimation of overall service quality. We accept reliability as the percentage of invocations which are not erroneous and executed within the expected QoS levels.

The need for a QoS model that covers necessary quality aspects of SOA also addressed by industry. OASIS introduced a QoS specification called Business QoS (bQoS) [16] in 2010.

3   Multi-objective Optimisation

In this section, we explain the necessary elements required to support replication of results for the multi-objective optimisation approach we propose here: the objective function and genetic operators. We also discuss the choice of our algorithm and parameter values selected for the genetic algorithm.

3.1   Objective Function

We introduce three different objective functions (Equation 1, 2 and 3) for the four QoS parameters we intend to include in the next version of ATAM: price, reliability, availability and response time. The reason for having three different functions instead of four is caused by the possibility of using the same formulation for both reliability and availability.

The objective function for the price parameter is straightforward. The function is the sum of the costs of all services required in a topology. In our approach we considered price as an objective rather than a constraint in order to allow the tester to explore all the possible solutions on the pareto-optimal front. The following is introduced as the objective function for minimising total cost of test data generation:

$$Minimize \sum_{i=1}^{n} p_{s_i} \tag{1}$$

where n is the total number of services in the selected topology and p_{s_i} is price of the ith service.

The objective function for the other three QoS characteristics are not as straightforward. The total reliability, availability or response time of a topology is as high as the combined value of each characteristic for all the services used in that topology. We introduced the concept of 'combined characteristic value' (referred to as combined value from this point on) because the reliability, availability and response time of an individual service is not solely defined by the behaviour of that service in isolation. For example in the case of reliability, the reliability of a service S also depends on the reliability of the services that generate the necessary input for S. Figure 1 illustrates an example solution and explains necessary concepts in combined value calculations.

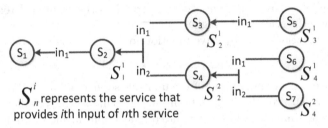

S_n^i represents the service that provides ith input of nth service

Fig. 1. Example topology scenario for generating realistic test data using services. In the figure, each node in the topology represents a service description. In this scenario service S_1 is the service that provides the required test input. The services on the other end, such as S_5, S_6 and S_7, are services that either require tester input or automatically generated input using a predefined algorithm. Each edge in the figure represents an input required by the service that is targeted by the edge. The source of the edge is the service that provides the required input. For example, in this scenario, the input for S_1 is provided by S_2.

The same formulation can be used in the calculation of combined values for reliability and availability. For response time however, we provide a third formulation. Due to space constraints we explain our formulation for reliability and availability focusing on the reliability only. The combined value for reliability (referred to as combined reliability (cr)) of a service is formulated as:

$$cr(S_n) = r_{S_n} \times ir(S_n)$$

where $cr(S_n)$ is the combined reliability and r_{S_n} is the reliability of the service S_n and $ir(S_n)$ is the reliability function that calculates the combined reliability of the services that provide inputs for S_n.

The reliability calculation for inputs $(ir(S_n))$ varies based on the number of services providing the input. This is because in our approach a service in the composition can get the required inputs in two possible manner:

case1 *From the tester or predefined algorithm*: In this case, the input reliability of the service is accepted 100% reliable (i.e. reliability score $= 1.0$). Services S_5, S_6 and S_7 in Figure 1 are examples to this case.

case2 *From some arbitrary number of services*: In this case, the input reliability of the service is equal to the lowest of the combined reliability of the services that provide its inputs. For example, service S_3 takes input from service S_5 while S_4 takes input from two services S_6 and S_7 as depicted in Figure 1.

In the light of these definitions, we formulated our input reliability function to suit these two cases as follows:

$$ir(S_n) = \begin{cases} 1.0 & \text{if } S_n \text{ is case 1} \\ MIN(cr(S_n^1), cr(S_n^2), \dots, cr(S_n^{in(S_n)})) & \text{if } S_n \text{ is case 2} \end{cases}$$

where S_n^i is the service providing ith input for service S_n and $in(S_n)$ is the total number of inputs service S_n has.

The reliability score of a composition is equal to the combined reliability of the first service (service at the highest level). In the light of the given combined reliability calculation, the objective function for maximising the total reliability is formulated as:

$$Maximise \; r_{S_1} \times ir(S_1) \tag{2}$$

Having services that require inputs from multiple services allows ATAM to invoke the services which provide the inputs in parallel. Due to this parallel invocation ability response time needs to be calculated in a similar fashion to reliability and availability.

We formulated combined response time as:

$$cres(S_n) = res_{S_n} + ires(S_n)$$

where $cres(S_n)$ is the combined response time and res_{S_n} is the response time of the service S_n and $ires(S_n)$ is the response time function that calculates the combined response time of the services that provide inputs for S_n.

There is also a minor difference in combined response time calculation. Even though the cases (case1 and case2) are also valid for response time the values for the cases are different. Function that suits these two cases for response time as follows:

$$ires(S_n) = \begin{cases} 0 & \text{if } S_n \text{ is case 1} \\ MAX(cres(S_n^1), cres(S_n^2), \dots, cres(S_n^{in(S_n)})) & \text{if } S_n \text{ is case 2} \end{cases}$$

where S_n^i is the service providing ith input for service S_n and $in(S_n)$ is the total number of inputs service S_n has.

We believe the testers use expected response time as a means to minimise the time it takes to generate test cases. As a result, the objective function needs to minimise the response time.

$$Minimise \; res_{S_1} + ires(S_1) \tag{3}$$

3.2   Representation and Genetic Operators

After discovering all possible topologies ATAM asks the tester to select one topology for test data generation. Then ATAM applies the optimisation process only to the selected topology using the service groups for this topology.

ATAM was implemented in a way to facilitate genetic operators (mutation and crossover) with service groups and topologies. As a result, in our multi-objective solution, genomes are represented as an array of integers as depicted in Figure 2. Each element in the array represents a 'service group' and values in each element represent a service that belongs to that group. The numbering of service groups and web services are based on the order of their discovery. For example, if the tester selects one of the topologies from Figure 2 the values of the elements in each genome represent the services which are in the places of $S_1, S_2, ..., S_9$ based on their order in the genome.

The initial population is generated using the number of services in each service group that form the selected topology. Each solution in the initial population is generated by randomly assigning a number to each array between 0 and the number of services in the service group which is represented by the element.

The mutation operator modifies the value in an element with a number between 0 and the number of services that the group contains. The crossover operator produces a new solution by combining two solutions into a new solution as depicted in Figure 2.

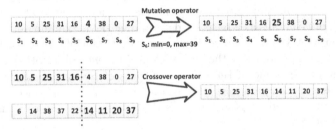

Fig. 2. The mutation and the crossover operators

3.3   Mutli-Objective Algorithm and Parameters

We selected the Non-dominated Sorting Genetic Algorithm-II (NSGA-II) as optimization technique for our problem due to its reported performance against other algorithms for similar problems [23]. We used a modified version of the popular ECJ framework [9] which provides a built-in NSGA-II algorithm, and integrated it to ATAM. After some tuning we found out that the ideal parameters that provide the most diverse solutions for our problem are: 35% mutation probability (for each gene) and one-point crossover with 50% crossover probability.

4   Case Studies

Not having real-world case studies is one of the major set backs in ScS testing research [5]. As expected, we were unable to find existing services with measures of suitable QoS values, thus we evaluated our approach by simulation using a synthetically generated case study. The only QoS characteristic that the authors have access to is price. As a result, we selected a group of existing commercial services with publicly available pricing details (presented in Table 1) which are

Table 1. Services used as a basis for the synthetically generated case study. The services and the given prices in this table are collected from Remote Methods website [19].

Service		Price (per query)			
Group	Description	Max	Company	Min	Company
1	Phone verification	$0.300	StrikeIron	Free	WebServiceMart
2	Traffic information	$0.300	MapPoint	Free	MapQuest
3	Geographic data	$0.165	Urban Mapping	$0.010	Urban Mapping
4	Bank info verification	$0.160	Unified	$0.090	Unified
5	IP to location	$0.020	StrikeIron	Free	IP2Location
6	Stock Quote	$0.020	XIgnite	$0.008	CDYNE
7	Historical financial data	$0.017	Eoddata	$0.007	XIgnite
8	Nutrition data	$0.010	CalorieKing	Free	MyNetDiary
9	Web search	$0.005	Google	Free	Bing

collected from the Remote Methods website [19]. Due to access restrictions on non-commercial users we were unable to verify the accuracy of the provided prices. We were also unable to verify continual service availability.

For 7 out of 9 groups in Table 1 we were able to identify at least two existing services with price details as a basis for determining the maximum and minimum prices. For groups 3 and 4 we were unable to find multiple services. However, for these two groups we used the prices from different subscription models of the same service. In some of the groups, outputs of free to use services might not be the same as the paid ones (due to the level of detail provided), however, we accepted these services as alternatives to the paid ones due to their similar service descriptions.

In our case study we focused on two QoS characteristics: price and reliability. The reason for choosing price is due to existence of real-world references and for choosing reliability is our better understanding of the concept compared to availability and response time.

At present, making a realistic projection on the relation between reliability and price is a challenging task due to lack of real-world reliability data. To compensate, we defined three price-reliability models (referred to as model) that construe the relation between price and reliability as illustrated in Figure 3.

We started generating our case study with the assumption that reliability and price are positively correlated. The reliability value for each service in each

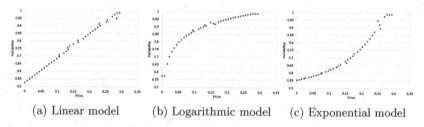

(a) Linear model   (b) Logarithmic model   (c) Exponential model

Fig. 3. Example reliability-price distribution of services in a group (Group size: 40 services) for all three models (with three very strong correlation).

group cover is between 0.50 to 0.99. The price value for each service in each group is assigned using the minimum and maximum prices given on the Table 1. We generated 9 service groups with 40 services in each group for all 3 models initially. Data for other group sizes (20 and 30) are generated by removing service entries from the initially generated groups.

Composition size in the experiments represents the number of groups (starting from Group 1) that are included in a given topology. For example, composition size 4 means a composition which includes the first 4 groups.

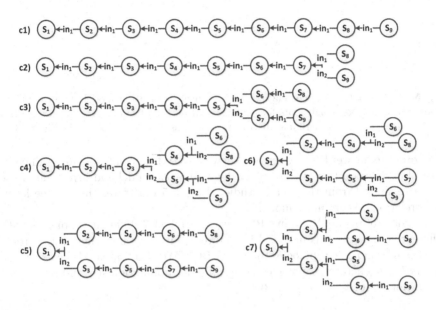

Fig. 4. Illustration of the topologies that are used in order to answer RQ2 and RQ4

5   Research Questions and Experimental Validation

In this section, we provide the four research questions we used to investigate our approach, our method of investigation, results from our experiments and our answers to the research questions.

5.1   Research Questions

We ask the following four questions:

RQ1 Is there a relation between the used price-reliability model and the discovered pareto front?

RQ2 How is discovered pareto front affected by different levels of price-reliability correlation in the models?

RQ3 What are the effects of parameters such as composition size, group size and topology complexity on the discovered pareto front?

RQ4 What are the effects of parameters on performance?

5.2   Method of Investigation

In order to answer RQ1, we generated pareto fronts for all three models with most of the possible combinations (allowed by our cases study) using different group sizes (20, 30 and 40) and composition sizes (3,4,5,7 and 9). Then we investigated the fronts generated from these groups.

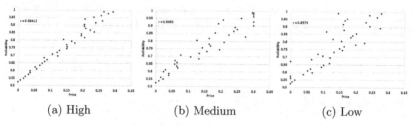

(a) High   (b) Medium   (c) Low

Fig. 5. Different levels of correlation for the linear correlation model. The calculated Pearson's correlation coefficients for each correlation level from low to high are 0.8574, 0.9459 and 0.98412. The original distribution for this group is depicted in Figure 3a.

In order to answer RQ2, we created new case studies by introducing 3 different level of correlation (low, medium and high) in each model and investigated the fronts generated from these case studies. Figure 5 illustrates these three levels of correlation on the linear model.

In order to answer RQ3 and RQ4, we developed 7 different topologies with different complexities as depicted in Figure 4. First, we investigated the fronts generated from these topologies to answer RQ3. We also compared the execution times of the generated topologies using different population size and number of generations in order to answer RQ4.

5.3   Results

The discovered pareto fronts for all models are depicted in Figure 6 and 9 (using the topology complexity (c1) in Figure 4). As can be observed from these figures, the discovered fronts are very closely related to the model used. Due to space constraints we only provide the fronts discovered using the logarithmic model from the original case study. We found out that the other models also comply with the trends observed from the logarithmic model. This similarity can be also observed in Figure 9.

The results suggest that the group size and composition size does have a minor effect on the overall form of the discovered pareto front. Composition size, as expected, causes a shift on the position of the discovered front. The results provide evidence for the fact that the observed shift is closely related with price of the added/eliminated service group in the composition. As depicted in Figure 6a, the biggest shift is observed between composition sizes 3 and 4 due to the relatively high price of the services in group 4. The effect of the reliability in this case is less observable due to the relatively low difference between reliability scores of services in each group.

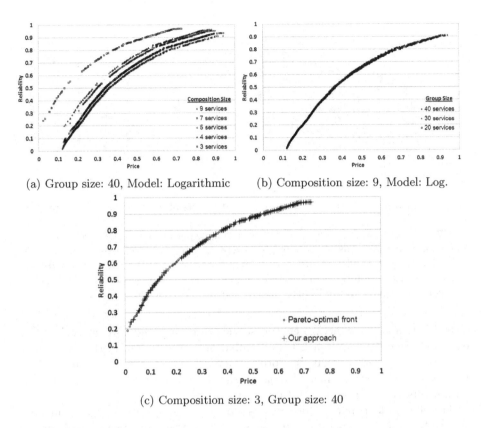

(a) Group size: 40, Model: Logarithmic   (b) Composition size: 9, Model: Log.

(c) Composition size: 3, Group size: 40

Fig. 6. The effects of the parameters used in our experiments on the generated pareto-front. Composition size represents the number of service groups that form a topology. Group size represents the number of web services in a service group. Model represents price-reliability relation. Figure 6c depicts the difference between the globally optimal front and the front discovered by our approach.

Search-based optimisation techniques are often applied to problems with large solution sets due to their effectiveness in such scenarios. In our problem, the search space is not fixed and it can vary greatly (based on the composition and group sizes). We investigated the performance of our approach in small search spaces in order to compare to known globally optimal fronts. For topologies small enough for exhaustive search we found that NSGA-II finds a pareto front very near the true pareto front as depicted in Figure 6c. The Euclidean distance between the optimal front and the discovered front is ~ 0.00024 price-reliability points. We also measured the average execution time of our approach and the time it takes to calculate the optimal pareto front in Figure 6c. Exhaustive calculation of the optimal front took 4.5 seconds where as our approach (using NSGA-II with population size 400 and 100 generations) took 0.8347 seconds.

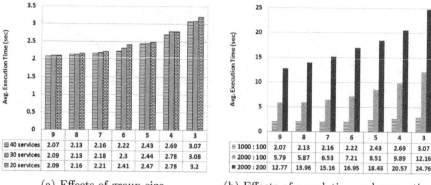

(a) Effects of group size   (b) Effects of population and generations

Fig. 7. The effects of the parameters used in our experiments on the execution time. The values on the x-axis of each graph represent the composition sizes. The average execution times for different group and composition sizes are provided in Figure 7a. The average execution times in Figure 7b are for different population sizes and number of generations (given in the format "population size : number of generations").

In SOA, a short interval between service discovery and invocation is often expected. Since our approach uses a global optimisation on all the discovered services it is important that the optimisation process should take minimal time to execute. This requirement led us to the investigation of the effects of the parameters of our approach on the execution time.

The effects of the parameters such as group size, composition size, population size and number of generations on the execution time are depicted in Figure 7. The results suggest that the effect of group size is negligible as depicted in Figure 7a. The performance difference between groups' sizes is only fraction of a second for all composition sizes ranging from 0.02s up to 0.13s, whereas for almost all group sizes there is about 1s difference in execution times between composition sizes 9 and 3.

One unexpected effect which is very noticeable in both graphs is the negative correlation between the composition size and the execution time. We believe that this negative correlation is caused by ECJ's internal mechanism that maintains crowding distance. In order to prove our hypothesis, we measured the execution of the same case study with different population sizes and different number of generations. The results from these experiments are depicted in Figure 7b.

During our experiments we also investigated the relation among composition size, number of generations and execution time. Our initial assumption was that higher the composition size and the number of generations longer it should take longer to execute. The results provided evidence for the validity of our assumption. When we increased the number of generations from 100 to 200 execution time increased 2.21 times for composition size 9 whereas 2.03 times for composition size 3.

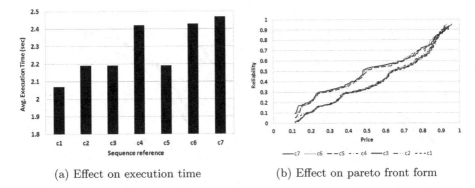

(a) Effect on execution time   (b) Effect on pareto front form

Fig. 8. The effects of topology complexity on pareto-front and execution time. The labels on the graphs ((c1) to (c7)) represent the results for the topologies in Figure 4. Pareto fronts in Figure 8b are presented in a different form (lines instead of individual solutions as points) than the previous figures in order to make these 7 pareto fronts more distinguishable.

As mentioned, the results up to this point in the paper are collected using the topology (c1) in Figure 4. The effects of complexity on execution time and pareto front are depicted in Figure 8. Investigation of the results in Figure 8a shows a positive correlation between the number of services that require multiple inputs and the execution time. As in the case of topology (c1) the lowest execution time achieved because this topology doesn't contain any services with multiple inputs. Whereas topologies (c4), (c6) and (c7) (topologies containing 2 multiple input services) have the highest execution times. Another important observation was that a topology's form has a negligible effect on execution times. For example, very similar topologies such as (c5) and (c7) have different execution times.

Complexity is found to have the highest effect among all parameters on the from of the discovered pareto front. This effect can clearly be observed in Figure 8b where the discovered fronts are separated into two distinct groups. The first group consists of topologies (c5), (c6) and (c7) and the second group consists of (c1), (c2), (c3) and (c4). The cause of this grouping is the similarities between the forms of the topologies in each group. The effect of complexity is almost negligible in higher reliability scores whereas, for the rest of the pareto front it is more visible. The fronts (c1) and (c7) in Figure 8b are a distance of ~ 0.073 price-reliability points apart.

The results suggest that correlation level in models have the highest effect on the form of the discovered pareto fronts as depicted in Figure 9. Comparison between the fronts of the original case study (Figure 6) and the modified ones shows that there is a positive correlation between the correlation level and the number of solutions on the discovered front. Correlation level has the same effect on all three models as depicted in Figure 9. We believe that this effect is caused by NSGA-II's tendency to generate solutions around the services with better QoS scores than expected in the model.

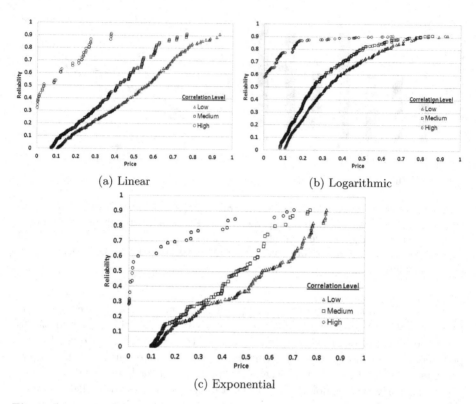

(a) Linear   (b) Logarithmic

(c) Exponential

Fig. 9. Pareto fronts discovered from all three models (Composition Size: 9, Group size: 40 services) with different levels of correlation. The original distribution for this group is depicted in Figure 6.

5.4   Answers to Research Questions

The results depicted in Figure 6 and 9 **answer RQ1** by providing evidence for the fact that the *cost-reliability relationship is replicated* in the discovered pareto fronts *when correlation is high*. However, as the price-reliability *correlation decreases, less smooth fronts with fewer solutions* are discovered.

The results also provide evidence for the fact that level of correlation in a model can effect the form of the discovered pareto fronts. The results, which **answer RQ2**, suggest a *negative correlation between the model correlation level and the number of solutions on the discovered front*.

In order to **answer RQ3** we have to provide a separate explanation for each parameter. The evidence from Figure 6b suggests that *group size has a negligible effect* on the discovered pareto front compared to composition size and topology complexity. As for the effects of the composition size, the results suggest that as one might hope, the *lower the composition size, the cheaper and more reliable the solution* will be (using the same set of services and topology). The results from

Figure 8b suggest that among all the three parameters we investigated *topology complexity has relatively highest impact.*

As for the **answer to RQ4** regarding the performance, we find that *composition size has the strongest effect*, with smaller topologies consuming more machine time; a curious effect we believe is due to the influence of crowding distance. The results suggest that *group size has a negligible effect.*

6   Conclusion and Future Work

In this paper, we presented a multi-objective solution to service-oriented test data generation. We focused on the cost of test data generation and the reliability of the test data sources as our primary objectives. We chose a widely used multi-objective algorithm NSGA-II and investigated the behaviour of our approach in possible different situations, drawn from different models and parameters for our problem and solution domain. The results provided evidence for robustness of our approach.

As for future work, we believe that our approach can be extended to be used in multi-objective service selection in service composition. The results from our experiments provide evidence for the robustness of our approach and to the possibility of adapting it to service selection.

References

1. Alshahwan, N., Harman, M.: Automated session data repair for web application regression testing. In: ICST 2008, pp. 298–307. IEEE, Lillehammer (2008)
2. Blanco, R., García-Fanjul, J., Tuya, J.: A first approach to test case generation for BPEL compositions of web services using scatter search. In: ICSTW 2009, pp. 131–140. IEEE, Denver (2009)
3. Bozkurt, M., Harman, M.: Automatically generating realistic test input from web services. In: SOSE 2011, pp. 13–24. IEEE, Irvine (2011)
4. Bozkurt, M., Harman, M.: Optimised realistic test input generation. Presented at the SSBSE 2011 (September 2011),
   http://www.ssbse.org/2011/fastabstracts/bozkurt.pdf
5. Bozkurt, M., Harman, M., Hassoun, Y.: Testing & verification in service-oriented architecture: A survey. STVR (to appear)
6. CBDI Forum, http://everware-cbdi.com/cbdi-forum
7. Conroy, K., Grechanik, M., Hellige, M., Liongosari, E., Xie, Q.: Automatic test generation from GUI applications for testing web services. In: ICSM 2007, pp. 345–354. IEEE, Paris (2007)
8. Di Penta, M., Canfora, G., Esposito, G., Mazza, V., Bruno, M.: Search-based testing of service level agreements. In: GECCO 2007, pp. 1090–1097. ACM, London (2007)
9. ECJ 20, http://cs.gmu.edu/~eclab/projects/ecj/
10. Elbaum, S., Rothermel, G., Karre, S., Fisher II, M.: Leveraging user-session data to support web application testing. IEEE Transactions on Software Engineering 31(3), 187–202 (2005)

11. Harman, M.: The current state and future of search based software engineering. In: Briand, L., Wolf, A. (eds.) FOSE 2007, Los Alamitos, CA, USA, pp. 342–357 (2007)
12. Lakhotia, K., Harman, M., McMinn, P.: A multi-objective approach to search-based test data generation. In: GECCO 2007, pp. 1098–1105. ACM, London (2007)
13. Luo, X., Ping, F., Chen, M.H.: Clustering and tailoring user session data for testing web applications. In: ICST 2009, pp. 336–345. IEEE, Denver (2009)
14. McMinn, P., Shahbaz, M., Stevenson, M.: Search-based test input generation for string data types using the results of web queries. In: ICST 2012, pp. 141–150. IEEE, Montreal (2012)
15. McMinn, P., Stevenson, M., Harman, M.: Reducing qualitative human oracle costs associated with automatically generated test data. In: STOV 2010, pp. 1–4. ACM, Trento (2010)
16. OASIS: SOA-EERP business quality of service (bQoS) (September 2009), http://docs.oasis-open.org/ns/soa-eerp/bqos/200903
17. Oster, N., Saglietti, F.: Automatic Test Data Generation by Multi-objective Optimisation. In: Górski, J. (ed.) SAFECOMP 2006. LNCS, vol. 4166, pp. 426–438. Springer, Heidelberg (2006)
18. Pinto, G.H.L., Vergilio, S.R.: A multi-objective genetic algorithm to test data generation. In: ICTAI 2010, vol. 1, pp. 129–134. IEEE, Arras (2010)
19. Remote Methods, http://www.remotemethods.com/
20. Sagarna, R., Yao, X.: Handling constraints for search based software test data generation. In: ICST 2008, pp. 232–240. IEEE, Lillehammer (2008)
21. Wan Ab. Rahman, W., Meziane, F.: Challenges to describe QoS requirements for web services quality prediction to support web services interoperability in electronic commerce. In: IBIMA 2008, vol. 4, pp. 50–58. International Business Information Management Association (IBIMA), Kuala Lumpur (2008)
22. Yoo, S., Harman, M.: Pareto efficient multi-objective test case selection. In: ISSTA 2007, pp. 140–150. ACM, London (2007)
23. Zhang, Y., Harman, M., Mansouri, S.A.: The multi-objective next release problem. In: GECCO 2007, London, UK, pp. 1129–1137 (July 2007)

Improving Software Security Using Search-Based Refactoring

Shadi Ghaith[1] and Mel Ó Cinnéide[2]

[1] IBM Dublin Software Lab., Ireland
sghaith@ie.ibm.com
[2] School of Computer Science,
University College Dublin, Ireland
mel.ocinneide@ucd.ie

Abstract. Security metrics have been proposed to assess the security of software applications based on the principles of "reduce attack surface" and "grant least privilege." While these metrics can help inform the developer in choosing designs that provide better security, they cannot on their own show exactly how to make an application more secure. Even if they could, the onerous task of updating the software to improve its security is left to the developer. In this paper we present an approach to *automated* improvement of software security based on search-based refactoring. We use the search-based refactoring platform, Code-Imp, to refactor the code in a fully-automated fashion. The fitness function used to guide the search is based on a number of software security metrics. The purpose is to improve the security of the software immediately prior to its release and deployment. To test the value of this approach we apply it to an industrial banking application that has a strong security dimension, namely *Wife*. The results show an average improvement of 27.5% in the metrics examined. A more detailed analysis reveals that 15.5% of metric improvement results in real improvement in program security, while the remaining 12% of metric improvement is attributable to hitherto undocumented weaknesses in the security metrics themselves.

1 Introduction

Software security is generally defined as the engineering of software so that continues to function correctly under malicious attack [15]. It includes matters such as ensuring that the software is free of vulnerabilities such as buffer overflow or unhandled errors, and that SQL queries are not formed with untrusted user input. In this paper we focus on software security defined as the exposure of *classified* data to the rest of the program [1]. Classified data is data whose release to an unauthorised user would constitute a breach of security. For example, a person's bank details are classified, while the address of particular bank branch would not be classified. If we reduce the extent to which classified data is exposed to the rest of the program, we can reduce the chance that a malicious attack on the deployed program will succeed in accessing classified data.

G. Fraser (Ed.): SSBSE 2012, LNCS 7515, pp. 121–135, 2012.

On the other hand, refactoring is defined as a process that improves software design without changing observable behavior [9]. While refactoring is normally carried out "by hand" using only the refactoring support provided by the IDE, more recent research has investigated the possibility of using a fully automated approach that relies on search-based refactoring [10,11,13,21]. In this approach, a search technique such as hill climbing, simulated annealing or a genetic algorithm is used to drive the refactoring process, which is guided by a software metric, or set of metrics, that the developer wishes to optimise.

In this paper we explore the possibility of using such an automated refactoring approach with the goal of automating the improvement of program security. In a sense, this use of automated refactoring is far more promising than using it to improve software design. Automated refactoring is liable to change the design of the refactored program radically. Even though the new design may be better in many ways, the developers have to invest time to understand it and this may prove to be more costly than the benefits accrued from the design improvements achieved. However, if automated refactoring is used to improve program security, then this drawback does not arise. Security only matters when the software is released and deployed and hence open to attack. Refactoring to improve security can therefore be applied as part of the final build process prior to the release and deployment of the software. The developers can continue to work with the original program whose design is familiar to them

The remainder of this paper is structured as follows. In section 2 we present an overview of the Alshammari et al. security metrics that we use in this paper. In section 3 we present the search-based refactoring platform, Code-Imp, and apply this to an industrial example in section 4. Related work is surveyed in section 5 and our conclusions and future work are presented in section 6.

2   Overview of Security Metrics

The security metrics we use in this paper are those defined by Alshammari et al. [1,3,4]. These metrics are based on the information flow within a program or software design and cover such areas as data encapsulation, cohesion, coupling, composition, extensibility, inheritance and design size. The formulae for these metrics are introduced based on the concept of classified (critical) members and classes. As defined above, a classified attribute is one whose release to an unauthorised user would constitute a breach of security. A classified method is one that interacts directly with a classified attribute. A class is said to be *critical* if it contains at least one classified attribute or method. Table 1 summarizes the Alshammari et al. security metrics that we use to guide the search-based refactoring process. In the remaining paragraphs of this section these metrics are introduced in more detail, based on descriptions by Alshammari et al. [1].

Table 1. Summary of the Alshammari et al. Security Metrics [1]

Metrics Group	Metrics Description
The cohesion-based metrics (CMAI, CAAI, CAIW, CMW)	measure the potential flow of classified attributes' values to accessor and mutator methods, designs with a large amount of classified flow.
The coupling-based metric (CCC)	measures interactions between classes and classified attributes, rewarding designs that minimise such interactions.
The composition based metric (CPCC)	penalises designs with critical classes higher in the class hierarchy, where they can be accessed by a large number of subclasses.
The extensibility-based metrics (CCE and CME)	rewards designs with fewer opportunities for extending critical classes or classified methods.
The inheritance-based metrics (CSP, CSI, CMI and CAI)	reward designs with fewer opportunities for inheriting from critical superclasses.
The design size-based metric (CDP)	rewards designs with a lower proportion of critical classes.
The data encapsulation-based metrics (CIDA, CCDA and COA)	assess the accessibility of classified attributes and methods.

Cohesion-Based Metrics:

- *Classified Mutator Attribute Interactions (CMAI)* is the ratio of the sum (for all classified attributes) of the number of mutator methods that may access classified attribute to the total number of possible interactions between the mutator methods and classified attributes. A mutator method is one that can set the value of an attribute.
- *Classified Accessor Attribute Interactions (CAAI)* is the ratio of the sum (for all classified attributes) of the number of accessor methods that may access classified attribute to the total number of possible interactions between the accessor methods and classified attributes. An accessor method is one that can return the value of an attribute.
- *Classified Attributes Interaction Weight (CAIW)* is the ratio of the sum (for all classified attributes) of the number of methods that may access the classified attribute to the sum (for all attributes) of the number of methods that may access the attribute.
- *Classified Methods Weight (CMW)* is the ratio of the number of classified methods to the total number of methods.

Coupling Metric:

- The *Critical Classes Coupling (CCC)* metric aims to find the degree of coupling between classes and classified attributes in a given design. It is the ratio of the number of links from classes to classified attributes defined in other classes to the total number of possible links from all classes to classified attributes defined in other classes.

Composition Metric:

– The *Composite-Part Critical Classes (CPCC)* metric is the ratio of the number of critical composed-part classes to the total number of critical classes.

Extensibility Metrics:

– The metric *Critical Classes Extensibility (CCE)* is defined as the ratio of the number of the non-finalised critical classes to the total number of critical classes.
– The metric *Critical Methods Extensibility (CME)* is defined as the ratio of the number of the non-finalised classified methods to the total number of classified methods.

Inheritance Metrics:

– *Critical Superclasses Proportion (CSP)* is the ratio of the number of critical super classes to the total number of critical classes in an inheritance hierarchy.
– *Critical Superclasses Inheritance (CSI)* is the ratio of the sum of classes that may inherit from each critical superclass to the number of possible inheritances from all critical classes in a class hierarchy.
– *Classified Methods Inheritance (CMI)* is the ratio of the number of classified methods that can be inherited in a hierarchy to the total number of classified methods in that hierarchy.
– *Classified Attributes Inheritance (CAI)* is the ratio of the number of classified attributes that can be inherited in a hierarchy to the total number of classified attributes in that hierarchy.

Design Size Metric:

– (CDP) Design size simply takes into account the size, i.e. the number of classes, in a given program.

Data Encapsulation (Accessibility) Metrics:

– *Classified Instance Data Accessibility (CIDA)* is ratio of classified instance public attributes to classified instance attributes.
– *Classified Class Data Accessibility (CCDA)* is ratio of classified class public attributes to classified class attributes.
– *Classified Operation Accessibility (COA)* is the ratio of classified public methods to classified methods.

3   Overview of Code-Imp and Refactorings

Code-Imp (Combinatorial Optimisation for Design Improvement) is a fully automated refactoring framework developed in order to facilitate experimentation in automatically improving the design of existing programs [11, 16, 19, 20]. It takes

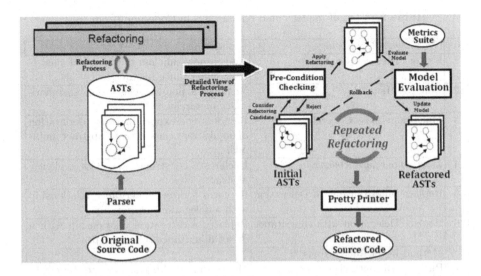

Fig. 1. Architecture of the Code-Imp automated refactoring framework (from [11])

Java version 6 source code as input and produces as output a refactored version of the program. Its output also comprises applied refactorings and metrics information gathered during the refactoring process.

Figure 1 depicts the architecture of Code-Imp. The right side of the figure shows the process of refactoring in detail. Code-Imp first extracts the initial ASTs (Abstract Syntax Trees) from the source code. Code-Imp then searches the ASTs for candidate refactorings. A refactoring is acceptable its pre- and post-conditions are satisfied and it complies with the demands of the search technique in use (in the case of a hill-climb, this means improving the quality of the design based on the metrics suite; in the case of, e.g. simulated annealing, a drop in quality may also be accepted). This process is repeated many times. After the final refactoring is applied, the ASTs are pretty printed to source code files. During the refactoring process, a rollback mechanism is supported by logging each change to the ASTs. The change history service makes it possible to perform a rollback at different levels of granularity. For example, at the finest level of granularity, individual refactorings can be reversed. At a coarser level of granularity, a composite refactoring such as a Pull Up Method (which also contains a Pull Up Field refactoring) can be reversed.

There are three aspects to the search-based refactoring process that takes place:

– the set of refactorings that can be applied;
– the type of search technique employed;
– the fitness function that directs the search.

Code-Imp currently supports 14 design-level refactorings categorized into three groups according to their scope as shown in Table 2. These are roughly based on refactorings from Fowler's catalogue [9], though they differ somewhat in the

Table 2. A list of implemented refactorings in Code-Imp (from [16])

No.	Class-Level Refactorings	Description
1	Extract Hierarchy	Adds a new subclass to a non-leaf class C in an inheritance hierarchy.
2	Collapse Hierarchy	Removes a non-leaf class from an inheritance hierarchy.
3	Make Superclass Concrete	Removes the explicit *abstract* declaration of an abstract class without abstract methods.
4	Make Superclass Abstract	Declares a constructorless class explicitly abstract.
5	Replace Inheritance with Delegation	Replaces a direct inheritance relationship with a delegation relationship.
6	Replace Delegation with Inheritance	Replaces a delegation relationship with a direct inheritance relationship.
	Method-Level Refactorings	
7	Push Down Method	Moves a method from a class to those subclasses that require it.
8	Pull Up Method	Moves a method from some class(es) to the immediate superclass.
9	Decrease Method Accessibility	Decreases the accessibility of a method from protected to private or from public to protected.
10	Increase Method Accessibility	Increases the accessibility of a method from protected to public or from private to protected.
	Field-Level Refactorings	
11	Push Down Field	Moves a field from a class to those subclasses that require it.
12	Pull Up Field	Moves a field from some class(es) to the immediate superclass.
13	Decrease Field Accessibility	Decreases the accessibility of a field from protected to private or from public to protected.
14	Increase Field Accessibility	Increases the accessibility of a field from protected to public or from private to protected.

details. Fowler's refactorings were designed specifically with the goal of design improvement in mind, whereas the goal of Code-Imp is to *explore* the design space.

The refactoring process is driven by a search technique; in this paper hill-climbing and simulated annealing are used. The simplest search technique is steepest-ascent hill-climbing, where the next refactoring to be applied is the one that produces the best improvement in the fitness function. In first-ascent hill-climbing, the first refactoring found to improve the fitness function is applied. Hill climbing suffers from not being able to escape from local optima. Simulated annealing avoids this drawback by accepting both positive and negative moves to escape local optima with the possibility of accepting negative moves decreasing over time.

The fitness function is a measure of how "good" the program is, so the fitness function used depends on the quality that we are trying to improve. When Code-Imp is used to improve software design, the fitness function is a combination of software quality metrics. In the work described in this paper, the fitness function is based on the security metrics described in section 2. These can be combined using either a weighted-sum approach or Pareto optimality [10].

4   Case Study

In order to test the ability of automated refactoring to improve program security, we chose a sample industrial application that has a strong security dimension to refactor using Code-Imp. The application we use is Wife 6.1, one of the most commonly used open source SWIFT[1] messaging applications built in Java. It has been used by variety of financial institutions and banks [12] and comprises 3,500 lines of code and eighty classes. This is a small application, but serves as a realistic test case for assessing if automated security improvement is possible. The Code-Imp search process is guided by a fitness function based on the security metrics by Alshammari et al. [1].

Our approach is as follows. In section 4.1 we describe how the metrics to be used in the study were determined. These metrics are then combined into a fitness function using Pareto optimality. The results of refactoring Wife using this Pareto-optimal fitness function and a variety of search techniques are presented in section 4.2. In section 4.3 we analyse more closely the effect of the refactoring process on the metrics and draw some conclusions about the metrics as well as the value of refactoring to improve security. The overall conclusions to be drawn from the case study are presented in section 4.4.

4.1   Initial Metric Assessment

The first step is to annotate the Wife source code to specify the classified attributes that will be used subsequently to derive the classified methods and classes as described above. In total, five fields were marked as classified including the Bank Identifier Code (BIC) of the message and the International Bank Account Number (IBAN) of the bank account. This annotation is shown in program 1.

We then use Code-Imp to refactor this program 16 times, on each occasion using just one of the metrics from table 1 to guide the refactoring process. Our goal is to find out which security metrics provide the best possibility to improve the security of the program. Steepest Ascent Hill-Climbing was used as the search technique, as it is simple and deterministic.

---

[1] SWIFT (Society for Worldwide Interbank Financial Telecommunication) is an industry-owned cooperative providing messaging services to financial institutions.

Program 1. The Classified Field Annotation

```
@Security(classified = true)
private String bic;

@Security(classified = true)
private String iban;
```

The results of this are presented in table 3. 12 of the metrics prove to be completely inert under refactoring. This is largely to do with the exact set of refactoring types that Code-Imp supports. If other refactoring types were supported, we could expect to see other metrics being affected. It is also partly dependent on the nature of the program to be refactored. Real applications, like Wife, contain certain patterns that are candidates to be targeted for refactoring, and those refactorings are likely to cause changes to the same set of security metrics and not to others. For example, if a program is not widely dependent on composition, then it is unlikely to see changes to composition metrics; the same applies to inheritance, and so on.

Table 3. Results for Running Individual Metrics against Wife Application

	Initial Value	Final Value	Percentage Change
CPCC	1.0	1.0	0.0%
CAIW	0.0037	0.0022	40.8%
CMW	0.0611	0.0611	0.0%
CSP	0.0	0.0	0.0%
CAI	0.0	0.0	0.0%
CCDA	0.0	0.0	0.0%
CMAI	0.0149	0.0149	0.0%
CCC	0.0421	0.0372	13.9%
CCE	1.0	1.0	0.0%
CIDA	0.0	0.0	0.0%
CME	1.0	1.0	0.0%
CMI	0.0	0.0	0.0%
CDP	0.1026	0.0909	13.6%
COA	0.8966	0.5172	42.3%
CAAI	0.0263	0.0263	0.0%
CMI	0.0	0.0	0.0%

The results of this experiment are clear: four metrics are improved by this process namely CAIW, CCC, CDP and COA, so we combine these in the next section to form a single fitness function.

4.2 Pareto Optimal Search

The problems associated with using weighted summation to combine ordinal-scale metrics are well documented [10]. To avoid these, we use a Pareto optimal approach to combining the four security metrics identified as promising in the previous section. In the Pareto-optimal approach, a refactoring is regarded as an improvement only if it improves at least one of the metrics and does not cause any of the others to deteriorate.

Code-Imp was run using all three search algorithms described in section 3. Steepest-Ascent Hill Climbing (HCS) was run once as it is deterministic, while First-Ascent Hill Climbing (HCF) and Simulated Annealing (SA) were each run 14 times. The small number of runs was due to the long execution time of the simulated annealing process. As this is a single-instance case study rather than a randomised experiment, the inability to use statistical inference to produce generalisable results is not critical.

The results are summarised in table 4. In the case of HCF and SA, the results presented are those from the best run.

The first observation is that HCS produces identical metric improvements here as it did when the individual metrics were used on their own. This indicates that in this example there was no conflict between the metrics, so in no case did Pareto optimality prevent a refactoring from being applied. This suggests that this set of metrics form a good combination.

HCF produces identical results to HCS, except for a minor improvement in CAIW. Simulated Annealing also produces a similar result to HCF and HCS, with the exception that the CAIW metric is improved dramatically. This is an example of simulated annealing escaping from a local optimum, and illustrates that better solutions many be found using such stochastic approaches. There was a heavy price to pay for this improvement in CAIW: while the refactoring sequences generated by the hill climbs were modest, (HCF 42 refactorings, HCS 57 refactorings), the best of the simulated annealing runs was 2194 refactorings in length. In both cases the improvement in CAIW is attributable to a weakness in this metric, as explained in Section 4.3.

Table 4. Security Metrics Enhancements for all Search Algorithms

	First Ascent Hill Climbing (HCF)	Steepest Ascent Hill Climbing (HCS)	Simulated Annealing (SA)
COA	42.3%	42.3%	42.3%
CAIW	41.2%	40.2%	72.6%
CCC	13.9%	13.9%	13.9%
CDP	13.6%	13.6%	13.6%

4.3 Qualitative Metrics Analysis

In this subsection we delve into the metric changes in greater detail. Our aim is to understand what refactorings caused the metrics to improve and to determine if this is a real improvement or simply an artifact of the search-based refactoring process. We focus on the results of the steepest-ascent hill climb using Pareto optimality, as shown in column three of table 4.

It is worth mentioning that these metrics are *ordinal*, i.e. using a scale of measurement in which data are listed in rank order but without fixed differences among the entries. This means that expressing the difference in terms of percentage may not be entirely meaningful. Nevertheless, we are assuming that the larger the difference in value, the more likely it is that a considerable improvement has taken place.

In the following paragraphs, we consider in turn each of the metrics that showed an improvement during the refactoring process.

Classified Operation Accessibility (COA): This is the ratio of classified public methods to classified methods, from the group of Data Encapsulation Metrics. The most significant change in security metrics is associated with this metric, as it undergoes a 42% enhancement from a value of 0.8966 to value of 0.5172. The change is mainly driven by the refactoring `Decrease Method Accessibility`. When applied to a public classified method, it obviously reduces the numerator and increases the denominator of the ratio, hence reducing (improving) the value of the metric. The improvement in this metric corresponds to a real improvement in program security, as it reduces the accessibility of security-critical fields.

Classified Attributes Interaction Weight (CAIW): This is the ratio of the sum (for all classified attributes) of the number of methods that may access the classified attribute to the sum (for all attributes) of the number of methods that may access the attribute, from the group of Cohesion-Based Metrics. This metric undergoes an improvement of 40% from an initial value of 0.0037 to a final value of 0.0022. Looking at the refactorings that cause the metric to improve it can be found that the refactoring `Increase Field Accessibility` is the largest contributor. This is surprising because increasing the accessibility of a field (e.g. from `protected` to `public`) would be expected to negatively impact security, if it impacts it at all. However what is happening is that reducing the accessibility of a non-classified attribute will enhance the ratio of the summation of the access to classified attributes compared to non-classified ones. This demonstrates a weakness in the CAIW metric as it detects an improvement in security although absolutely no extra protection has been achieved for the security-critical data. The other main contributor to this metric enhancement is the `Pull Up Method` refactoring which reduces the access to classified attributes in the subclass and limits them to those of the superclass. In other words moving a method to a superclass will limit its access to classified methods in subclass. This aspect of improvement in the CAIW metric represents a real improvement in security.

Classes Design Proportion (CDP): The is the ratio of the number of critical classes to the total number of classes, from the group of Design Size Metrics. This metric is enhanced by 13.6% from an initial value of 0.1026 to an improved value of 0.0909. All improvements were caused by the `Extract Hierarchy` refactoring and were mainly due to the fact that dividing a non-classified class into two classes will generate two non-classified classes and hence enhance the ratio of classified to non-classified classes. As with CAIW, this "improvement" is meaningless and demonstrates a weakness in the metric as absolutely no extra protection to classified data is achieved as a result of the metric improvement. On the other hand, in the cases where the `Extract Hierarchy` refactoring divided a classified class into one classified and one non-classified class, the improvement in CDP was real. Breaking a critical class into two classes with only one having access to classified data means that data is rendered accessible to fewer methods and is therefore more secure.

Critical Classes Coupling (CCC): This is the ratio of the number of links from classes to classified attributes defined in other classes to the total number of possible links from all classes to classified attributes defined in other classes, from the group of Coupling-Based Metrics. The behaviour of CCC is very similar to that of CDP both in terms of the percentage of enhancement (13.9%) and the effect of the `Extract Hierarchy` refactoring on its value. It is interesting to note note that every time CDP was affected by an `Extract Hierarchy` refactoring, CCC was affected by a proportional value. Both the metric weakness and the recommendation for the CDP metric above are applicable.

4.4 Case Study Conclusions

The key research question we wish to investigate is whether automated refactoring can be used to improve program security. 12 out of the 16 Alshammari et al. metrics we tested proved to be inert under the refactorings we applied, but this is a consequence of the refactoring suite we use. For the four metrics that were affected by the refactoring, in each case some of the metric improvement corresponded to real improvement in program security, as detailed in the *Real* column of Table 5. This is a positive result in terms of the ability of the search-based refactoring approach to improve security.

Table 5. Security Metrics Enhancements for Steepest-Ascent Hill Climbing

	Improvement	Real	Artificial
COA	42.3%	42.3%	0.0%
CAIW	40.2%	14.5%	25.7%
CCC	13.9%	3.5%	10.4%
CDP	13.6%	2.3%	11.3%
Average	27.5%	15.6%	11.9%

As a by-product of our analysis, we made several other interesting discoveries about the metrics we used to guide the refactoring process. These were derived from where the improvement in the security metrics was found not to represent a true improvement in security, as detailed in the *Artificial* column of Table 5.

CAIW is a poorly formed metric that rewards the increasing of the accessibility of a non-classified attribute, even though this obviously has no impact on program security, and in fact is an example of poor object-oriented design. CDP was also found to be poorly-formed as it rewards the splitting of non-classified classes which again has no impact on program security, and is again is an example of poor object-oriented design. The ability of search-based refactoring to pinpoint metric weaknesses like this was also demonstrated in previous work that used the QMOOD metrics suite [5] to guide search-based refactoring [20].

5   Related Work

This work merges two research strands: software security metrics and search-based refactoring.

The traditional approach to measuring software security is to measure the number of security-related bugs found in the system or the number of the system is reported in security bulletins [8]. More recent work has focussed on measuring properties of the software design that are related to security [1,8].

As well as proposing the metrics suite using in this paper [1,4], Alshammari et al. developed a hierarchical approach to assessing security [2]. They chose two principles to measure the security of designs from the perspective of information flow: *grant least privilege* [6] which means that each program component will have access to *only* the parts that it legitimately requires, and *reduce attack surface* [7] which means to reduce the amount of code running with an unauthenticated user. In a follow-up study that is more closely related to our work [3], Alshammari et al. looked at the effect on their security metrics of applying refactorings to program designs. While this can provide static insight into the properties of the refactorings, it does not automate the improvement of the program as our approach does.

Smith and Thober [22] try to expose information flow security by using code refactoring to partition a system into high security and low security components, where high security components can take high or low security input but cannot send output to low security components unless this output is investigated and approved (i.e. declassified). The program needs to be analysed first to identify high and low security components, and then refactored to isolate the high security components. Metrics are not used to measure code security, instead a manual (or partly-automated) partitioning is used to isolate code with various security levels. Full automation is impossible, as both partitioning and declassification need to be performed by developers knowledgeable in the code. By contrast, our approach requires minimal developer input.

The main application of search-based refactoring has been to automate the improvement of a program's design. O'Keeffe and Ó Cinnéide [20] propose an automated design improvement approach to improve software flexibility, understandability, and reusability based on the QMOOD quality model [5]. They report a significant improvement in understandability and minimal improvement in flexibility, however the the QMOOD reusability function was found to be unsuitable for automated design improvement. Other work by the same authors [18] investigates if a program can be automatically refactored to make its metrics profile more similar to that of another program. This approach can be used to improve design quality when a sample program has some desirable features, such as ease of maintenance, and it is desired to achieve these features in another program.

Seng et al. [21] propose an approach for improving design quality by moving methods between classes in a program. They use a genetic algorithm with a fitness function based on coupling, cohesion, complexity and stability to produce a desirable sequence of move method refactorings. Improving design quality by moving methods among classes was also investigated by Harman and Tratt [10]; their key contribution is the use a Pareto optimal approach to make combination of metrics easier. Jensen and Cheng [13] use refactoring in a genetic programming environment to improve the quality of the design in terms of the QMOOD quality model. Their approach was found to be capable of introducing design pattern through the refactoring process, which helps to change the design radically. Kilic et al. explore the use of a variety of population-based approaches to search-based parallel refactoring, finding that local beam search could find the best solutions [14]. Search-based refactoring has also been used to improve other aspects of software, e.g. to improve program testability [17]. The work presented in this paper is the first attempt to use search-based refactoring to improve software security.

6   Conclusions and Future Work

This paper builds on previous work that shows that refactoring can have a substantial effect on security metrics when applied to a software design [3]. We extend this work by using the search-based refactoring platform, Code-Imp, guided by security metrics to test if the security of source code can be improved in an automated fashion. In our study of an industrial software application, Wife, we achieved an overall real improvement of 15.5% in the metrics affected by the refactoring. This improvement is obtained by a fully-automated search-based refactoring process that requires no developer input other than the annotation of the classified fields. This security gain at such little cost indicates the value and potential of the approach.

Previously search-based refactoring has been used mainly to refactor a program in order to improve its design [10,11,19]. This creates the difficult problem of explaining the new design to the developer. In using search-based refactoring to improve security, such problems do not arise. The process is applied to a program/library just prior to it being released; the developers continue to work with the original version of the code.

As summarised in section 4.4, we also demonstrated that some of the Alshammari et al. security metrics are poorly formed and require reworking in order to more truly reflect program security. These are issues to be addressed by security researchers.

Future work involves extending Code-Imp with new refactorings that can impact the inert security metrics, e.g. `Extract Method`, as well as adding refactorings that have a specific security dimension, e.g. `Make Class Final`. One weakness of the security metrics employed in this paper is that they do not define clearly the type of attack they protect the program from. If security metrics were to specify this attack profile, then security test cases could be created that exploit the security vulnerabilities injected into a sample program. Then it could be tested if our refactoring approach could actually fix the security vulnerabilities. This would a be a more robust demonstration of construct validity than the examination of the changes brought about by refactoring that we performed in this paper.

While our experiences in refactoring Wife have demonstrated the potential value of search-based refactoring to improve program security, further experiments with larger applications, extra refactorings and more refined metrics are required to fully explore the potential of this approach.

Acknowledgment. This work was supported, in part, by Science Foundation Ireland grant 10/CE/I1855. The authors would like to thank Iman Hemati Moghadam for his help in performing the experiments with the Code-Imp refactoring platform. Shadi Ghaith gratefully acknowledges the support of IBM in this work.

References

1. Alshammari, B., Fidge, C., Corney, D.: Security metrics for object-oriented class designs. In: Proceedings of the International Conference on Quality Software, pp. 11–20. IEEE (2009)
2. Alshammari, B., Fidge, C., Corney, D.: A hierarchical security assessment model for object-oriented programs. In: Proceedings of the International Conference on Quality Software, pp. 218–227. IEEE, Los Alamitos (2011)
3. Alshammari, B., Fidge, C.J., Corney, D.: Assessing the impact of refactoring on security-critical object-oriented designs. In: Han, J., Thu, T.D. (eds.) Proceedings of the Asia Pacific Software Engineering Conference, pp. 186–195. IEEE Computer Society (2010)
4. Alshammari, B., Fidge, C.J., Corney, D.: Security metrics for object-oriented designs. In: Nobel, J., Fidge, C.J. (eds.) The 21st Australian Software Engineering Conference, pp. 55–64. IEEE, Hyatt Regency (2010)
5. Bansiya, J., Davis, C.: A hierarchical model for object-oriented design quality assessment. IEEE Transactions on Software Engineering 28, 4–17 (2002)
6. Bishop, M.A.: The Art and Science of Computer Security. Addison-Wesley Longman Publishing Co., Inc., Boston (2002)
7. Blackwell, C.: A security architecture to protect against the insider threat from damage, fraud and theft. In: Proceedings of the 5th Annual Workshop on Cyber Security and Information Intelligence Research, CSIIRW 2009, pp. 45:1–45:4. ACM, New York (2009)
8. Chowdhury, I., Chan, B., Zulkernine, M.: Security metrics for source code structures. In: Proceedings of the Fourth International Workshop on Software Engineering for Secure Systems, SESS 2008, pp. 57–64. ACM, New York (2008)
9. Fowler, M.: Refactoring: Improving the Design of Existing Code. Addison-Wesley, Boston (1999)
10. Harman, M., Tratt, L.: Pareto optimal search based refactoring at the design level. In: Proceedings of the 9th Annual Conference on Genetic and Evolutionary Computation, GECCO 2007, pp. 1106–1113. ACM, New York (2007)
11. Hemati Moghadam, I., Ó Cinnéide, M.: Code-Imp: a tool for automated search-based refactoring. In: Proceedings of the 4th Workshop on Refactoring Tools, WRT 2011, pp. 41–44. ACM, New York (2011)
12. Wife swift application. In: Wife Swift Application. Prowide Open Source SWIFT (2012), http://www.prowidesoftware.com

13. Jensen, A., Cheng, B.: On the use of genetic programming for automated refactoring and the introduction of design patterns. In: Proceedings of the Conference on Genetic and Evolutionary Computation, pp. 1341–1348. ACM (July 2010)
14. Kilic, H., Koc, E., Cereci, I.: Search-Based Parallel Refactoring Using Population-Based Direct Approaches. In: Cohen, M.B., Ó Cinnéide, M. (eds.) SSBSE 2011. LNCS, vol. 6956, pp. 271–272. Springer, Heidelberg (2011)
15. McGraw, G.: Software Security: Building Security In. Addison-Wesley Professional (2006)
16. Hemati Moghadam, I., Ó Cinnéide, M.: Automated refactoring using design differencing. In: Proceedings of the European Conference on Software Maintenance and Reengineering, pp. 43–52. IEEE Computer Society (2012)
17. Ó Cinnéide, M., Boyle, D., Hemati Moghadam, I.: Automated refactoring for testability. In: Proceedings of the International Conference on Software Testing, Verification and Validation Workshops (March 2011)
18. O'Keeffe, M., Ó Cinnéide, M.: Automated design improvement by example. In: Proceeding of the Conference on New Trends in Software Methodologies, Tools and Techniques, pp. 315–329 (2007)
19. O'Keeffe, M., Ó Cinnéide, M.: Search-based refactoring: an empirical study. Journal of Software Maintenance and Evolution 20(5), 345–364 (2008)
20. O'Keeffe, M., Ó Cinnéide, M.: Search-based refactoring for software maintenance. Journal of Systems and Software 81(4), 502–516 (2008)
21. Seng, O., Stammel, J., Burkhart, D.: Search-based determination of refactorings for improving the class structure of object-oriented systems. In: GECCO 2012, pp. 1909–1916. ACM, Seattle (2006)
22. Smith, S.F., Thober, M.: Refactoring programs to secure information flows. In: Proceedings of the Workshop on Programming Languages and Analysis for Security, PLAS 2006, pp. 75–84. ACM, New York (2006)

Combining Search-Based and Adaptive Random Testing Strategies for Environment Model-Based Testing of Real-Time Embedded Systems

Muhammad Zohaib Iqbal[1,2], Andrea Arcuri[1], and Lionel Briand[1,3]

[1] Certus Center for V & V, Simula Research Laboratory, P.O. Box 134, Lysaker, Norway
[2] Department of Informatics, University of Oslo, Norway
[3] SnT Center, University of Luxembourg, Luxembourg
{zohaib,arcuri}@simula.no, lionel.briand@uni.lu

Abstract. Effective system testing of real-time embedded systems (RTES) requires a fully automated approach. One such black-box system testing approach is to use environment models to automatically generate test cases and test oracles along with an environment simulator to enable early testing of RTES. In this paper, we propose a hybrid strategy, which combines (1+1) Evolutionary Algorithm (EA) and Adaptive Random Testing (ART), to improve the overall performance of system testing that is obtained when using each single strategy in isolation. An empirical study is carried out on a number of artificial problems and one industrial case study. The novel strategy shows significant overall improvement in terms of fault detection compared to individual performances of both (1+1) EA and ART.

1 Introduction

Real-time embedded systems (RTES) are widely used in critical domains where high system dependability is required. These systems typically work in environments comprising of large numbers of interacting components. The interactions with the environment are typically bounded by time constraints. Missing these time deadlines, or missing them too often for soft real-time systems, can lead to serious failures leading to threats to human life or the environment. There is usually a great number and variety of stimuli from the RTES environment with differing patterns of arrival times. Therefore, the number of possible test cases is usually very large if not infinite. Testing all possible sequences of stimuli is not feasible. Hence, systematic automated testing strategies that have high fault revealing power are essential for effective testing of industry scale RTES. The system testing of RTES requires interactions with the actual environment. Since, the cost of testing in actual environments tends to be high, environment simulators are typically used for this purpose.

In our earlier work, we proposed an automated system testing approach for RTES software based on environment models [6, 11]. The models are developed according to a specific strategy using the Unified Modeling Language (UML) [19], the Modeling and Analysis of Real-Time Embedded Systems (MARTE) profile [18] and our

G. Fraser (Ed.): SSBSE 2012, LNCS 7515, pp. 136–151, 2012.
© Springer-Verlag Berlin Heidelberg 2012

proposed profile [14]. These models of the environment were used to automatically generate an environment simulator [12], test cases, and obtain test oracle [6, 11].

In our context, a test case is a sequence of stimuli generated by the environment that is sent to the RTES. A test case can also include changes of state in the environment that can affect the RTES behavior. For example, with a certain probability, some hardware components might break, and that affects the expected and actual behavior of the RTES. A test case can contain information regarding when and in which order to trigger such changes. So, at a higher level, a test case in our context can be considered as a setting specifying the occurrence of all these environment events in the simulator. Explicit "error" states in the models represent states of the environment that are only reached when RTES is faulty. Error states act as the oracle of the test cases, i.e., a test case is successful in triggering a fault in the RTES if any of these error states is reached during testing.

In previous work, we investigated several testing strategies to generate test cases. We used random testing (RT) [7] as baseline, and then considered two different approaches: Search-based Testing (SBT) [13] and Adaptive Random Testing (ART) [6]. For SBT, an *order function* was defined that utilizes the information in environment models to guide the search toward the error states. In contrast, with ART, test cases are rewarded based on their *diversity*. The results indicated that, apart from the failure rate of the system under test (SUT), the effectiveness of a testing algorithm also depends on the characteristics of the environment models. For problems where the environment model is easier to cover or where the failure rate of the RTES is high, even RT outperforms SBT. However, for more complex problems, SBT showed much better performance than RT. This raised the need for a strategy that combines the individual benefits of the two strategies and utilizes adaptive mechanisms based on the feedback from executed test cases.

In this paper, we extend our previous work by devising such a hybrid strategy that aims at combining the best search technique, i.e., (1+1) Evolutionary Algorithm (EA) in our experiments and ART (which is the algorithm that gave best results in our earlier experiments in [11]) in order to achieve better overall results in terms of fault detection. We defined two different strategies for combining these algorithms, but due to space constraints, in this paper, we only discuss the strategy that showed the best results. The hybrid strategy (HS) discussed here starts with running (1+1) EA and switches to ART when (1+1) EA stops yielding fitter test cases. The decision of when to switch (referred to as *configuration*) can have significant impact on the performance of the strategy and one main objective of this paper is to empirically investigate different configuration options. The other combination strategy started by running ART and later switched to (1+1) EA if consecutive test cases generated through ART showed better fitness compared to previously executed test cases. It did show improvements over the individual algorithms, but fared worse than HS.

We evaluate the fault detection effectiveness of HS by performing a series of experiments on 13 artificial problems and an industrial case study. The RTES of the artificial problems were based on the specifications of two industrial case studies. Their environment models were developed in a way to vary possible modeling characteristics so as to understand their effect on the performance of the test strategies. We could not have covered such variations in environment models with one or even a few industrial case studies, hence the motivation to develop artificial cases. The industrial case study used is of a marine seismic acquisition system, which was developed by a company leading in this industry sector. For all these cases, we compared the performance of HS (with best configuration) with that of ART, (1+1) EA, and RT. The results suggest that in terms of success rates (number of times an algorithm found a fault within a given test budget), for the problems where RT/ART showed better performance over (1+1) EA, HS results are similar to ART/RT and for the problems where (1+1) EA was better, HS results are similar to those of (1+1) EA, thus suggesting that HS combines the strength of both algorithms.

The rest of the paper is organized as follows. Section 2 discusses the related work, while Section 3 provides an introduction to the earlier proposed environment model-based system testing methodology that we improve in this paper. Section 4 describes the proposed hybrid strategy, whereas Section 5 reports on the empirical study carried out for evaluation purposes. Finally, Section 6 concludes the paper.

2   Related Work

Depending on the goals, testing of RTES can be performed at different levels: model-in-the-loop, hardware-in-the-loop, processor-in-the-loop, and software-in-the-loop [9]. Our approach falls in the software-in-the-loop testing category, in which the embedded software is tested on the development platform with a simulated environment. The only variation is that, rather than simulating the hardware platform, we use an adapter for the hardware platform that forwards the signals from the SUT to the simulated environment. This approach is especially helpful when the software is to be deployed on multiple hardware platforms or the target hardware platform is stable.

There are only a few works in literature that discuss RTES testing based on environment models rather than system models. Auguston *et al.* [8] discusses the modeling of environment behaviors for testing of RTES using an event grammar. The behavioral models contain details about the interactions with the SUT and possible hazardous situations in the environment. Heisel *et al.* [10] propose the use of a requirement model and an environment model along with the model of the SUT for testing. Adjir *et al.* [1] discuss a technique for testing RTES based on the system model and assumptions in the environment using Labeled Prioritized Timed Petri Nets. Larsen *et al.* [15] propose an approach for online RTES testing based on time automata to model the SUT and environmental constraints. Iqbal *et al.* [14] propose an environment modeling methodology based on UML and MARTE for black-box system testing. Fault detection effectiveness of testing strategies based on these models was evaluated and reported in [13], including RT/ART [6], GA, and (1+1) EA. The results indicate that SBT show significantly better performance over RT for a number of cases and significantly worse performance than RT for a number of other cases.

There has been some work to combine SBT with RT. Andrews *et al.* propose the use of GA to tune parameters for random unit testing [4]. An evolutionary ART algorithm that uses the ART distance function as a fitness function for GA is proposed in

[21]. In [20], the authors propose a search-based ART algorithm by using a variant of ART distance function as the fitness function for Hill Climbing to optimize the results of ART when the input domains are more than two dimensional.

The work presented here improves the work on environment model-based testing presented in [13] by combining the strengths of both ART and (1+1) Evolutionary Algorithm. Approaches discussed in the literature for combining ART/RT with SBT are restricted to improving ART or tuning RT by using search techniques. In contrast, here we want to use (1+1) EA to generate test cases that exploit the characteristics of environment models as well as benefit from the test diversity generated by ART, thus combining the two approaches.

3   Environment Model-Based Testing

In this section, we discuss in more details the various components of our environment model-based testing approach.

3.1   Environment Modeling and Simulation

For RTES system testing, software engineers familiar with the application domain would typically be responsible for developing the environment models. The environment models consist of a domain model and several behavioral models. The domain model, represented as a class diagram, captures the structural details of the RTES environment, such as the environment components, their relationships, and their characteristics. The behavior of the environment components is captured by state machines. These models are developed, based on our earlier proposed methodology by using UML, MARTE, and our proposed profile for environment modeling [14]. These models not only include the nominal functional behavior of the environment components (e.g., booting of a component) but also include their robustness (failure) behavior (e.g., break down of a sensor). The latter are modeled as "failure" states in the environment models. The behavioral models also capture what we call "error" states. These are the states of the environment that should never be reached if the SUT is implemented correctly. Therefore, error states act as oracles for the test cases. Java is used as an action language and OCL (Object Constraint Language) is used to specify constraints and guards. An important feature of these environment models is that they capture the non-determinism in the environment, which is a common characteristic for most RTES environments. Non-determinism may include, for example, failures of components, or user commands. Each environment component can have a number of non-deterministic choices whose exact values are selected at the time of testing.

Using model to text transformations, the environment models are automatically transformed into environment simulators implemented in Java. The transformations follow specific rules that we discussed in detail in [12]. During simulation a number of instances can be created for each environment component, which can interact with each others and the SUT (for example multiple instances of a sensor component). The generated simulators communicate with the SUT through a communication layer (e.g., TCP layer), which is written by software engineers. They are also linked with the test framework that provides the appropriate values for each simulation execution. The choice of Java as target language is based on actual requirements of our industrial partner, where the RTES under study only involves soft real-time constraints.

3.2   Testing RTES Based on Environment Models

In our context, a test case execution is akin to executing the environment simulator. During the simulation, values are required for the non-deterministic choices in the environment models. A test case, in our context, can be seen as a test data matrix, where each row provides a series of values for a non-deterministic choice of the environment component (the number of rows is equal to the number of non-deterministic choices). Each time a non-deterministic choice needs to be made, a value from the corresponding matrix row is selected.

To calculate the distance between two test data matrices m_1 and m_2 for ART we use the function $dis(m_1, m_2) = \sum_r \sum_c abs(m_1[r,c] - m_2[r,c])/ |D(r)|$, where r and c represent the rows and columns of the matrices. In other words, we sum the absolute difference of each variable weighted by the cardinality of the domain of that variable. Often, these variables represent the time in timeout transitions. Therefore, ART rewards diversity for the values of non-deterministic choices. The results of the first experiments we conducted showed that RT/ART perform better than SBT [6].

For search-based testing, rather than using a fitness function, we use an *order* function. An order function is used to determine whether one solution is better than another, without having the problem of defining a precise numerical score (this is often difficult when several objectives need to be combined and tight budget constraints do not allow a full multi-objective approach). The new order function h can be seen as an extension of the fitness function developed for model-based testing based on system specifications [16]. The original fitness function uses the so-called "approach level" and normalized "branch distance" to evaluate the fitness of a test case. For environment model-based testing, we introduced the concept of "time distance" with a look-ahead branch distance and the concept of "time in risky states" [13].

In our context, the goal is to minimize the order function h, which heuristically evaluates how far a test case is from reaching an error state. If a test case with test data m is executed and an error state of the environment model is reached, then $h(m) = 0$. The approach level (A) refers to the minimum number of transitions in the state machine that are required to reach the error state from the closest executed state. Fig. 1 shows a dummy example state machine to elaborate the concept. The state named *Error* is the error state. Events *e1*, *e2*, and *e3* are signal events, whereas events *after* "*t, s*", *after* "*t1, ms*", and *after* "*t2, ms*" are time events with *t*, *t1*, and *t2* as the time values and *ms* and *s* as time units referring to milliseconds and seconds. If the desired state is *Error* and the closest executed state was *State5*, then the approach level is 1.

We use branch distance heuristic (B) to heuristically score the evaluation of the guards on the outgoing transitions from the closest executed state. In previous work [2], we have defined a specific branch distance function for OCL expressions that is reused here for calculating the branch distance. In the dummy state machine in Fig. 1 we need to solve the guard "*y > 0*" so that whenever *e4* is triggered, then the simulation can transition to *Error*.

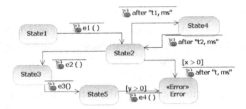

Fig. 1. A dummy state machine to explain search heuristics

The third important part of the order function is the time distance (T), which comes into play when there are timeout transitions in the environment models. For example, in Fig. 1, the transition from *State2* to *Error* is a timeout transition. If a transition should be taken after z time units, but it is not, we calculate the maximum consecutive time c the component stayed in the source state of this transition (e.g., *State2* in the dummy example). To guide the search, we use the following heuristic: $T = z - c$, where $c \leq z$. For transitions other than time transitions, we initially decided to calculate branch distance after an event is triggered. As investigated in our earlier work [13], this is not suitable for time transitions and therefore the concept of a look-ahead branch distance *(LB)* was introduced. *LB* represents the branch distance of OCL guard on a time transition when it is not fired (i.e., the timeout did not occur). Because OCL evaluations are free from side-effects [2], this approach is feasible in our context.

The fourth important part of the order function is "time in risky states" *(TIR)*. *TIR* favors the test cases that spent more time in the state adjacent to the error state (i.e., the *risky* state). For example, for the state machine shown in Fig. 1, this heuristic will favor the test cases that spend more time in the risky states *State2* or *State5*. For instance in *State2*, it is possible to increase the value of *t1* in the time event *after "t1, ms"*, which will increase the time spent in this state. *TIR* is less important than the other three heuristics and is only used when the other heuristics fail to guide the search. The order function h using the four previously described heuristics, given two test data matrices m_1 and m_2 as input, is defined as:

$$h(m_1,m_2)= \begin{cases} v(m_1, m_2) & \text{if } v(m_1, m_2) \mathrel{!=} 0 \\ 1 & \text{if } v(m_1, m_2) = 0 \text{ and } \mathrm{TIR}_{\mathrm{sum}}(m_1) > \mathrm{TIR}_{\mathrm{sum}}(m_2) \\ 0 & \text{if } v(m_1, m_2) = 0 \text{ and } \mathrm{TIR}_{\mathrm{sum}}(m_1) = \mathrm{TIR}_{\mathrm{sum}}(m_2) \\ -1 & \text{otherwise} \end{cases} \quad (1)$$

$$v(m_1,m_2)= \begin{cases} 1 & \text{if } A_{\mathrm{min}}(m_1) < A_{\mathrm{min}}(m_2) \text{ or } (A_{\mathrm{min}}(m_1)=A_{\mathrm{min}}(m_2) \text{ and } B_{\mathrm{min}}(m_1) < B_{\mathrm{min}}(m_2)) \\ & \text{or } (A_{\mathrm{min}}(m_1)=A_{\mathrm{min}}(m2) \text{ and } B_{\mathrm{min}}(m_1)= B_{\mathrm{min}}(m_2) \text{ and } \mathrm{ITD}_{\mathrm{min}}(m_1,m_2) = 1) \\ 0 & \text{if } A_{\mathrm{min}}(m_1)=A_{\mathrm{min}}(m_2) \text{ and } B_{\mathrm{min}}(m_1)= B_{\mathrm{min}}(m_2) \text{ and } \mathrm{ITD}_{\mathrm{min}}(m_1, m_2)=0) \\ -1 & \text{otherwise} \end{cases}$$

$$\mathrm{ITD}_e(m_1,m_2) = \begin{cases} 1 & \text{if } LB_e(m_1) < LB_e(m_2) \text{ or } (LB_e(m_1) = LB_e(m_2) \text{ and } T_e(m_1) < T_e(m_2)) \\ 0 & \text{if } (LB_e(m_1) = LB_e(m_2) \text{ and } T_e(m_1) = T_e(m_2)) \\ -1 & \text{otherwise} \end{cases}$$

where for an error state e, $A_{min}(m)$ represents the minimum approach level over all error states, $B_{min}(m)$ represents the minimum branch distance, T_e represents the time distance, LB_e is the look-ahead branch distance for an error state e, and $TIR_{sum}(m)$ is the sum of time spent in risky states for all error states for test data matrix m.

The results, based on our extensive experiments evaluating various heuristics [13], suggested that (1+1) EA with the order function in (1) gave best results in cases where the approach to a risky state was non-trivial (i.e., simulation cannot reach a risky state in <5 random test cases). But in cases where the approach was easy, RT outperformed evolutionary algorithms.

4   Hybrid Strategy by Combining Search-Based and Adaptive Random Testing

In this section we present our proposed hybrid strategy (HS) that combines (1+1) EA and ART to improve the overall fault detection effectiveness of our testing approach. As discussed earlier (Section 3), previous studies showed that, in some cases, RT/ART could perform better than SBT. The difference between their performances was mostly significant and at times even extreme. In [11] and [13], we identified two possible reasons for this behavior. First of all, for the problems with high failure rates, randomized algorithms were found to be much better than SBT [11]. For high failure rates, there is no need for search, as solutions are anyway found quickly. Secondly, the performance of the algorithms also depended on the properties of environment models, and in particular how easy is it to traverse the models in order to reach the error states. In other words, by combining ART and (1+1) EA, we hope to achieve consistently good results, regardless of the properties of the SUT or its environment.

In the environment models, there are transitions on paths leading to error states that depend only on the behavior of the SUT (i.e., they can only be triggered when the SUT behaves in a certain way). Transition from a risky state to an error state is one such example as it is only triggered when the SUT behaves in an erroneous way. Another example can be when a guard on a transition depends on a specific response from the SUT. To execute this behavior of SUT, the overall environment needs to behave in a particular way. This particular behavior of the environment that is required to trigger SUT behavior cannot be determined before simulation, since for practical reasons discussed earlier the design of the SUT is not visible. Hence, the information of what should be executed in the environment to trigger this behavior is not available in the environment models. The fitness function for SBT (which exploits the environment models to guide the search towards error states) in this case does not give enough gradient to generate fitter test cases (i.e., a search plateau). In these cases maximizing the diversity of the environment behavior (e.g., by using entirely different values for the test data matrix, irrespective of their effect on the fitness) appears to be a better option, thus favoring RT/ART. This can explain the scenarios where RT/ART show better results than (1+1) EA.

On the other hand, if in the environment models, there are transitions on the path to error states which are triggered by specific behaviors of the environment (e.g., a transition triggered as a result of a specific non-deterministic event in the environment, such as a failure of an environment component) or time transitions, then fitness function for SBT is specifically designed to deal with these cases and are more suitable for such cases than RT. For example, in the fitness function, the time distance heuristic is defined specifically for time transitions and favors test cases that are closer to executing the transitions (i.e., with a value of c closer to z, see Section 3.2). OCL constraints in guards that are independent of SUT behavior but dependent on the state of environment components (e.g., a constraint requiring a sensor to be broken), can be solved by directly changing the values of these components' attributes. For such constraints, our previous results showed that SBT are an order of magnitude better than RT [2].

HS combines ART, which showed best results in our initial experiments [11], with our proposed SBT strategy that showed best performance [13], i.e., (1+1) EA with improved time distance and the "time in risky state" heuristic (ITD-TIR). The strategy is designed to combine the strengths of both (1+1) EA and ART. This strategy starts by applying (1+1) EA. If (1+1) EA does not find fitter test cases after running n number of test cases, the testing algorithm is switched to ART. All the test cases that were executed so far are now used for distance calculations in ART. Fig. 2 shows the pseudo-code for HS. The idea behind switching from (1+1) EA to ART is that there is not enough time for a random walk to get out of a fitness plateau. And so, in this scenario, applying ART can yield better results. Running system test cases is very time consuming, so only few fitness evaluations are feasible within reasonable time (e.g., 1000 test cases can already take several hours). Therefore, in case of fitness plateau, it is reasonable to switch strategy, and rather reward diversity instead of the fitness value. Though the choice of n is arbitrary it can have significant consequences on the performance of this strategy. A too small value of n will result in an early switch to ART. If the given problem matches the case where (1+1) EA performs better, then the performance of HS will be affected. Similarly, if n is too large then the remaining testing budget might not be sufficient for ART to perform well.

Algorithm	**HybridStrategy(mx, n, w)**
Input	mx: # of maximum fitness evaluations, n: # of consecutive test cases with no improved fitness, w: # of random test-cases to generate for comparison in ART
Declare	Y: set of executed test cases = { }, W: set of randomly generated test cases = { }
	ev: # of fitness evaluations performed = 0, z: # of consecutive test cases with no improved fitness found so far = 0, T_c: a random test case, T_m: mutated test case,
	T_w: a test case from W, T_e: test case from W selected according to ART criteria,
	D_w: minimum distance of test case T_w with all the test cases in Y, d: maximum value of D_w obtained over W

1. **begin**
2.   Generate a random test case T_c
3.   Execute T_c and evaluate whether environment error state is reached, Add T_c to Y
4.   **while** environment error state not reached *AND* $ev <= mx$ *AND* $z <= n$
5.   Mutate T_c to get T_m, Execute T_m and evaluate whether environment error state is reached
6.   Add T_m to Y,   Increment ev
7.   **if** *fitness(T_m)* $>=$ *fitness(T_c)* **then** $T_c = T_m$, $z = 1$
8.   **else** Increment z
9.   **while** environment error state not reached *AND* $ev <= mx$
10.   Sample w random test cases and add them to W, $d = 0$
11.   **for each** $T_w \in W$
12.   Calculate D_w
13.   **if** $D_w > d$ **then** $d = D_w$, $T_e = T_w$
14.   Execute T_e and evaluate whether environment error state is reached
15.   Add T_e to Y, Increment ev
16. **end**

Fig. 2. Pseudo code of the proposed hybrid strategy (HS)

5   Empirical Study

The objective of this empirical study is to evaluate the fault detection effectiveness of the proposed hybrid strategy.

5.1   Case Study

To enable experimentation with diverse environment models and RTES, we developed 13 different artificial RTES that were inspired by two industrial cases we have been involved with [11] and one case study discussed in the literature [22]. Since, there are no benchmark RTES available to researchers, we specifically designed these artificial problems to conduct our experiments (called P1 – P13). The goal while developing the models of these RTES was to vary various characteristics of the environment models (e.g., guarded time transitions, loops) that were expected to have an impact on the test heuristics. Nine of these artificial problems were inspired by a marine seismic acquisition system developed by one of our industrial partners. They covered various subsets of the environment of that RTES. Three problems were inspired by the behavior of another industrial RTES (automated recycling machine) developed by another industrial partner. The thirteenth artificial problem was inspired by the train control gate system described in [22].

These RTES are multithreaded, written in Java and they communicate with their environments through TCP. Each of the artificial problems had one error state in their environment models and non-trivial faults were introduced by hand in each of them. We could have rather seeded the faults in a systematic way, as for example by using a mutation testing tool [3] but opted for a different procedure since the SUTs are highly multi-threaded and use a high number of network features (e.g., opening and reading/writing from TCP sockets), features that are not handled by current mutation testing tools. Furthermore, our testing is taking place at the system level, and though small modifications made by a mutation testing tool might be representative of faults at the unit level, it is unlikely to be the case at the system level for RTES. On the other hand, the faults that we manually seeded came from our experience with the industrial RTES and from the feedback of our industry partners.

The industrial case study we also report on (called IC) is a very large and complex seismic acquisition system (mentioned above) that interacts with many sensors and actuators. The timing deadlines on the environment are in the order of hundreds of milliseconds. The company that provided the system is a market leader in its field. For confidentiality reasons we cannot provide full details of the system. The SUT consists of two processes running in parallel, requiring a high performance, dedicated machine to run. For the industrial case study, we did not seed any fault and the goal was to find the real fault that we uncovered earlier [6].

5.2   Experiment

In this paper, we want to answer the following research questions:

RQ1. Which configuration is best in terms of fault detection for the proposed hybrid strategy (HS)?

RQ2. How the fault detection of the best HS configuration compares with the performance of ART, (1+1) EA, and RT for (a) the artificial problems (P1-13) and (b) the industrial case study (IC)?

To answer these research questions, we have conducted two distinct sets of experiments, one for the artificial problems (to answer RQ1 and RQ2a) and one for the industrial RTES (to answer RQ2b). For test case representation in these experiments we used a dynamic representation with a length equal to 10 for the test cases (which correspond to each row of the test data matrix m). In our earlier experiments this setting showed the best results [11]. For (1+1) EA we calculated the mutation rate as $1/k$, where k is the number of total elements in a test data matrix. This strategy is widely used for SBT and was initially suggested in [17]. We used the fitness function that performed best in our previous experiments [13], as discussed in Section 4: Improved Time Distance with Time in Risky State (ITD-TIR).

To answer RQ1, we used 12 different values for the number of test cases which fitness should be considered before switching from (1+1) EA to ART: $n \in \{10, 20, 50, 60, 70, 80, 90, 100, 200, 300, 400, 500\}$. We ran these 12 configurations on each of the 13 artificial problems. To answer RQ2a, we selected the configuration of HS that gave the best result in terms of fault detection for the 13 artificial problems. We compared this configuration with the results of (1+1) EA, ART, and RT on these problems. RT was used as a comparison baseline.

For the artificial problems, the execution time of each test case was fixed to 10 seconds and we stopped each algorithm after 1000 sampled test cases or as soon as we reached any of the error states. The choice of running each test case for 10 seconds was based on the properties of the RTES and the environment models. The objective was to allow enough time for the test cases to reach an error state. We ran each of the strategies 20 times on each artificial problem with different random seeds. The total number of sampled test cases was 1,561,390, which required around 180 days of CPU resources. Therefore, we performed these experiments on a cluster of computers.

Table 1. Success Rates (SR) for 12 configurations of HS on the 13 problems

Configurations → / Problems ↓	10	20	50	60	70	80	90	100	200	300	400	500
P1	0.5	0.75	0.95	1	1	1	1	1	1	1	1	1
P2	0.85	0.95	1	1	1	1	1	1	0.9	1	1	1
P3	1	1	1	1	1	1	1	1	0.9	0.8	0.6	0.5
P4	0.05	0.2	0.8	0.85	0.7	0.75	0.9	0.9	1	1	0.9	1
P5	0.85	1	1	1	1	1	1	1	1	1	1	1
P6	0	0.15	0.45	0.4	0.45	0.5	0.45	0.6	0.7	0.7	0.5	0.6
P7	0.3	0.4	0.8	0.8	0.85	0.95	0.8	0.8	0.8	0.8	0.8	1
P8	1	1	1	1	1	1	1	1	1	1	0.95	1
P9	0.05	0.05	0.45	0.55	0.55	0.35	0.6	0.4	0.8	0.45	0.5	0.55
P10	1	1	1	1	0.95	0.85	1	0.95	0.65	0.55	0.4	0.45
P11	1	1	1	0.95	0.95	0.9	1	0.9	0.65	0.05	0.1	0.4
P12	1	1	1	1	0.95	1	1	1	0.9	0.9	0.75	0.65
P13	1	1	1	1	1	1	1	1	0.9	0.7	0.95	0.85
Average SR	0.66	0.73	0.88	0.89	0.88	0.87	0.9	0.89	0.86	0.77	0.73	0.77
Average Rank	6.38	6.73	5.19	5.77	5.23	6.31	6.50	6.19	6.73	8.46	7.73	6.69

To answer RQ2b, we carried out experiments on the described seismic acquisition system. We run each test case for 60 seconds, where 1000 test case executions (fitness evaluations) can take more than 16 hours. This choice has been made based on the properties of the RTES and discussions with the actual testers. Due to the large amount of resources required, we only ran the configuration that on average gave best results for the artificial problems (i.e., $n=50$) and compared its fault detection rate with that of (1+1) EA, ART, and RT. We carried out 39 runs for each of these four test strategies. The total number of sampled test cases was 55,283, which required over 55 days of computation on a single, high-performance, dedicated machine.

To analyze the results, we used the guidelines described in [5] which recommends a number of statistical procedures to assess randomized algorithms. First we calculated the success rates of each algorithm: the number of times it was successful in reaching the error state out of the total number of runs. These success rates are then compared using the Fisher Exact test, quantifying the effect size using an odds ratio (ψ) with a 0.5 correction. When the differences between the success rates of two algorithms were not significant, we then looked at the average number of test cases that each of the algorithms executed to reach the error state. We used the Mann-Whitney U-test and quantified the effect size with the Vargha-Delaney A_{12} statistics. The significance level for these statistical tests was set to 0.05.

5.3   Results and Discussion

Table 1 provides the success rates (in terms of fault detection) for various HS configurations. The last row of the table shows the average ranking of each configuration based on the statistical differences among them. Configurations that are statistically equivalent (i.e., p-values above 0.05) are assigned a similar ranking. This is done by assigning scores based on pairwise comparisons of configurations. Whenever a configuration is better than the other and the difference is statistically significant, its score is increased (for details, see [5]). Then, based on the final scores, each configuration is assigned ranks ranging from 1 (best configuration) to 12 (worst configuration). In case of ties, ranks are averaged. As the success rates and average rankings indicate, using a very low (< 50) or very high value (>=200) of n results in a degraded performance for HS. With a low value of n, HS makes the switch from (1+1) EA to ART too early, which does not give sufficient time for (1+1) EA to converge and hence running HS becomes similar to only running ART. In cases where ART performs well, such configurations of HS also perform well (see Table 2 for the performance of ART on artificial problems). For instance, for $n = 10$, the average success rate is 66% and average ranking is 6.38. Similarly, when HS switches too late, it does not give enough time to ART (given the upper bound of 1000 iterations) and hence running HS is similar to running (1+1) EA in such cases. These configurations perform well in cases where (1+1) EA performs well (Table 2) and poor otherwise. The best results are provided for values between *50* and *100* and the differences in results in this range are not significant. Though the results are not fully consistent across all problems, configuration $n = 50$ has the best average rank across all problems and is always very close to the maximum success rates. We can hence answer RQ1 by stating that, overall, $n=50$ shows the best results for HS and therefore this configuration can be used when applying HS on new problems.

For RQ2a we compared the best HS configuration ($n = 50$) with RT, ART, and (1+1)EA. Table 2 shows the corresponding success rates of these algorithms and Table 3 shows a comparison of HS with the other three algorithms based on statistical tests. The statistics for the situations where HS is significantly better are bold-faced and are italicized where it is significantly worse. Table cells with a '-' denote no significant differences. P-values obtained as a result of Fisher Exact test on the success rates are denoted as p and odds ratio as ψ. In cases where there is no statistical difference in success rates, the number of iterations is considered and the p-values of the Mann-Whitney U-test are denoted as $it\text{-}p$ and corresponding effect sizes by A_{12}.

When compared to (1+1) EA, HS showed better fault detection performance in four of the artificial problems (P3, P10 – P12) and had similar results otherwise. These are the problems where (1+1) EA, when ran in isolation, showed poor results when compared to RT and ART (as visible from Table 2). For example in the case of P11, (1+1) EA was not able to find the a in any of the runs. On the other hand it is 100% for HS, RT, and ART, which means that these strategies were able to find a fault in every run. Hence, HS shows significant improvement over (1+1) EA.

When compared to RT, HS showed significantly better results in terms of success rates for six artificial problems (P1, P4, P5, P6, P7, and P9) and had similar results for all the other problems. Similarly with ART, in terms of success rates, HS showed better results for six artificial problems (P1, P2, P4, P6, P7, and P9) and had similar results for the rest. P1, P4, P6, P7, and P9 are the problems where ART and RT showed poor results when compared to (1+1) EA (Table 2). For example in the cases of P4, P6, and P9, the success rate of both RT and ART is 0, but that of (1+1) EA and HS is 1 and 0.8, respectively. Hence, in terms of success rates, HS shows significantly better results when compared to RT and ART. However, in terms of number of iterations required to detect the fault, HS is significantly worse than RT in four problems (P8, P10, P12, and P13) and significantly worse than ART in six problems (P3, P8, P10, P11, P12, and P13). But, for all these problems, the success rate of HS, RT, and ART is 1, which means that whenever these algorithms run they find the fault (within the budget of 1000 test cases). Therefore, we can answer RQ2a by stating that HS shows overall significantly better performance than ART, RT, and (1+1) EA in terms of fault detection, but was slower than RT/ART in finding faults for problems where these two algorithms perform better than (1+1) EA. But since the success rate of HS is 100%, and therefore the first run is expected to reach the error state, this difference in execution time has limited practical impact.

For RQ2b we compared the performance of the best configuration of HS ($n = 50$) with that of ART, RT, and (1+1) EA on the industrial case study. The last row of Table 3 shows a comparison of the results of the four strategies on this case study (IC) and the last column of Table 2 shows the corresponding success rates. The results are similar to that obtained for those artificial problems where RT and ART perform better than (1+1) EA. HS outperformed (1+1) EA. When compared with the results of ART and RT, there is no significant difference though (100% success rate). These results are consistent with RQ2a and we can therefore answer RQ2 by stating that, overall, HS shows significantly better results when compared to (1+1) EA, RT, and ART. However, as for RQ2a, for problems where ART performed much better than (1+1) EA, though the success rates of HS and ART are similar, ART find the faults faster than HS.

HS starts with (1+1) EA and switches only when *fifty* consecutive test cases do not show better fitness. Fitness evaluations make HS slower than ART/RT but its effectiveness considerably improves over ART/RT for the problems where they showed poor results. In the light of these results, we can conclude that when applying our testing approach, using HS seems to be the most practical choice as its performance, unlike that of (1+1) EA, ART, and RT, is not drastically affected by the properties of the SUT and its environment models. As a result, testers can apply this strategy in confidence, knowing it will perform well in most circumstances.

Table 2. Success Rates of HS (best configuration), RT, ART, and (1+1) EA

	P1	P2	P3	P4	P5	P6	P7	P8	P9	P10	P11	P12	P13	Avg.	IC
HS	0.95	1	1	0.8	1	0.45	0.8	1	0.45	1	1	1	1	0.88	1
ART	0.4	0.75	1	0	0.95	0	0.15	1	0	1	1	1	1	0.63	1
EA	1	1	0.5	1	1	0.7	0.85	1	0.35	0.45	0	0.7	0.95	0.73	0.74
RT	0.45	1	1	0	0.65	0	0.2	1	0	1	1	1	1	0.64	0.97

Table 3. Comparison of best HS configuration with RT, ART, & (1+1)EA

Problem	HS vs. (1+1) EA	HS vs. RT	HS vs. ART
P1	-	p = 0.0012, ψ =15.74	p = 0.0004, ψ =19.12
P2	-	it-p = 0.0065, A_{12} = 0.25	p = 0.047, ψ =14.55
P3	p = 0.0004, ψ = 41.00		it-p = 0.013, A_{12} = 0.73
P4	-	p = 1.5e-07, ψ = 150.33	p = 1.5e-07, ψ = 150.33
P5	-	p = 0.0083, ψ = 22.78	-
P6	-	p = 0.0012, ψ = 33.87	p = 0.0012, ψ = 33.87
P7	-	p = 0.0004, ψ = 13.44	p = 8.7e-05, ψ = 18.33
P8	-	it-p = 0.009, A_{12} = 0.74	it-p = 0.0004, A_{12} = 0.825
P9	-	p = 0.0012, ψ = 33.87	p = 0.0012, ψ = 33.87
P10	p = 0.0001, ψ = 49.63	it-p = 0.0006, A_{12} = 0.81	it-p = 0.0002, A_{12} = 0.85
P11	p = 1.4e-11, ψ = 1681.00		it-p = 0.0032, A_{12} = 0.77
P12	p = 0.02, ψ = 18.38	it-p = 0.0016, A_{12} = 0.79	it-p = 0.0008, A_{12} = 0.81
P13	-	it-p = 0.0199, A_{12} = 0.71	it-p = 0.021, A_{12} = 0.71
IC	p = 0.0004, ψ = 28.83	-	it-p = 0.015, A_{12} = 0.66

5.4   Threats to Validity

Although the artificial problems that we developed were based on industrial RTES and are not trivial to test (they are multithreaded and hundreds of lines long), these artificial problems may not be representative of complex RTES. To reduce this threat, we used artificial problems inspired by actual RTES and intentionally varied the properties of their environments in ways that could affect the testing strategies.

A typical problem when testing RTES is the accurate simulation of time. To be on the safe side, to evaluate whether our results are reliable, we selected a set of experiments and ran them again with exactly the same random seeds. We obtained equivalent results with a small variance of a few milliseconds, which in our context did not affect the testing results.

Another possible threat to validity could be the faults that were manually seeded in the artificial problems. The faults that we seeded came from our experience with the industrial RTES and from the feedback of our industry partners, so they are representative of real faults. We could not use a more systematic fault seeding techniques due to the reasons mentioned in Section 5.1.

6   Conclusion

In this paper, we proposed a hybrid strategy (HS) that combines (1+1) Evolutionary Algorithm (EA) and Adaptive Random Testing (ART) for black-box automated system testing of real-time embedded systems (RTES). The strategy was developed to combine the benefits of both algorithms, since their individual results varied greatly depending on the failure rate of the system under test and properties of its environment. The ultimate goal was to obtain a strategy with consistently good results. The proposed strategy starts with running (1+1) EA and switches to ART when the (1+1) EA search stops yielding fitter test cases. We empirically investigated when to switch to ART and identified an optimal setting for HS. Results indicate that switching too early or too late than the identified setting has a negative impact on the performance of the strategy. Based on the experiments, when using HS in practice, we propose switching to ART after (1+1) EA generates 50 consecutive test cases that do not improve fitness. We evaluated the proposed strategy and compared its performance with that of running (1+1) EA and ART individually. We also use random testing (RT) as a comparison baseline. The empirical evaluation uses an industrial case study and 13 artificial problems that were developed based on two industrial case studies belonging to different domains. The models of these artificial problems were developed in order to vary their characteristics, thus potentially affecting the performance of the evaluated testing strategies. Overall, the results indicate that HS shows significantly better performance in terms of fault detection (an overall 88% success rate for artificial problems and 100% for the industrial case study) than the other three algorithms (for artificial problems: ART: 63%, RT: 64%, and (1+1) EA: 74% and for the industrial case study: ART: 100%, RT, 97%, (1+1) EA: 74%). Unlike the other strategies, variations in environment properties do not have a drastic impact on the performance of HS and it is therefore the most practical approach, showing consistently good results for different problems.

Acknowledgements. The work is supported by supported by the Norwegian Research Council and was produced as part of the ITEA 2 VERDE project. L. Briand was supported by a FNR PEARL grant. We are thankful to our industrial partners for their support.

References

1. Adjir, N., De Saqui-Sannes, P., Rahmouni, K.M.: Testing Real-Time Systems Using TINA. In: Núñez, M., Baker, P., Merayo, M.G. (eds.) TESTCOM 2009. LNCS, vol. 5826, pp. 1–15. Springer, Heidelberg (2009)
2. Ali, S., Iqbal, M.Z., Arcuri, A., Briand, L.: A Search-based OCL Constraint Solver for Model-based Test Data Generation. In: 11th International Conference on Quality Software, pp. 41–50. IEEE (2011)
3. Andrews, J., Briand, L., Labiche, Y., Namin, A.: Using mutation analysis for assessing and comparing testing coverage criteria. IEEE Transactions on Software Engineering 32, 608–624 (2006)
4. Andrews, J.H., Menzies, T., Li, F.C.H.: Genetic algorithms for randomized unit testing. IEEE Transactions on Software Engineering 37, 80–94 (2011)
5. Arcuri, A., Briand, L.: A Practical Guide for Using Statistical Tests to Assess Randomized Algorithms in Software Engineering. In: 33rd International Conference on Software Engineering (ICSE), pp. 1–10 (2011)
6. Arcuri, A., Iqbal, M.Z., Briand, L.: Black-Box System Testing of Real-Time Embedded Systems Using Random and Search-Based Testing. In: Petrenko, A., Simão, A., Maldonado, J.C. (eds.) ICTSS 2010. LNCS, vol. 6435, pp. 95–110. Springer, Heidelberg (2010)
7. Arcuri, A., Iqbal, M.Z., Briand, L.: Random Testing: Theoretical Results and Practical Implications. IEEE Transactions on Software Engineering 38, 258–277 (2012)
8. Auguston, M., Michael, J.B., Shing, M.: Environment behavior models for automation of testing and assessment of system safety. Information and Software Technology 48, 971–980 (2006)
9. Broekman, B.M., Notenboom, E.: Testing Embedded Software. Addison-Wesley Co., Inc. (2003)
10. Heisel, M., Hatebur, D., Santen, T., Seifert, D.: Testing Against Requirements Using UML Environment Models. In: Fachgruppentreffen Requirements Engineering und Test, Analyse & Verifikation, pp. 28–31. GI (2008)
11. Iqbal, M.Z., Arcuri, A., Briand, L.: Automated System Testing of Real-Time Embedded Systems Based on Environment Models. Simula Research Laboratory, Technical Report (2011-19) (2011)
12. Iqbal, M.Z., Arcuri, A., Briand, L.: Code Generation from UML/MARTE/OCL Environment Models to Support Automated System Testing of Real-Time Embedded Software. Simula Research Laboratory, Technical Report (2011-04) (2011)
13. Iqbal, M.Z., Arcuri, A., Briand, L.: Empirical Investigation of Search Algorithms for Environment Model-Based Testing of Real-Time Embedded Software. In: Proceedings of the International Symposium on Software Testing and Analysis (ISSTA), pp. 199–209. ACM (2012)
14. Iqbal, M.Z., Arcuri, A., Briand, L.: Environment Modeling with UML/MARTE to Support Black-Box System Testing for Real-Time Embedded Systems: Methodology and Industrial Case Studies. In: Petriu, D.C., Rouquette, N., Haugen, Ø. (eds.) MODELS 2010, Part I. LNCS, vol. 6394, pp. 286–300. Springer, Heidelberg (2010)
15. Grabowski, J., Nielsen, B. (eds.): FATES 2004. LNCS, vol. 3395. Springer, Heidelberg (2005)
16. Lefticaru, R., Ipate, F.: Functional search-based testing from state machines. In: Proceedings of the International Conference on Software Testing, Verification, and Validation, pp. 525–528. IEEE Computer Society (2008)

17. Mühlenbein, H.: How genetic algorithms really work: I. mutation and hillclimbing. In: Parallel Problem Solving from Nature, vol. 2, pp. 15–25 (1992)
18. OMG: Modeling and Analysis of Real-time and Embedded systems (MARTE), Version 1.0 (2009), http://www.omg.org/spec/MARTE/1.0/
19. OMG: Unified Modeling Language Superstructure, Version 2.3 (2010), http://www.omg.org/spec/UML/2.3/
20. Schneckenburger, C., Schweiggert, F.: Investigating the dimensionality problem of Adaptive Random Testing incorporating a local search technique. In: IEEE International Conference on Software Testing Verification and Validation Workshop (ICSTW 2008), pp. 241–250 (2008)
21. Tappenden, A.F., Miller, J.: A novel evolutionary approach for adaptive random testing. IEEE Transactions on Reliability 58, 619–633 (2009)
22. Zheng, M., Alagar, V., Ormandjieva, O.: Automated generation of test suites from formal specifications of real-time reactive systems. The Journal of Systems & Software 81, 286–304 (2008)

Testing of Concurrent Programs
Using Genetic Algorithms

Vendula Hrubá[1], Bohuslav Křena[1], Zdeněk Letko[1], Shmuel Ur[2], and Tomáš Vojnar[1]

[1] IT4Innovations Centre of Excellence, FIT, Brno University of Technology, Czech Republic
{ihruba,krena,iletko,vojnar}@fit.vutbr.cz
[2] Shmuel Ur Innovations, Ltd.
shmuel.ur@gmail.com

Abstract. Noise injection disturbs the scheduling of program threads in order to increase the probability that more of their different legal interleavings occur during the testing process. However, there exist many different types of noise heuristics with many different parameters that are not easy to set such that noise injection is really efficient. In this paper, we propose a new way of using genetic algorithms to search for suitable types of noise heuristics and their parameters. This task is formalized as the test and noise configuration search problem in the paper, followed by a discussion of how to represent instances of this problem for genetic algorithms, which objectives functions to use, as well as parameter tuning of genetic algorithms when solving the problem. The proposed approach is evaluated on a set of benchmarks, showing that it provides significantly better results than the so far preferred random noise injection.

1 Introduction

The arrival of multi-core processors into common computers accelerated development of software with multi-threaded design. Multi-threaded programming is, however, significantly more demanding and offers much more space for errors. Moreover, errors in concurrency are often very difficult to discover and localise due to the non-deterministic nature of multi-threaded computation. This situation stimulates research efforts devoted to all sorts of methods for testing, analysis, and verification.

Formal methods of verification, such as, e.g., model checking [11], aim at precise program verification. Unfortunately, these precise approaches do not scale well for complex software. This is one of the main reasons why heuristic approaches such as lightweight static analyses, testing, and dynamic analyses are still very popular.

When dealing with concurrent programs, testing and dynamic analysis that rely on executing the program under test and evaluating the witnessed run suffer from the problem of non-deterministic scheduling of program threads. Due to this problem, a single execution of a program is insufficient to determine correctness of the program even for the particular input data used in the execution. Moreover, even if the program has been executed many times with the given input without spotting any failure, it is still possible that its future execution with the same input will produce an incorrect result.

One way to address this problem is to use *deterministic testing* that can be viewed as model checking bounded in various ways (e.g., in the number of context switches) [16,19]. This technique attempts to systematically test all interleaving scenarios up to some bound,

G. Fraser (Ed.): SSBSE 2012, LNCS 7515, pp. 152–167, 2012.

which is quite demanding (especially for long runs) because one needs to track which scheduling scenarios have been witnessed and systematically force new ones.

A lightweight alternative to the above is to use *noise injection techniques* [7] based on heuristically disturbing the scheduling of program threads in hope of observing so far unseen scheduling scenarios. Although this approach cannot prove correctness of a program even under some bounds on its behaviour, it was demonstrated in [7,14] that it can rapidly increase the probability of spotting concurrency errors without introducing any false alarms.

Noise injection can be implemented by instrumenting the program under test by noise generation code that influences the execution of selected threads at selected program locations. Noise injection can use different *noise seeding heuristics* given by the *type of noise* (e.g., noise based on injecting calls of $yield()$, calls of $wait()$, halting one thread till other threads can continue, etc.), the *strength of the noise* (e.g., how many times the yield operation should be called when injected at a certain location, for how long a thread should wait, etc.), and the *frequency of the noise* (how often some noise is generated at a particular location). Moreover, various *noise placement heuristics* can be used, including the use of a fixed set of program locations at which some concurrency-relevant actions appear (such as accesses to shared memory, synchronisation, etc.), using a randomly selected subset of such a set, or some more involved heuristics driven by the so-far obtained coverage of the program behaviour [14].

Our previous work on noise injection [14] shows that there is no silver bullet among the many existing noise injection heuristics. Results provided by them depend on the tested program as well as on the run-time environment (the type and number of processors and the actual workload are usually the most significant factors). Actually, some configurations can decrease the probability of an error manifestation. This is helpful for run-time healing of errors, but it is highly undesirable for detecting them. Moreover, the number of possible settings of the noise injection (and also of the test itself) together with the considerable time needed to run a test in order to evaluate the efficiency of a certain noise configuration makes exhaustive searching for suitable noise configurations impractical. This is exactly the case where *metaheuristic search techniques* [17] can help.

Genetic algorithms [17] are metaheuristic search techniques which try to find the best solutions by sampling the search space. They start with an initial set (called a *generation*) of possible solutions (also called *individuals*). Each individual is evaluated and assigned a value called a *fitness* representing the suitability of the solution it represents. The next generation of individuals is typically obtained by a stochastic recombination (called a *crossover*) and *mutation* of individuals selected according to their fitness.

In this paper, we propose a new way of using genetic algorithms to search for suitable types of noise heuristics and their parameters. We formalize this task as the *test and noise configuration search problem* (the TNCS problem). Then, we show how to represent instances of this problem for genetic algorithms, and we discuss which basic objective functions may be useful as building blocks of complex objective functions suitable in the given context. We also discuss parameter tuning of genetic algorithms when solving the TNCS problem. Next, we instantiate the framework by a concrete combined objective function suitable especially (but, as our experiments show, not only) for data race detection. Finally, we evaluate the proposed approach on a set of benchmarks, showing that it provides significantly better results than the so far preferred random noise injection.

Plan of the paper. The rest of the paper is organised as follows. In Section 2, we discuss the related work. In Section 3, we formulate the TNCS problem and discuss the basic objective functions suitable in its context. In Section 4, we propose how to utilise genetic algorithms to solve the TNCS problem. Section 5 focuses on setting parameters of genetic algorithms for the TNCS problem. In Section 6, we propose a concrete combined objective function for use with the TNCS problem, and we provide experimental evidence on how the proposed approach can improve the testing process. Finally, Section 7 provides concluding remarks and comments on the possible future work.

2 Related Work

Most existing works in the area of search-based testing of concurrent programs focus on applying various metaheuristic techniques to control the state space exploration within the *guided model checking* approach [11]. The basic idea is to explore areas of the state space that are more likely to contain concurrency errors even when the entire state space will not be explored. Metaheuristic algorithms that have been applied within the guided model checking approach for finding deadlocks and/or assertion violations include simulated annealing [6], genetic algorithms [11,3], the partial swarm optimisation (PSO) [6], and the ant colony optimisation (ACO) [1,2]. An advantage of this approach is that the underlying model checking offers a well-defined state space and a high degree of determinism. On the other hand, the approach shares limitations of model checking in terms of scalability and cost of modelling of the environment. In our approach, we focus on testing and dynamic analyses which are able to handle much larger real programs.

In [9], genetic algorithms are applied within the process of debugging of concurrent programs based on repeated testing with noise injection. Genetic algorithms are used in order to find a noise configuration which causes concurrency-related bugs to occur with a high probability while preferring settings with noise concentrated to a minimal number of locations (which is motivated by concentrating on the problem of debugging). Compared to [9], we do not search for which concrete locations should be noised with which noise. Instead, we search for which noise seeding and noise placement heuristics (or which combinations of these heuristics) with which parameters can provide good results for a particular test and environment. This allows us to use a simpler representation of individuals and to support much larger test cases with plenty of possible locations to be noised. Moreover, we propose new objective functions (based, e.g., on results that can be obtained through various dynamic analyses) which allow us to focus not only on debugging but also on testing. Finally, compared with the initial results presented in [9], we present a more thorough experimental evaluation.

The problem of increasing the probability of an error manifestation within the debugging process is targeted in [4,18] too. However, these works do not consider metaheuristic search algorithms. In [4], program locations are first statically classified according to their suitability for noise injection. Then, a probabilistic algorithm is used to find a subset of program locations that increase the error manifestation ratio. In [18], a machine learning feature selection algorithm is used to identify a subset of program locations where to inject noise. In this case, the test is executed many times, and program locations where the noise was injected in each execution are collected together with information whether the error got manifested.

Finally, in [15], we presented our preliminary results with a steepest ascending search algorithm. The experiments proved our concept but showed that the local search technique is not suitable for the given setting. The used algorithm showed a tendency to find a local optimum only. Therefore, in this work, we focus on global search techniques, namely, genetic algorithms.

3   Testing of Concurrent Programs as a Search Problem

In this section, we present our proposal of how search techniques can be combined with noise-based testing of concurrent programs by identifying suitable combinations of noise injection heuristics as well as their parameters. In particular, we formulate the proposed use of search techniques via the so-called test and noise configuration search (TNCS) problem. Subsequently, we discuss several objective functions that can be useful when dealing with various instances of the TNCS problem (typically as building blocks of more complex combined objective functions as we illustrate in Section 6).

3.1   The Test and Noise Configuration Search Problem

As we have mentioned already in the introduction, there are two main issues that must be solved when using noise injection. First, one needs to determine program locations where to insert noise. Heuristics which target this problem are called *noise placement heuristics*. Second, one needs to determine which *noise seeding heuristics*, i.e., which way of disturbing thread scheduling, should be used. Moreover, most types of the heuristics are adjustable by one or more parameters influencing their behaviour and efficiency (e.g., noise seeding heuristics are often parametrised by their strength). Further, one can combine several noise placement and seeding techniques within one execution. Indeed, our results presented in [14] show that such a combination provides in many cases better results than using a single heuristics. Finally, it is usually the case that there exist multiple test cases for a given program that can also be parametric.

With respect to the above, we formulate the *test and noise configuration search problem* (the TNCS problem) as the problem of selecting test cases and their parameters together with types and parameters of noise placement and noise seeding heuristics that are suitable for a certain test objective.

Formally, let $Type_P$ be a set of available types of noise placement heuristics each of which we assume to be parametrised by a vector of parameters. Let $Param_P$ be a set of all possible vectors of parameters. Further, let $P \subseteq Type_P \times Param_P$ be a set of all allowed combinations of types of noise placement heuristics and their parameters. Analogically, we can introduce sets $Type_S$, $Param_S$, and S for noise seeding heuristics. Next, let $C \subseteq 2^{P \times S}$ contain all the sets of noise placement and noise seeding heuristics that have the property that they can be used together within a single test run. We denote elements of C as *noise configurations*. Further, like for the noise placement and noise seeding heuristics, let $Type_T$ be a set of test cases, $Param_T$ a set of vectors of their parameters, and $T \subseteq Type_T \times Param_T$ a set of all allowed combinations of test cases and their parameters. We let $TC = T \times C$ be the set of *test configurations*.

Now, the TNCS problem can be expressed as searching for a test configuration from TC suitable wrt. some given objective function. One can also consider the natural generalisation of the TNCS problem to searching for a set of test configurations, i.e., a member of 2^{TC}, suitable wrt. some given objective function.

3.2   Objective Functions for the Context of the TNCS Problem

We next suggest several possible objective functions that can be useful in various instances of the TNCS problem, typically combined into more complex objective functions as we illustrate in Section 6.

First, an objective function that can often be found useful is to minimise the impact of noise injection on the *time of execution* of a test case. The more noise is injected into the execution the slower the execution typically is. The slowdown can be unwelcome especially when the time for testing is limited. Then, due to the slowdown, less executions of a test case and/or less test cases will be considered which may in turn negate the aim of using noise injection to improve the quality of testing. The time aspect is also important when a program under test needs to meet certain throughput or response time requirements that could be broken by an excessive use of noise.

Next, since the primary goal of testing is to find errors, a natural objective function is to maximise the *number of errors* that occur (and are detected by the test harness) when executing tests with a certain configuration. Once some test configuration is found suitable wrt. the number of errors it allows one to observe, one could think that this configuration is not useful any more since the errors were already detected. However, this test configuration can be used for further testing in hope that it will allow one to discover even more errors (recall that due to the non-determinism of scheduling, not all errors will show up in a single run or a set of runs). Moreover, one can also think of using this test configuration in regression testing or when testing similar applications.

Another sensible objective function, tightly related to the above, is to monitor test executions under particular test configurations by some *dynamic analyser* and to maximise the number of warnings about dangerous behaviour of the program under test that get reported. Test configurations delivering good results in this case can subsequently be used for more extensive testing in hope of finding a real error even though an actual error was not seen during evaluation of the test configuration. The reliability of this approach of course depends on the precision of the chosen analyser. A high ratio of false positives and/or negatives makes this objective function unreliable.

A further possibility is to use a suitable *coverage metric* allowing one to judge how much of the possible behaviour of the program under test has been covered (and hence how likely it is that some undesired behaviour was omitted) and to look for test configurations maximising the obtained coverage. Concurrency-related metrics based on dynamic analyses which we presented in [13] can be especially useful here. These metrics are not based on simply counting the number of produced warnings, but on much finer measures. Some of them are based on monitoring events that make the internal state of a dynamic analyser change, e.g., the *HBPair* metrics based on the happens-before relations, and some express how many internal states a certain dynamic analyser reached, e.g., the *GolidLockSC* metrics based on monitoring the internal states of the GoldiLock analyser [8]. Of course, there are many other existing coverage metrics which can be considered. For instance, the synchronisation coverage [5] (*Synchro*) which measures how well the various synchronisation mechanisms used in the program under test are tested (by measuring how many different scenarios of the use of the synchronisation mechanisms were witnessed). A drawback of many concurrency coverage metrics is that it is often impossible to compute what the full coverage is; this is, however, not a problem here since we are interested in relative comparisons of the coverage achieved through different test configurations.

Fitness of a test configuration $tc \in TC$ wrt. the above objective functions has typically to be evaluated by a *repeated execution* of the test case encoded in tc with the test parameters and noise configuration that are also a part of tc. Recall that the noise configuration can contain multiple types of noise heuristics. We assume all of them to be used in each testing run, which is consistent with our definition of noise configurations that allows only those combinations of noise heuristics that can indeed be used together. Further note that the repeated execution makes sense due to the non-determinism of thread scheduling. The evaluation of individual test runs must of course be combined, which can be done, e.g., by computing the *average evaluation* or by computing a *cumulative evaluation* across all the performed executions.

In addition, it is also possible to define some simple objective functions directly on the test configurations. For instance, one can minimise/maximise the number of enabled heuristics, volume or frequency of noise to be injected, etc. Such objective functions are typically not sensible alone, but can make sense when combined with other objective functions.

4   A Genetic Approach to the TNCS Problem

In this section, we present our proposal of using a genetic approach to solve the TNCS problem. The approach is presented on a concretization of the TNCS problem for the context of using the IBM Concurrency testing tool (ConTest) [7] (with some extensions) for noise-based testing. We therefore start by presenting the concrete set of noise configurations considered. Subsequently, we present how one can apply the genetic approach in this concrete setting.

4.1   ConTest-Based Noise Configurations

We consider noise injection heuristics implemented in ConTest extended by our plug-in implementing a coverage-based noise placement heuristics [14]. Hence, we consider three noise placement heuristics: the *random* heuristics which picks program locations randomly, the *sharedVar* heuristics which focuses on accesses to shared variables, and our *coverage-based* heuristics [14] which focuses on accesses near a previously detected thread context switch. The *sharedVar* heuristics has two parameters modifying its behaviour with 5 valid combinations of its values. The *coverage-based* heuristics is controlled by 2 parameters with 3 valid combinations of values. All these noise placement heuristics inject noise at selected places with a given probability. The probability is set globally for all enabled noise placement heuristics by a *noiseFreq* setting from the range 0 (never) to 1000 (always). The *random* heuristics is enabled by default when $noiseFreq > 0$. The *random* heuristics can be suppressed by one parameter of the *sharedVar* heuristics which explicitly disables the combination of these two heuristics.

The considered infrastructure offers 6 basic and 2 advanced noise seeding techniques. The basic techniques cannot be combined, but any basic technique can be combined with one or both advanced techniques. The basic heuristics are: *yield*, *sleep*, *wait*, *busyWait*, *synchYield*, and *mixed*. The *yield* and *sleep* techniques inject calls of the `yield()` and `sleep()` functions. The *wait* and *synchYield* techniques lock a special monitor and then either wait for some time or call `yield()`. The *busyWait* technique inserts code that just loops for some time. The *mixed* technique randomly chooses one of the five other techniques at each noise injection location. The *haltOneThread* technique occasionally stops one thread until any other thread cannot run. Finally, the *timeoutTamper* heuristics randomly reduces the time-outs used in the program under test in calls of `sleep()` (to ensure that they are not used for synchronisation).

4.2   Individuals, Their Encoding, and Genetic Operations on Them

In order to utilise a genetic algorithm to solve the TNCS problem with the considered set of noise configurations, we let the particular test configurations play the role of *individuals*. We encode the test configurations as *vectors of integers*. The test configuration is either reduced to solely a noise configuration (when a single test case without parameters is considered), or it consists of the noise configuration extended by one or more specific entries controlling the test case settings. We, however, concentrate here on the noise configurations only, which form vectors of numbers in the range $(0, 0, 0, 0, 0, 0)$– $(1000, 5, 3, 6, 2, 2)$. Here, the first entry controls the *noiseFreq* setting, the next two control the *sharedVar* and *coverage-based* noise placement heuristics. The last three entries control the setting of the basic and advanced noise seeding heuristics. Each entry in the vector is annotated by a flag saying whether there exists an ordering on the values of the entry. We call entries whose values are ordered as *ordinal entries*.

We consider the standard one-point, two-point, and uniform element-wise (any-point) *crossover operators* [17] in the form they are implemented in the ECJ library [20]. *Mutation* is also done on an element-wise basis, and it handles ordinal and non-ordinal entries differently. Non-ordinal entries are set to a randomly chosen value from the particular range (including the current value). Ordinal entries (e.g., entries encoding the strength of noise or the parameter controlling the number of threads the test should use) are handled using the standard Gaussian mutation [17] (with the standard deviation set to 10 % of the possible range or minimal value 2). Finally, we consider standard proportional and tournament-based fitness selection operators [17] as they are implemented in the ECJ library.

5   Parameters of Genetic Algorithms and the TNCS Problem

Genetic algorithms are adjustable through a number of parameters influencing the efficiency of the search process. The way these parameters should be set to obtain a high efficiency usually depends on the considered problem. In this section, we provide our findings on how to set the parameters of a standard genetic algorithm when solving the TNCS problem. We focus mainly on the following questions: How to set up the breeding infrastructure, i.e., which standard selection and crossover operators should be used, how to set up their parameters, which value of mutation probability provides good results, and whether elitism or random generation of individuals can help. We also target the question whether it is better to run a few big generations or instead more small generations in case the time for testing is limited.

5.1 Experiments for Finding Suitable Parameters of Genetic Algorithms

We conducted all our experiments aimed at finding a suitable setting of the parameters of genetic algorithms on one selected case study only. This is mostly due to the high time consumption of evaluating each test configuration through multiple test executions. In particular, we used the *Crawler* test case which is based on a part of an older version of a major IBM production software. The crawler creates a set of threads waiting for a connection. If a connection simulated by the testing environment is established, a worker thread serves it. There is an error in a method that is called when the crawler shuts down. The error causes an unhandled exception. The trickiness of the error can be seen from the very low probability of its manifestation (approximately 0.0006 when no noise is used). The case study has 19 classes and 1.2 kLOC. There is a single test case available with no parameters (making test configurations equal to noise configurations).

We conducted our experiments on multiple machines, all having Intel i5 661 processors, running 64-bit Linux and Java 6. We used our infrastructure SearchBestie [15] and IBM ConTest to evaluate test configurations and the ECJ library [20] to implement the genetic algorithms. We narrowed the search space down by sampling the *noiseFreq* parameter by ten, i.e., by reducing its possible values to 0, 10, ..., 1 000.

With the aim of observing as many behaviours differing in their various important concurrency-related aspects as possible, we considered an objective function maximising the obtained coverage under three different concurrency coverage metrics, namely, *Synchro*, *Avio** and *HBPair** [13]. This objective function covers three different aspects of concurrency behaviour: interleaving of accesses from different threads to shared memory locations via *Avio**, successful synchronisation of program threads inducing a happens-before relation via *HBPair**, and information about whether the implemented synchronisation does something helpful via *Synchro*. We used results of approximately 1 million randomly noised executions to estimate the 100 % achievable coverage (denoted as *max* below) for each of the metrics and set up the following fitness function:

$$\frac{1}{3} * \left(\frac{Avio^*}{Avio^*_{max}} + \frac{HBPair}{HBPair^*_{max}} + \frac{Synchro}{Synchro_{max}} \right)$$

The evaluation of each test configuration consisted of 5 executions of the test case with the noise parameters encoded in the test configuration. The value of the fitness function was then computed using the accumulated coverage of all the five executions.

We fixed the number of evaluated individuals to 2 000 in each experiment. According to our experiments, this value is sufficient to reach saturation of the selected coverage metrics in the Crawler case study. We set the size of the considered populations and number of generations as follows (population/generation size): 200/10, 80/25, 40/50, 20/100, and 10/200. We considered the breeding infrastructure to consist of two selection operators which select individuals for the crossover operator. The output of the crossover operator was mutated using the mutation operator described in Section 4.2.

We performed two sets of experiments. In the first one, we considered the standard fitness proportional selection operators, four different standard crossover operators (*one-point*, *two-point*, and *any-point* with the probability of mutating each element of the vector set to 0.1 and 0.25), and four different probabilities of applying the mutation operator (0.01, 0.1, 0.25, and 0.5). All experiments were repeated 10 times, therefore in the first set of experiments we performed 8 000 000 executions of the test (5 × 2 000 × 5 × 4 × 4 × 10). In the second set of experiments, we fixed the considered size of the population to 40 and the number of generations to 50, the crossover operator to *any-point* with probability 0.1, and the mutation probability to 0.01. We then studied the influence of different selection operators, elitism which puts into the next generation a number of individuals (0, 2, 4) without breeding, and a random creation of a few individuals (0, 2, 4) that are put into the following generation. In the second set of experiments, we fixed the number of evaluated individuals in each experiment to 1 600 individuals and performed 720 000 executions (5 × 1 600 × 3 × 3 × 10). Last but not least, we considered the fitness proportional and tournament selection operators (with the size of the tournament being 2 or 4) and their combinations. In the third set of experiments we evaluated 1 200 individuals in each experiment and performed 480 000 executions of the test (5 × 1 200 × 4 × 2 × 10).

From each experiment, we collected various data concerning the generated populations including, in particular, the following two statistics: (1) The average fitness value in each generation *aver* and (2) the best individual fitness in each generation *best*. Our goal was then to identify parameters of the genetic algorithms under which the best test configuration out of all discovered test configurations is found, and it is found as quickly as possible. For that, we used the *best* and *aver* statistics. The results of the experiments are summarised below with some more technical details available in [12].

5.2   Results of Experiments with the Parameters of Genetic Algorithms

The values of the *best* and *aver* statistics that we obtained from the first set of experiments presented above show that small populations combined with the *any-point* crossover and mutation set to 0.01 are able to find the best individual (i.e., the best test configuration) out of all the encountered ones quite fast (within a few generations). Very small populations (10 and rarely also 20) are, however, sometimes not able to find the best individual and get stuck in a local optimum. On the other hand, in larger populations, it takes much longer to arrive to the best individual. The *any-point* crossover operator exceeded the other two operators.

The best individuals obtained by the genetic algorithm in our experiments had fitness higher than 0.5, and they therefore covered more than 50 % of the concurrent behaviour as defined by our fitness function. The overall best individuals achieved fitness 0.64. The average fitness of the final population was in the worst case 0.35 only, which is quite similar to fitness 0.33 that we achieved by randomly generating individuals to evaluate. The highest average fitness was close to the maximum fitness of 0.64, which represents a situation when nearly all individuals in the generation were the same.

In the second set of experiments from the above, we clearly saw the positive effect of elitism (set to 10 % of the population). The selection operators seem to affect the results only a little. The best results seem to be provided by a combination of the tournament selection operator (with the size of the tournament set to a high value) and the fitness proportional selection operator.

Based on the results summarised above and in more details presented in [12], we found as suitable the following setting of the parameters of genetic algorithms for the considered concretization of the TNCS problem: Size of population 20, two different selection operators (tournament among 4 individuals and fitness proportional), the *any-point* crossover with a higher probability (0.25), a low mutation probability (0.01), and two elites (that is 10 % of the population). This parameter setting is used in the experiments presented in the next section.

6   A Concrete Application of the Proposed Approach

In this section, we first propose a complex objective function for the TNCS problem that carefully combines the above discussed basic objective functions, finally leading to a concrete application of genetic algorithms for improving the process of testing of concurrent programs. In particular, the stress is on looking for data races, but as our experiments show, the approach helps in finding other kinds of concurrency-related errors too. Next, we present a collection of benchmarks and results of experiments with them which illustrate the efficiency of our approach.

6.1   A GoldiLocks-Based Objective Function

Based on our experience with different concurrency coverage metrics and dynamic error detectors, we have decided to build our concrete objective function on maximizing the coverage obtained under the concurrency coverage metric *GoldiLockSC* [13] based on the GoldiLocks algorithm [8], together with maximizing the number of actual warnings produced by this algorithm. We have chosen the GoldiLocks algorithm for our objective function because it has a low ratio of false positives, and it is able to continue in the analysis even after an error is detected. Moreover, our results indicate that the concurrency coverage metric *GoldiLockSC* has multiple positive properties. In particular, the coverage under this metric usually grows smoothly (i.e., with a minimum of shoulders) and does not stabilise too early (i.e., before most behaviours relevant from the point of view of data race detection are examined). Further, based on the discussion presented in Section 3.2, we also reflect in our objective function an intention to minimize the execution time and to maximize the number of detected errors.

In summary, we thus aim at (1) maximizing coverage under the concurrency coverage metric *GoldiLockSC* [13], (2) maximizing the number of warnings produced by GoldiLocks, (3) maximizing the number of detected real errors due to data races, and (4) minimizing the execution time. The different basic objectives are combined using a system of weights assigned to them.

To be more precise, the *GoldiLockSC* metric counts the encountered internal states of the GoldiLocks algorithm (here, SC stands for the optimised version of the algorithm with the so-called *short circuits*, i.e., cheap checks done before the full algorithm is used). We weight the different coverage tasks of this metric as well as the error manifestation according to their severity. In particular, the coverage tasks of the *GoldiLockSC* metric are tuples $(ploc, state)$ where $ploc$ identifies the program location at which some shared memory location is accessed, and $state \in \{SCT, SCL, LS, E\}$ denotes the internal state of the GoldiLocks algorithm. We divide the tasks into three categories according to severity of their $state$. The SCT state represents a situation where the first short circuit check of GoldiLocks (checking whether a variable is accessed by a single thread only) proves correctness of the given access. This situation is common for sequentially executed code, and so we assign it weight 1. The SCL and LS states mean that the first check does not succeed, but it is possible to use further heuristic short circuit checks (SCL) or use the full algorithm (LS) to infer a lock (or locks) whose locking proves correctness of the access. We assign such tasks with weight 5. Finally, the E state means that the algorithm detected a data race and produced a warning message. We weight such tasks with 10. We denote the weighted coverage as $WGoldiLockSC$.

A GoldiLocks warning has the form of a tuple $(var, ploc_1, ploc_2)$ where var identifies a shared variable, and $ploc_1, ploc_2$ represent two program locations between which a data race was detected. Sometimes, a single coverage task with $state = E$ produced at $ploc_1$ leads to several warnings differing in the $ploc_2$ or var values. We denote by $GLwarn$ the number of different warnings issued during the test execution, and we give them the weight of 1 000.

Finally, as we have already mentioned, we also aim at maximizing the number of detected error manifestations ($error$) and minimizing the execution time ($time$). Error manifestations are detected by looking for unhandled exceptions. They are given a very high weight of 10 000. With respect to all the described objectives, we then define the fitness function as follows (expecting the time to be measured in miliseconds):

$$\frac{WGoldiLockSC + 1000 * GLwarn + 10000 * error}{time}$$

6.2   Case Studies

We concentrate primarily on data race detection, but we also try to apply our genetic approach to case studies containing other kinds of concurrency errors (and, as we will show, we obtain quite positive results even in such cases). In particular, we evaluate our approach on 5 test cases containing concurrency-related errors. The test cases are listed in Table 1. In the table, the *kLOC* column shows the size of the considered test case, and the *Param* column indicates the number of its parameters and the number of possible values of each parameter (e.g., 3, 3 means that the test takes two parameters, each with three possible values).

The *Animator*, *Crawler*, and *FTPServer* test cases contain a data race which leads to unhandled exceptions. The *Airlines* case study contains a high level atomicity violation that is detected by a final check at the end of the execution which throws an unhandled exception. Finally, the *Rover* test case contains a deadlock and an atomicity violation which leads to an unhandled exception [10].

We admit that the described case studies are not very large, and one could surely found much bigger ones. Let us, however, stress that the reason why we did not consider truly large benchmarks is *not* a bad scalability of our approach, but rather the large number of experiments that we did with the various parameter settings which in summary take a lot of time even on smaller benchmarks.

The *Airlines* and *Animator* test cases were run on Intel Core2 6600 machines, the *Rover* test case on a machine with an Intel i5-2500 processor, and the *FTPServer* test case on a machine with two Intel X5355 processors. In case of the *Crawler* test case, two different hardware environments were used. The first (denoted simply as *Crawler* in Table 1) used a machine with an Intel i5 661 processor, while the second (denoted *Crawler**) was executed on a machine with four AMD Opteron 8389 processors. These two options were used on purpose in order to study how our approach works in different hardware environments. All mentioned computers ran 64bit Linux and Java version 6. A more detailed description of the test cases and their parameters can be found in [12].

6.3 Experimental Results

To evaluate the efficiency of our approach when using the GoldiLocks-based objective function, we again used the infrastructure described in Section 5. We use the setting of parameters of genetic algorithms inferred in Section 5. Although this setting was inferred for a different objective function and using sampled values of the *noiseFreq* parameter only, we suppose that it represents a good option even for other experiments with our genetic algorithm. Indeed, the objective function used in Section 5 was designed to be rather general in order to cover a lot of different concurrent behaviours. Moreover, we analysed the correlation between the values of the fitness function of Section 5 and the *GoldiLocksSC* metric used in the GoldiLocks-based objective function on the performed experiments and realized that the correlation is high. After all, the combination of *HBPair** and *Avio** focuses on the same events as the GoldiLocks algorithm.

In the experiments, we allowed the elite individuals to be re-evaluated in the following generations. This is motivated by the fact that a few executions of an individual (5 in our case) are often not sufficient to prove whether the configuration can make a concurrency error to manifest. Indeed, tricky concurrency-related errors manifest very rarely even if a suitable noise heuristics is used [14]. The reevaluation of elites therefore gives the most promising individuals another chance to spot an error. This setting is a compromise between a high number of executions needed to evaluate every individual more times and the available time we have.

We compare our genetic approach with the random approach to the choice of noise heuristics and their parameters. In the random approach, we randomly select 2 000 test and noise configurations and let our infrastructure evaluate them in the same way we evaluate individuals in the genetic approach. Table 1 summarises our results. The table is based on average results obtained from 10 executions of the genetic and random approach. It is divided into three parts. In the left part (*Test case*), the test cases are identified, and their size and information about their parameters are provided.

Table 1. An experimental comparison of the proposed genetic approach with the random approach to setting test and noise parameters

Test case			Best configuration			Search process		
Name	kLOC	Params	Gen.	Error	Time	Error	Error*	Time
Airlines	0.3	5,5,10	15	3.0 / 1.7	3.8 / 2.5	3.2	8.8	3.0
Animator	1.5	–	25	21.8 / 10.9	1.1 / 1.3	4.3	5.4	1.3
Crawler	1.2	–	22	– / –	1.3 / 1.5	0.3	1.1	3.3
Crawler*	1.2	–	25	– / –	1.1 / 1.1	0.4	1.0	2.8
FTPServer	12.2	10	14	1.2 / 1.0	3.8 / 4.7	0.9	1.7	1.9
Rover	5.4	7	3	★	33.7 / 19.4	3.2	8.8	3.0

An Evaluation of the Best Individuals. The middle part of Table 1 (*Best configuration*) contains three columns which compare the best individual obtained by our genetic approach and found by the random approach. The *Gen.* column contains the number of generation (denoted as gen below) within which we discovered the average best individual according to the considered fitness function. The numbers indicate that we are able to find the best individual according to the considered fitness function within the first quarter of the considered generations. This motivates our future work to design a suitable termination condition for our specific testing process.

The *Error* column of the *Best individual* section of Table 1 compares the ability of the best individual to detect an error. The column contains two values (x_1/y_1). The first value x_1 is computed as the fraction of the average number of errors found by the best individual computed by the genetic algorithm and the average number of errors discovered by the best individual found by the random generation provided that an equivalent number of executions is provided to the random approach (this number is computed as gen times the size of the population which is 20). The second number y_1 is computed as the fraction of the average number of errors found by the best individual computed by the genetic algorithm and the average number of errors discovered by the best individual found randomly in 2000 evaluations. The –/– value represents a situation where none of the best individuals was able to detect the error within the allowed 5 executions. The ★ symbol means that the genetically obtained best individual did not spot any error while the best individual found by the random generation did (we discuss this situation in more detail below).

Similarly, the *Time* column of the *Best configuration* section of Table 1 compares average times needed to evaluate the best individual obtained by our approach and the best individual found by the random approach. Again, two values are presented (x_2/y_2). The first value x_2 is computed as the average time needed by the best individual found by the random approach if only $gen * 20$ evaluations are considered, divided by the average time the genetically found best individual needed. The second value y_2 shows the average time needed by the best individual found by the random generation when it was provided with 2000 evaluations, divided by the average time needed by the genetically found best individual.

The values that are higher than 1 in the *Error* and *Time* columns of the *Best individual* section of Table 1 represent how many times our approach outperforms the random approach. In general, one can see that the best individual found by our genetic approach has a higher probability to spot a concurrency error, and it also need less time to do so. Even if we let the random approach perform 2000 evaluations, our best individual is still better. Exceptions to this are the *Rover* and *Crawler* test cases. In the *Crawler* test case, the error manifests with a very low probability. The best individuals in both cases were not successful in spotting the error (note, however, that the error was discovered during the search process as discussed below). In the *Rover* test case, the best individual found by the genetic algorithm was not able to detect an error and some of the best individuals found by the random approach did detect the error (as again discussed below, the error was discovered during the search process too). This results from the fact that the genetic approach converged to an individual that allows a very fast evaluation (over 30 times faster than the best configuration found by the random generation). This, however, lowered the quality of the found configuration from the point of view of error detection, indicating that as a part of our future research, we may think of further adjusting the fitness function such that this phenomenon is suppressed.

An Evaluation of The Search Process. The right part of Table 1 (*Search process*) provides a different point of view on our results. In this case, we are not interested in just one best individual learned genetically or by random generation that is assumed to be subsequently used in debugging or regression testing. Instead, we focus on the results obtained during the search process itself. The genetic algorithm is hence considered here to play a role of heuristics that directly controls which test and noise configurations should be used during a testing process with a limited number of evaluations that can be done (2000 in our case).

This part of the table contains three columns which compare the genetic and random approaches wrt. their successes in finding errors and wrt. the time needed to perform the 2000 evaluations. The first column (*Error*) compares the average number of errors spot during the search process and the average number of errors spot during the evaluation of 2000 randomly chosen configurations of the test and noise heuristics. The *Error** column compares the average number of errors detected by our genetic approach with the average number of errors spot by the random approach when the random approach is provided with the same amount of time as the genetic approach. Finally, the *Time* column compares the average total time needed by the random approach in 2000 evaluations and the average time needed by our genetic approach. Again, the values higher than 1 in all the columns represents how many times our approach outperforms the random approach.

The cumulative results presented in the *Error* and *Error** columns show that our approach mostly outperforms the random approach. The exceptions in the *Error* column reflect the already above mentioned preference of the execution time in our fitness function, which is further highlighted by the *Time* column. For instance, in the worst case (the *Crawler* test case), our genetic approach is more than 3 times faster but in total discovers three times less errors. On the other hand, in the best cases (the *Airlines* and *Rover*), we found three times more errors in three times shorter time. To give some idea about the needed time in total numbers, the average time needed to evaluate 2000 random individuals took on average 32 hours (whereas the genetic approach needed just 10.5 hours), and the average time needed to evaluate 2000 random individuals of our biggest test case *FTPServer* took 101 hours (whereas the genetic approach needed on average just 53 hours).

Overall, our results show that our approach outperforms the random approach. They also indicate that we should probably partially reconsider our fitness function that puts sometimes too much stress on the execution time, which can in some cases (demonstrated in the *Crawler* test case) be counter-productive.

Another positive fact is that our objective function helps to improve the testing process even for test cases that do not contain a data race. This can be attributed to that our fitness favours configurations within which the synchronisation occurs more often and therefore is tested more. The results obtained from our experiment with the *Crawler* test case evaluated using two different hardware configurations indicate that the genetic approach is able to reflect the environment and focus on the noise heuristics and their parameters which provide better results for the considered environment.

7   Conclusions and Future Work

In this work, we have formulated the test and noise configuration search (TNCS) problem, and we have proposed a way how to use genetic algorithms to solve it. We have performed experiments aimed at choosing suitable parameters of genetic algorithms to be used when solving the problem. We have instantiated the framework for the case of noise injection techniques implemented in the ConTest tool and its extensions and proposed a complex objective function suitable when aiming at data race detection (but successful even when looking for other kinds of bugs). We have performed experiments on a set of benchmark programs showing that our approach significantly outperforms the commonly used approach of randomly selecting noise configurations.

The proposed approach can be further improved, e.g., by development of termination conditions which would help one to determine when to stop the search process. Other interesting subjects for future work include development of improved objective functions.

Acknowledgement. This work was supported by the Czech Science Foundation (projects P103/10/0306 and 102/09/H042), the Czech Ministry of Education (projects COST OC10009 and MSM 0021630528), the EU/Czech IT4Innovations Centre of Excellence project CZ.1.05/1.1.00/02.0070, and the internal BUT projects FIT-S-11-1 and FIT-S-12-1.

References

1. Alba, E., Chicano, F.: Finding Safety Errors with ACO. In: Proc. of GECCO 2007. ACM Press (2007)
2. Alba, E., Chicano, F.: Searching for Liveness Property Violations in Concurrent Systems with ACO. In: Proc. of GECCO 2008, pp. 1727–1734. ACM Press (2008)
3. Alba, E., Chicano, F., Ferreira, M., Gomez-Pulido, J.: Finding Deadlocks in Large Concurrent Java Programs Using Genetic Algorithms. In: Proc. of GECCO 2008. ACM Press (2008)
4. Ben-Asher, Y., Eytani, Y., Farchi, E., Ur, S.: Noise Makers Need To Know Where To Be Silent–Producing Schedules That Find Bugs. In: Proc. of ISOLA 2006. IEEE CS (2006)

5. Bron, A., Farchi, E., Magid, Y., Nir, Y., Ur, S.: Applications of Synchronization Coverage. In: Proc. of PPoPP 2005. ACM Press (2005)
6. Chicano, F., Ferrer, J., Alba, E.: Elementary Landscape Decomposition of the Test Suite Minimization Problem. In: Cohen, M.B., Ó Cinnéide, M. (eds.) SSBSE 2011. LNCS, vol. 6956, pp. 48–63. Springer, Heidelberg (2011)
7. Edelstein, O., Farchi, E., Goldin, E., Nir, Y., Ratsaby, G., Ur, S.: Framework for Testing Multi-Threaded Java Programs. Concurrency and Computation: Practice and Experience 15(3-5), 485–499 (2003)
8. Elmas, T., Qadeer, S., Tasiran, S.: Goldilocks: A Race and Transaction-aware Java Runtime. In: Proc. of PLDI 2007. ACM Press (2007)
9. Eytani, Y.: Concurrent Java Test Generation as a Search Problem. ENTCS 144, 57–72 (2006)
10. Giannakopoulou, D., Pasareanu, C.S., Lowry, M., Washington, R.: Lifecycle Verification of the NASA Ames K9 Rover Executive. In: Proc. of ICAPS 2005 (2005)
11. Godefroid, P., Khurshid, S.: Exploring Very Large State Spaces Using Genetic Algorithms. STTT 6(2), 117–127 (2004)
12. Hrubá, V., Křena, B., Letko, Z., Vojnar, T.: Testing of Concurrent Programs Using Genetic Algorithms. Technical report FIT-TR-2012-01, BUT (2012)
13. Křena, B., Letko, Z., Vojnar, T.: Coverage Metrics for Saturation-Based and Search-Based Testing of Concurrent Software. In: Khurshid, S., Sen, K. (eds.) RV 2011. LNCS, vol. 7186, pp. 177–192. Springer, Heidelberg (2012)
14. Křena, B., Letko, Z., Vojnar, T.: Noise Injection Heuristics for Concurrency Testing. In: Kotásek, Z., Bouda, J., Černá, I., Sekanina, L., Vojnar, T., Antoš, D. (eds.) MEMICS 2011. LNCS, vol. 7119, pp. 123–135. Springer, Heidelberg (2012)
15. Křena, B., Letko, Z., Vojnar, T., Ur, S.: A Platform for Search-based Testing of Concurrent Software. In: Proc. of PADTAD 2010. ACM Press (2010)
16. Musuvathi, M., Qadeer, S., Ball, T.: Chess: A Systematic Testing Tool for Concurrent Software. Technical Report MSR-TR-2007-149, Microsoft Research (2007)
17. Talbi, E.-G.: Metaheuristics: From Design to Implementation. Wiley Publishing (2009)
18. Tzoref, R., Ur, S., Yom-Tov, E.: Instrumenting Where It Hurts: An Automatic Concurrent Debugging Technique. In: Proc. of ISSTA 2007. ACM Press (2007)
19. Šimša, J., Bryant, R., Gibson, G.: dBug: Systematic Testing of Unmodified Distributed and Multi-threaded Systems. In: Groce, A., Musuvathi, M. (eds.) SPIN Workshops 2011. LNCS, vol. 6823, pp. 188–193. Springer, Heidelberg (2011)
20. White, D.: Software Review: The ECJ Toolkit. Genetic Programming and Evolvable Machines 13, 65–67 (2012)

Reverse Engineering Feature Models with Evolutionary Algorithms: An Exploratory Study

Roberto Erick Lopez-Herrejon[1], José A. Galindo[2], David Benavides[2],
Sergio Segura[2], and Alexander Egyed[1]

[1] Institute for Systems Engineering and Automation
Johannes Kepler University Linz, Austria
{roberto.lopez,alexander.egyed}@jku.at
[2] Department of Computer Languages and Systems
University of Seville, Spain
{jagalindo,benavides,sergiosegura}@us.es

Abstract. Successful software evolves, more and more commonly, from
a single system to a set of system variants tailored to meet the simil-
iar and yet different functionality required by the distinct clients and
users. Software Product Line Engineering (SPLE) is a software development paradigm that has proven effective for coping with this scenario. At
the core of SPLE is variability modeling which employs Feature Models
(FMs) as the de facto standard to represent the combinations of features
that distinguish the systems variants. Reverse engineering FMs consist in
constructing a feature model from a set of products descriptions. This research area is becoming increasingly active within the SPLE community,
where the problem has been addressed with different perspectives and
approaches ranging from analysis of configuration scripts, use of propositional logic or natural language techniques, to ad hoc algorithms. In
this paper, we explore the feasibility of using Evolutionary Algorithms
(EAs) to synthesize FMs from the feature sets that describe the system
variants. We analyzed 59 representative case studies of different characteristics and complexity. Our exploratory study found that FMs that
denote proper supersets of the desired feature sets can be obtained with a
small number of generations. However, reducing the differences between
these two sets with an effective and scalable fitness function remains an
open question. We believe that this work is a first step towards leveraging
the extensive wealth of Search-Based Software Engineering techniques to
address this and other variability management challenges.

1 Introduction

Successful software evolves not only to adapt to emerging development technologies but also to meet the clients and users functionality demands. For instance, it
is not uncommon to find academic, professional, or community variants of commercial and open source applications such as editing, modelling, programming
or many other development tools, where each variant provides different functionality. Thus the distinction between variants is described in terms of their
features, that we define as increments in program functionality [25].

G. Fraser (Ed.): SSBSE 2012, LNCS 7515, pp. 168–182, 2012.

In practice, the most common scenario starts with a first system and forks a new independent development branch everytime a new variant with different feature combinations is required. Unfortunately, this approach does not scale as the number of features and their combinations increases even slightly [13]. *Software Product Line Engineering (SPLE)* is an emerging software development paradigm that advocates a disciplined yet flexible approach to maximize reuse and customization in all the software artifacts used throughout the entire development cycle [6, 13, 14, 17]. The driving goal of SPLE is to create *Software Product Lines (SPLs)* that realize the different software system variants in an effective and efficient manner. However, evolving SPLs from existing and invidividually-developed system variants is not an easy endeavor. A crucial requirement is accurately capturing the variability present in SPLs and representing it with *Feature Models (FMs)* [6, 12], the de facto standard for variability modeling. Current research has focused on extracting FMs from configuration scripts [22], propositional logic expressions [7], natural language [24], and ad hoc algorithms [3, 11].

Previous work from some of the authors has shown *Evolutionary Algorithms (EAs)* as an attractive alternative to synthesize FMs that are hard to analyze [18, 19, 21]. In this paper, we explore the feasibility of using EAs to reverse engineer FMs from the feature sets that describe the system variants and thus help coping with the evolution scenario – from system variants to SPLs – described above. Our study analyzed 59 representative feature sets from publicly available case studies of different sizes and complexity. For the implementation of our approach, we used a specific instantiation of the evolutionary algorithm ETHOM [21], integrated into the open source tool BeTTy [19].

We devised two fitness functions. With them we identified a trade-off between accuracy of the obtained feature model (the required feature sets vs of the obtained feature sets) and number of generations. That is, proper supersets of the desired feature sets can be obtained with a small number of generations. However, these supersets contain a large surplus of feature sets. In contrast, to reduce such surplus does require more generations. We believe our work is a first step towards leveraging the extensive wealth of Search-Based Software Engineering techniques to address many pressing variability management challenges.

2   Feature Models and Running Example

Feature models are the de facto standard to model the common and variable features of an SPL and their relationships [12], and thus represent the set of feature combinations that the products of the SPL can have. Features are depicted as labeled boxes and are connected with lines to other features with which they relate, collectively forming a tree-like structure.

A feature can be classified as: *mandatory* if it is part of a program whenever its parent feature is also part, and *optional* if it may or may not be part of a program whenever its parent feature is part. Mandatory features are denoted with filled circles while optional features are denoted with empty circles both at the child end of the feature relations. Features can be grouped into: *or* relation whereby one or more features of the group can be selected, and *alternative* relation where exactly one feature can be selected. These relations are depicted as filled arcs and empty arcs respectively.

Besides the parent-child relations, features can also relate across different branches of the feature model in the so called *Cross-Tree Constraints (CTC)* [5]. The typical examples of this kind of relations are: *i) requires* relation whereby if a feature A is selected a feature B must also be selected, and *ii) excludes* relation whereby if a feature A is selected then feature B *must not* be selected. In a feature model, these latter relations are depicted with doted single-arrow lines and doted double-arrow lines respectively. Additionally, more general cross-tree constraints can be expressed using propositional logic [5].

Figure 1 shows the feature model of our running example, a hypothetical SPL of Video On Demand systems. The root feature of a SPL is always included in all programs, in this case the root feature is VOD. Our SPL also has feature Play which is mandatory, in this case it is included in all programs because its parent feature VOD is always included. Feature Record is optional, thus it may be present or not in our product line members, the same holds for feature PPV (Pay Per View). Feature Display is also mandatory and like features Play and OS (Operating System) they are included in all our programs because their parent, the root, is always included. Features TV and Mobile constitute an alternative relation, meaning that our programs can have either one of them but only one. Similarly, features Aerial and Cable, and features Std (Standard) and Smart are respectively in alternative relations. Notice as well, requires relations between features Smart and Advanced, meaning that if a product contains feature Smart it must also contain feature Advanced. The same is the case between Smart and PPV, and between Cable and PPV. Finally, features Aerial and PPV are in an excludes relation meaning that both cannot be present in the same product. Next we provide a more formal definition of feature sets and their relation with products[1].

Definition 1. *A feature set is a 2-tuple [sel,$\overline{sel}$] where sel and $\overline{sel}$ are respectively the set of selected and not-selected features of a product. Let FL be the list of features of a feature model, such that sel, $\overline{sel}$ ⊆ FL, sel ∩ $\overline{sel}$ = ∅, and sel ∪ $\overline{sel}$ = FL.*

Definition 2. *A product is valid if the selected and not-selected features adhere to all the contraints imposed by the feature model.*

For example, the feature set fs=[{VOD, Play, Display, TV, Aerial, OS, Kernel}, {Record, Cable, Mobile, Std, Smart, Advanced, PPV}] is valid. As another example, a feature set with features TV and Mobile is not valid because it violates the constraint of the or relation which establishes that these two features cannot appear selected together in the same configuration.

---

[1] Adapted from [5] where feature sets are referred to as configurations.

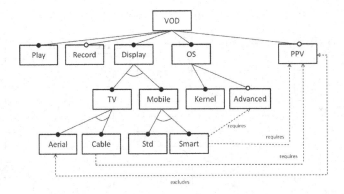

Fig. 1. Video On Demand SPL Feature Model

Table 1 depicts the feature sets of our running example VOD SPL of Figure 1. The ticks represent the selected features whereas empty entries represent not selected features. The column headers are abbreviations of the feature names.

We reiterate that in this paper we address the problem of reverse engineering feature models out of a lists of feature sets such as that in Table 1. It should be stressed though that a list of feature sets can be denoted by different feature models. In other words, the mapping from feature sets to feature models is one-to-many. This characteristic makes EAs specially attractive as different potential feature models can be analyzed and ranked according to distinct criteria. Next we present a short description the EAs infrastructure we employed.

3   Evolutionary Algorithms with ETHOM

We relied on the EA tool *ETHOM (Evolutionary algoriTHm for Optimized feature Models)* to implement our approach [21]. In this section, we present a basic overview of the main characteristics of ETHOM, and in next section we describe how it was applied for reverse engineering feature models.

3.1   Encoding

One of the most salient characteristics of ETHOM is its encoding of individuals (i.e chromosomes) which is specifically tailored to represent feature models. A feature model is thus represented as an array divided in two parts, one for the tree structure of the feature model and one for its cross-tree constraints. The enconding of the feature model of our running example in Figure 1 is shown in Figure 2.

The chromosomes of the structural part of the tree are tuples of the form <PR,CN> where:

- PR denotes the type of relationship a feature has with its parent. It can be M for mandatory, Op for optional, Alt for alternative, and Or for or relation.

Table 1. Feature Sets of VOD Software Product Line

FSet	VOD	Play	Rec	Disp	OS	TV	Mob	Sm	Std	Ker	Adv	Aer	Cab	PPV
FS1	✓	✓	✓	✓	✓	✓				✓			✓	✓
FS2	✓	✓		✓	✓	✓				✓			✓	✓
FS3	✓	✓	✓	✓	✓	✓				✓		✓		
FS4	✓	✓		✓	✓	✓				✓		✓		
FS5	✓	✓	✓	✓	✓		✓		✓	✓				✓
FS6	✓	✓	✓	✓	✓		✓		✓	✓				
FS7	✓	✓		✓	✓		✓		✓	✓				
FS8	✓	✓		✓	✓		✓		✓	✓				✓
FS9	✓	✓	✓	✓	✓	✓				✓	✓		✓	✓
FS10	✓	✓		✓	✓	✓				✓	✓		✓	✓
FS11	✓	✓	✓	✓	✓	✓				✓	✓	✓		
FS12	✓	✓		✓	✓	✓				✓	✓	✓		
FS13	✓	✓	✓	✓	✓		✓		✓	✓	✓			✓
FS14	✓	✓	✓	✓	✓		✓		✓	✓	✓			
FS15	✓	✓		✓	✓		✓		✓	✓	✓			✓
FS16	✓	✓		✓	✓		✓		✓	✓	✓			
FS17	✓	✓	✓	✓	✓		✓	✓		✓	✓			✓
FS18	✓	✓		✓	✓		✓	✓		✓	✓			✓

- CN denotes the number of children of the feature.

In addition, the order of these tuples is determined by a Depth-First Traversal (DFT) starting from the root. It should also be noted that the root of the feature model is not encoded. For example, the chromosome at entry with DFT value 6, corresponds to feature Mobile that is an alternative (PR value is Alt) feature of its parent (feature Display), and has two children (CN value is 2), namely Std and Smart. As another example, chromosome at entry with DFT value 11. This chromosome encodes feature Advanced, with an optional relation with its parent feature OS (PR value is Op), and with no children (CN value is 0).

The chromosomes of the cross-tree constraints are tuples of the form <TC,O,D> where:

- TC encodes the type of cross-tree constraint. An R value denotes a requires constraint whereas an E value denotes an excludes constraint.
- O encodes the origin feature of the cross-tree constraint represented with the corresponding DFT value.
- D encodes the destination feature of the cross-tree constraint represented with the corresponding DFT value.

For example, the tuple <E,4,12> encodes the excludes cross-tree constraint between features Aerial (DFT value 4) and PPV (DFT value 12). As another example, the tuple <R,8,11> encode a requires cross-tree constraint between features Smart (DFT value 8) and Advanced (DFT value 11).

Tree	DFT	0	1	2	3	4	5	6	7	8	9	10	11	12
		M,0	Op,0	M,2	Alt,2	Alt,0	Alt,0	Alt,2	Alt,0	Alt,0	M,2	M,0	Op,0	Op,0

CTC | E,4,12 | R,5,12 | R,8,12 | R,8,11 |

Feature DFT order
Play=0, Record=1, Display=2, TV=3, Aerial=4
Cable=5, Mobile=6, Std=7, Smart=8, OS=9
Kernel=10, Advanced=11, PPV=12

Fig. 2. ETHOM Encoding of Video On Demand feature model

3.2   Initial Population and Selection

There are different alternatives in the literature to randomly generate feature models [19,23]. ETHOM uses the following configuration parameters:

– Population size.
– Number of features.
– Percentage of cross-tree constraints.
– Maximum branching factor, defined as the maximum number of subfeatures of a feature, considering all the types of relationships.
– Percentage of mandatory relations.
– Percentage of optional relations.
– Percentage of **Alternative** relations.
– Percentage of **Or** relations.

There are multiple alternatives to implement selection in the EA literature [8]. The current version of ETHOM provides roulette wheel and tournament selection [21].

3.3   Crossover

There are multiple alternatives to implement crossover in the EA literature [8]. The current version of ETHOM provides one-point and uniform crossover [21]. Fig. 3 depicts an example of the application of one-point crossover in ETHOM. The process starts by selecting two parent chromosomes to be combined. For each array in the chromosomes, the tree and CTC arrays, a random point is chosen (so-called crossover point). Finally, the offspring is created by copying the content of the arrays from the beginning to the crossover point from one parent and the rest from the other one. Notice that the characteristics of this encoding guarantee a fixed size for the individuals.

3.4   Mutation

ETHOM defines four mutation operators that are applied with the probability set in the configuration. The mutation operators available are:

– Operator 1. Changes randomly a relation between two features from one kind to any other kind. For example, from mandatory (M) to optional (Op) or from Op to **Alternative** (**Alt**).
– Operator 2. Changes the number of children CN, to a number selected from 0 to a maximum branching factor parameter set up in ETHOM.

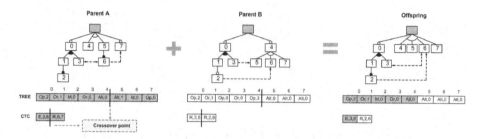

Fig. 3. Example of one-point crossover in ETHOM [21]

- Operator 3. Changes the type of cross-tree constraint, from excludes to requires and vice versa.
- Operator 4. Changes either the origin or destination feature (with equal probability) of a cross-tree constraint. It is checked that the resulting CTC does not have the same origin and destination values.

It should be noted that with application of cross-over and mutation operators there is possibility of creating feature models that are not semantically correct. ETHOM provides mechanisms for their identification and repair. For further details, please consult [21].

4   Applying ETHOM for Extracting Feature Models

In this section, we describe the concrete instantiation of ETHOM developed for reverse engineering feature models and how it was used in the experimental setting of our exploratory study.

4.1   ETHOM Configuration Parameters and Fitness Functions

Based on our previous experience using ETHOM to generate hard-to-analyze feature models, we set up ETHOM remaining configuration parameters as shown in Table 2. Notice that, as mentioned in Section 3.4, infeasible individuals (i.e. semantically incorrect feature models), are replaced.

A crucial decision in our study was selecting an adequate fitness function. Overall, our goal is to obtain features models that produce exactly the set of products desired. Unfortunately, existing work on formal analysis of feature models indicates that finding the relations between feature models and their related product specifications (e.g. logic representation) can be a hard and expensive computational task [5, 7, 23]. For this reason, we decided to analyze two fitness functions, one that focuses on obtaining the desired feature sets disregarding any surplus of feature sets and one that does penalize surplus. We argue that studying both alternatives could provide some insights as to whether the extra cost of computing the penalty would yield a faster (i.e. less number of generations) and more accurate result.

Table 2. ETHOM configuration parameters

Parameter	Value selected
Selection strategy	Roulette-wheel
Crossover strategy	One-point
Crossover probability	0.7
Mutation probability	0.01
Initial population size	100
Infeasible individuals	Replace
Maximum generations	25

Relaxed Fitness Function. The relaxed fitness function FFRelaxed maximizes the number of desired feature sets contained in a feature model disregarding any surplus feature sets that the model could denote. FFRelaxed is defined as follows:

$$FFRelaxed(sfs, fm) = |\{fs : sfs \mid validFor(fs, fm)\}|$$

Where sfs is the set of desired feature sets, fs is an individual feature set in sfs, fm a feature model, and validFor(fs,fm) is a function that receives a feature model and determines if a given feature set is contained in the set of feature sets represented by the feature model fm[2].

FFRelaxed is maximized to have as many feature sets from sfs as possible. Thus its maximum is the size of the set of desired feature sets sfs. However this function has a shortcomming, namely, that a feature model can contain more feature sets besides those in sfs. An ideal solution would then include all the feature sets in sfs and no more additional feature sets. However, reaching the maximum would not guarantee this. For instance, consider the feature sets of Table 1 and the feature model shown in Figure 1. If we apply FFRelaxed, we will get a value of 18 which is the maximum in this case since this is the number of feature sets we want the feature model to include and in this case this is exactly the set of feature sets represented by the feature model. However, if we change, for instance, the relationship between the root feature (VOD) and the feature Play from mandatory to optional, the number of feature sets represented by the feature model quickly raises from 18 to 36 but the value of the fitness function would remain the same since all the 18 desired feature sets are included but more are also included. The opposite could also eventually happen, i.e. that the feature model includes less feature sets than the ones expected. For instance, imagine that we add an excludes relationship between features Record and Mobile in the feature model of Figure 1. The total number of feature sets represented by the model will be reduced from 18 to 13 (feature sets FS5, FS6, FS13, FS14, and FS17 would not be included). If we apply now FFRelaxed, the result will be 13 not reaching the maximum of the fitness function in this case.

---

[2] For more details on the implementation of function valid please refer to [5].

Strict Fitness Function. The strict fitness function FFStrict penalizes having more feature sets than those desired. FFStrict is maximized and its values can range from 0 up to the concrete number of feature sets we want the feature model to have. It is defined as follows:

$$FFStrict(sfs, fm) == \begin{cases} 0 & : \#products(fm) \neq |sfs| \\ FFRelaxed(sfs, fm) & : \#products(fm) = |sfs| \end{cases}$$

Where sfs is the set of desired feature sets and #products(fm) is a function that receives a feature model and returns the number of feature sets it represents [3]. FFStrict returns 0 in the case the number of feature sets represented by the feature model is not equal to the number of desired feature sets and the value of applying FFRelaxed otherwise. A potential shortcomming of this function is that it equally penalizes all feature models that have more feature sets than the number desired, irrespective of how big the differences are.

4.2 Experimental Setting

In order to assess our approach we analyzed case studies from the SPLOT website [10], a publicly available repository of feature models that contains both academic and real-life examples. We selected 59 representative feature models based on their number of features and number of products. The number of products ranged from 1 to 896 and the number of features from 9 to 27. We chose these thresholds so that the fitness functions analysed yield results within a reasonable amount of time using the available tooling support.

Figure 4 sketches the overall control flow for each selected feature model. First, we used FAMA [1], a tool for the analysis of feature models, to generate the feature sets represented by a given feature model. ETHOM follows a standard EA control flow. Using information gathered from FAMA (e.g. number of features and feature names) the initial population is created. The evaluation takes into account the feature sets of the products we are looking for and a fitness function, either FFRelaxed or FFStrict. If the EA finds an individual with the feature sets desired, it records the number of generations along with time elapsed to obtain them and stops. Otherwise, ETHOM then proceeds with the standard selection, crossover and mutation steps before continuing with the evaluation of the new generation. As shown in Table 2, there is also a stop criteria according to the maximum number of generations, 25 generations in our study.

4.3 Evaluation Results and Analysis

We performed 10 runs for each of the two fitness functions for each of the 59 feature models we selected in order to get average values. The evaluation of both functions started with the same initial populations. All the experiments were performed on a Intel Xeon E5620 ©with 16 cores running at 2.40GHz.

---

[3] Implemented based on function *number of products* defined in [5].

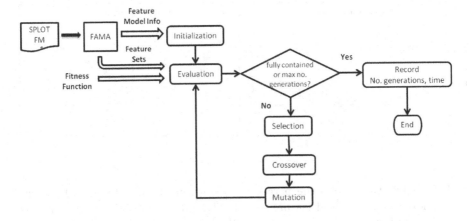

Fig. 4. Experiment Control Flow

This machine has 25 GB of shared RAM with other 5 machines inside a cloud, and was running CentOS 5 and Sun Java 1.6. Note that for our evaluation only one core was used per model. Execution time was measured using Java utilities.

Figure 5 summarizes the results obtained for `FFRelaxed`. In total, the algorithm reached the maximum of the fitness function in 94,64% of all the runs with an average execution time of 11 minutes. The average number of generations to reach a maximum was 5 with a standard deviation of 3,39. We found that there was only a model where the fitness function did not reach the maximum (within 25 generations) in any of the runs, shown in histogram of Figure 5(a). The percentage of cases where the maximum of the fitness function was reached within the first 2 generations was around 30%. This would suggest that our algorithm is performing quite effectively. However, we found that the number of denoted feature sets in the evolved feature models was far from the expected number of feature sets. To gauge at this difference, we defined the following surplus metric:

$$Surplus(sfs, fm) = \frac{\#products(fm) - |sfs|}{|sfs|} \times 100 \qquad (1)$$

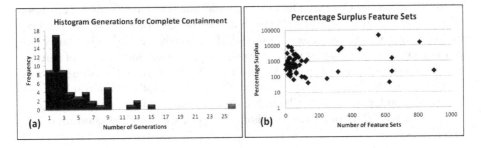

Fig. 5. FFRelaxed Results

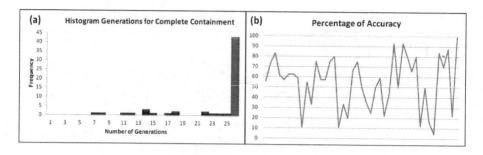

Fig. 6. FFStrict Results

This metric thus shows the percentage of increment or reduction of feature sets denoted by an evolved feature model. The differences obtained in our runs are shown in Figure 5(b). Please do note the logarithmic scale of the percentage surplus axis. On average, the value was 2401,24% of surplus feature sets with respect to the value expected. We did not find any feature model with the best score of the fitness function with less number of feature sets than sfs. We did not find any feature model with exactly the same number of feature sets either. With these data, we conjecture that our algorithm using FFRelaxed would get similar results to those of a random search because the number of generations to get a maximum of the fitness function is small and the surplus of the resulting feature model is high. It might also suggest that the fitness function is not guiding too much the algorithm and thus using random search of feature models would offer similar results. To confirm this conjecture, a proper and dedicated experiment comparing both approaches is called for, which is part of our future work.

Figure 6 summarizes our findings for FFStrict. With this fitness function 16 out of the 59 feature models reached a maximum within the limit of 25 generations, with an average of number of generations of 16,66 and a standard deviation of 5,56, see histogram in Figure 6(a). This would suggest that our algorithm is performing worse; however, it is important to highlight the accuracy obtained by the 43 feature models that did not reach the maximum. The accuracy here was defined as follows:

$$Accuracy(sfs) = bestFitnessAchieved(sfs)/|sfs| \times 100 \qquad (2)$$

Where sfs is the set of desired feature sets, and function bestFitnessAchieved collects the best fitness value obtained in all the generations of the 10 runs evaluated for a sfs (i.e. each feature model of our study). Figure 6(b) shows the accuracy percentages (y-axis) plotted arbitrarily by increasing number of feature sets of the corresponding feature model (x-axis). On average, the accuracy was 54,81% but with a wide standard deviation value of 25,4%. This result shows that even though penalty of not having the exact number of expected feature sets may appear harsh, it contributed to hone in the search to yield good accuracy percentages.

5 Related Work

There is extensive and increasing literature in reverse engineering mostly from source code. The novelty of our work lies on the application of EAs for reverse engineering of variability. In this section, we shortly describe those works that focus on reverse engineering feature models from specifications and those that employ search-based techniques.

Recent work by Haslinger et al. proposes an ad hoc algorithm to reverse engineer feature models from feature sets [11]. It works by identifying occurrence patterns in the selected and not selected features that are mapped to parent-child relations of feature models. Currently, they do not address general feature models that can contain any type of cross-tree constraints. The main distinction with our work is that only one feature model can be reversed engineered, whereas in our approach we could provide different feature models (if they exist) as alternatives for the designers to choose from.

Work by Czarnecki and Wasowski study reverse engineering of feature models but from a set of propositional logic formulas [7]. They provide an ad hoc algorithm that can potentially extract from a single propositional logic formula multiple feature models but that tries to preserve the original formulas and reduce redundancies. Subsequent work by She et al. highlighted the limitations of this approach, namely problems selecting the parents of features and incompleteness [22]. They proposed a heuristic to address these two issues that complements dependency information with textual feature description. In contrast with our work, their starting point are configuration and documentation files.

Closer to our work is Acher et al. that also tackle the reverse engineering of feature models from feature sets [3]. The salient difference between our approaches is that their work maps each feature set into a feature model which are later merged in to a single feature model. This mapping and merge operation rely on propositional logic techniques and tools which can be computationally expensive. A more detailed comparison and analysis of the advantages and disadvantages of both approaches is part of our future work.

Finally, our exploratory study was inspired by and builds upon the previous work on BeTTy, a benchmark and testing framework for feature model analysis [19]. This framework is geared to generate feature models that are hard to analyze, be it in execution time or memory footprint.

6 FutureWork

We argue that our exploratory study has opened up several research venues on the application of SBSE techniques for variability management. The following are some areas we plan to pursue as future work.

Improvement of fitness functions. The cornerstone of our work is devising an adequate fitness function that contains the set of products required, as tight as possible, but still remains scalable. A possibility is to experiment with functions that consider feature model metrics [4].

Parameter landscape analysis. Currently, ETHOM is equally configured for all the feature models we used in our runs. A detailed analysis of the parameter landscape of the problem is duly called for. We plan to experiment with ETHOM's configuration parameters. The ultimate goal is to see whether any particular parameter configurations can yield better results and how they would scale for larger feature models and feature sets.

Variability-aware mutation operators. Currently ETHOM blindly applies mutation operators without any considerations of any potential variability implications, that is, how it could impact the set of products denoted by a feature model. We want to extend the set of mutation operators so that they consider the impact they may have on variability. We could then set up different probabilities so that they could be applied distinctly perhaps depending on the nature of the required set of products. Integrating the work on analysis of feature model changes is a starting point [20, 23].

Quality of feature models. So far the emphasis of our work has been on obtaining a feature model that denotes the required set of feature sets. However, as we mentioned, more than one feature model can denote the same set of feature sets. The question is now, towards which equivalent feature model should the search be directed to? We believe that quality metrics for feature models [4] as well as quantification of developers feedback could also be integrated [2] to help answer this question.

Novel applications. Software Product Line Engineering covers the entire development life cycle [17], from early design to deployment and maintenance. Thus, there are plenty of areas where SBSE could be potentially applied. A salient one is testing, where EA approaches could be tailored to consider variability implications in their search. A solid first step in that direction is the work by Segura et al. [20]. As another example, the area of fixing inconsistencies in models with variability [15,16]. These new applications will in turn require more general feature model and problem encodings and effective fitness functions.

Comparative studies. As mentioned throughout the paper, we plan to perform a comparative study of basic search techniques (e.g. hill climbing), other EAs, and feature model composition (e.g. [3]) with our approach.

7   Conclusions

In this paper we explored the feasibility of EAs for reverse engineering feature models from feature sets. We devised two fitness functions that respectively focused on: *i)* getting the desired feature sets while disregarding any surplus (FFRelaxed), *ii)* getting the desired number of feature sets and then on the desired feature sets (FFStrict).

With these two functions we were able to identified a trade-off between accuracy of the obtained feature model (the required feature sets vs of the obtained feature sets) and number of generations. That is, proper supersets of the the desired feature sets can be obtained with a small number of generations. However, these supersets contain a large surplus of feature sets. In contrast, reducing such surplus does require more generations but still can yield good accuracy results. Despite this encouraging results, devising a fitness function that can reduce, if not eliminate, this trade-off is still an open question. We hope that this work has highlighted some of the the many potential areas where SBSE techniques can help tackle many open challenges in the realm of variability management.

The sources of our exploratory study can be downloaded from:
http://www.lsi.us.es/~dbc/material/ssbse2012

Acknowledgments. This research is partially funded by the Austrian Science Fund (FWF) project P21321-N15 and Lise Meitner Fellowship M1421-N15, and Marie Curie Actions - Intra-European Fellowship (IEF) project number 254965. It is also supported by the European Commission (FEDER) and Spanish Government under CICYT project SETI (TIN2009-07366) and by the Andalusian Government under THEOS project and Talentia scholarship. We would also like to thank Jules White and the MAGNUM research group for its support running the required experiments.

References

1. FAMA Tool Suite (2012), http://www.isa.us.es/fama/
2. Acher, M., Cleve, A., Collet, P., Merle, P., Duchien, L., Lahire, P.: Reverse Engineering Architectural Feature Models. In: Crnkovic, I., Gruhn, V., Book, M. (eds.) ECSA 2011. LNCS, vol. 6903, pp. 220–235. Springer, Heidelberg (2011)
3. Acher, M., Cleve, A., Perrouin, G., Heymans, P., Vanbeneden, C., Collet, P., Lahire, P.: On extracting feature models from product descriptions. In: Eisenecker et al. [9], pp. 45–54
4. Bagheri, E., Gasevic, D.: Assessing the maintainability of software product line feature models using structural metrics. Software Quality Journal (2010)
5. Benavides, D., Segura, S., Cortés, A.R.: Automated analysis of feature models 20 years later: A literature review. Inf. Syst. 35(6), 615–636 (2010)
6. Czarnecki, K., Eisenecker, U.: Generative Programming: Methods, Tools, and Applications. Addison-Wesley (2000)
7. Czarnecki, K., Wasowski, A.: Feature diagrams and logics: There and back again. In: SPLC, pp. 23–34. IEEE Computer Society (2007)
8. Eiben, A., Smith, J.: Introduction to Evolutionary Computing. Springer (2003)
9. Eisenecker, U.W., Apel, S., Gnesi, S. (eds.): Proceedings of Sixth International Workshop on Variability Modelling of Software-Intensive Systems, Leipzig, Germany, January 25-27. ACM (2012)
10. Generative Software Development Lab. Computer Systems Group, University of Waterloo, C.: Software Product Line Online Tools(SPLOT) (2012), http://www.splot-research.org/
11. Haslinger, E.N., Lopez-Herrejon, R.E., Egyed, A.: Reverse engineering feature models from programs' feature sets. In: Pinzger, M., Poshyvanyk, D., Buckley, J. (eds.) WCRE, pp. 308–312. IEEE Computer Society (2011)
12. Kang, K., Cohen, S., Hess, J., Novak, W., Peterson, A.: Feature-Oriented Domain Analysis (FODA) Feasibility Study. Tech. Rep. CMU/SEI-90-TR-21, Software Engineering Institute, Carnegie Mellon University (1990)

13. Krueger, C.W.: Easing the Transition to Software Mass Customization. In: van der Linden, F.J. (ed.) PFE 2002. LNCS, vol. 2290, pp. 282–293. Springer, Heidelberg (2002)
14. van der Linden, F.J., Schmid, K., Rommes, E.: Software Product Lines in Action: The Best Industrial Practice in Product Line Engineering. Springer (2007)
15. Lopez-Herrejon, R.E., Egyed, A.: Fast abstract. Searching the variability space to fix model inconsistencies: A preliminary assessment. In: Third International Symposium Search Based Software Engineering SSBSE 2011 (2011), http://www.ssbse.org/2011/fastabstracts/lopex-herrejon.pdf
16. Lopez-Herrejon, R.E., Egyed, A.: Towards fixing inconsistencies in models with variability. In: Eisenecker et al. [9], pp. 93–100
17. Pohl, K., Bockle, G., van der Linden, F.J.: Software Product Line Engineering: Foundations, Principles and Techniques. Springer (2005)
18. Segura, S., Parejo, J., Hierons, R., Benavides, D., Ruiz-Cortés, A.: Automated generation of hard feature models using evolutionary algorithms. Journal Submission (under review, 2012)
19. Segura, S., Galindo, J., Benavides, D., Parejo, J.A., Cortés, A.R.: BeTTy: benchmarking and testing on the automated analysis of feature models. In: Eisenecker et al.[9], pp. 63–71
20. Segura, S., Hierons, R.M., Benavides, D., Cortés, A.R.: Automated metamorphic testing on the analyses of feature models. Information & Software Technology 53(3), 245–258 (2011)
21. Segura, S., Parejo, J.A., Hierons, R.M., Benavides, D., Ruiz-Cortés, A.: ETHOM: An Evolutionary Algorithm for Optimized Feature Models Generation. Tech. Rep. ISA-2012-TR-01, Applied Software Engineering Research Group. Department of Computing Languages and Systems. University of Sevilla, ETSII. Avda. de la Reina Mercedes s/n (February 2012)
22. She, S., Lotufo, R., Berger, T., Wasowski, A., Czarnecki, K.: Reverse engineering feature models. In: Taylor, R.N., Gall, H., Medvidovic, N. (eds.) ICSE, pp. 461–470. ACM (2011)
23. Thüm, T., Batory, D.S., Kästner, C.: Reasoning about edits to feature models. In: ICSE, pp. 254–264. IEEE (2009)
24. Weston, N., Chitchyan, R., Rashid, A.: A framework for constructing semantically composable feature models from natural language requirements. In: Muthig, D., McGregor, J.D. (eds.) SPLC. ACM International Conference Proceeding Series, vol. 446, pp. 211–220. ACM (2009)
25. Zave, P.: Faq sheet on feature interaction, http://www.research.att.com/~pamela/faq.html

Searching for Pareto-optimal
Randomised Algorithms

Alan G. Millard[1], David R. White[2], and John A. Clark[1]

[1] Department of Computer Science, University of York, UK
{millard,jac}@cs.york.ac.uk
[2] School of Computing Science, University of Glasgow, UK
david.r.white@glasgow.ac.uk

Abstract. Randomised algorithms traditionally make stochastic decisions based on the result of sampling from a uniform probability distribution, such as the toss of a fair coin. In this paper, we relax this constraint, and investigate the potential benefits of allowing randomised algorithms to use non-uniform probability distributions. We show that the choice of probability distribution influences the non-functional properties of such algorithms, providing an avenue of optimisation to satisfy non-functional requirements. We use Multi-Objective Optimisation techniques in conjunction with Genetic Algorithms to investigate the possibility of trading-off non-functional properties, by searching the space of probability distributions. Using a randomised self-stabilising token circulation algorithm as a case study, we show that it is possible to find solutions that result in Pareto-optimal trade-offs between non-functional properties, such as self-stabilisation time, service time, and fairness.

1 Introduction

Search-Based Software Engineering (SBSE) research has traditionally focused on using metaheuristic search techniques to tackle problems related to deterministic algorithms [10]. Somewhat surprisingly, problems associated with randomised algorithms have received little attention, despite tremendous growth in the area over the past twenty five years. A *randomised algorithm* is one that makes stochastic choices during its execution, based on the outcome of a random number generator [20]. Since Rabin's seminal paper [22] randomised algorithms have gained universal acceptance, mostly due to their speed and simplicity. In fact, many randomised algorithms, such as random-pivot Quicksort, provide significant efficiency gains over the best known deterministic solutions, and are easier to implement than their deterministic counterparts [20]. Consequently, randomised algorithms are now becoming more commonplace, and have seen widespread adoption in practical applications.

To date, randomised algorithm research has typically placed emphasis on formally proving the correctness of the algorithms. Whilst functionality is clearly important, we are not only interested in provable functional correctness, but also in eliciting guarantees about non-functional properties. The SBSE community

G. Fraser (Ed.): SSBSE 2012, LNCS 7515, pp. 183–197, 2012.

has recognised that the satisfaction of non-functional requirements is an important consideration in algorithm design, and the problem is starting to attract greater attention [1]. In practice, we find that the non-functional properties of randomised algorithms can be influenced through the choice of probability distributions upon which they base their stochastic decisions. This is particularly interesting in the context of randomised distributed algorithms, where the decisions made by each processor in the system may affect the state of others.

The non-functional properties of randomised distributed algorithms are an emergent result of local interactions between processors. Thus, the choice of probability distribution to be used for each processor has consequences in terms of the emergent non-functional properties, and we wish to explore the effect of different choices. However, due to the complex interplay between the probabilistic behaviour of networked processors, the relationship between the emergent non-functional properties of the system and the probability distributions used are unlikely to be intuitive, and therefore difficult to predict *a priori*. The space of possible solutions is also infinitely large for continuous probability distributions, exacerbating the problem. The search for probability distributions that satisfy multiple objective criteria is therefore an appropriate application of SBSE.

Instead of searching for entirely novel algorithms, we seek to optimise pre-existing algorithms for some target criteria, by searching over the space of probability distributions. We have chosen to examine the problem of self-stabilising token circulation in a ring of networked processors as a case study, because simple randomised distributed algorithms already exist to solve this problem. We use Genetic Algorithms (GAs) in conjunction with Multi-Objective Optimisation (MOO) techniques, to find sets of solutions to this problem that approximate underlying Pareto fronts for pairs of conflicting non-functional objectives. However, the focus of this paper is not upon finding optimal solutions for this case study, we are simply interested in the discovery of trade-offs, primarily as an indication of whether similar trade-offs might exist for other randomised algorithms.

1.1   Contributions

The main contributions of this paper are to demonstrate that:

- GAs can be used in conjunction with MOO to explore the trade-offs between non-functional objectives associated with pre-existing randomised algorithms, by searching over the space of probability distributions.
- Such trade-offs exist for the randomised self-stabilising token circulation protocol used as a case study, and potentially for other randomised algorithms.
- The characteristics of these trade-offs vary with problem size (in this case, the number of processors in the system).

2   Related Work

In order to verify properties of randomised algorithms, a probabilistic model checking tool is required. Model checking is often used to ensure the correctness of a system, by constructing a finite state model of the system and then exhaustively checking this model against some specification [6]. Our approach uses the Probabilistic Symbolic Model Checker (PRISM) [18] language to describe a system of interest, the non-functional properties of which can then be measured using the PRISM model checker. The benefit of model checking over traditional testing, is that we are able to formally prove properties about solutions found through search, by checking every possible execution of an algorithm's code, rather than exploring only a sample of all possible execution paths.

Johnson [14] has previously used model checking to guide the synthesis of finite state automata with Genetic Programming. Each individual was evaluated against a formal specification given in temporal logic, and assigned a fitness value directly proportional to the number of functional properties satisfied. In an attempt to smooth the fitness landscape, Johnson included properties that check for partially correct programs. However, this method still suffers from discontinuities in the fitness landscape, making it hard for evolution to make gradual improvements towards the target specification.

Katz and Peled [16] have also used model checking to generate fitness values, for the rediscovery of classical two-process mutual exclusion algorithms using Genetic Programming, and later the synthesis of novel algorithms for mutual exclusion [15]. Of greater relevance to our work, is their synthesis of leader election protocols for unidirectional token rings [17], however only deterministic algorithms were evolved. The fitness landscape was smoothed by analysing the graph generated during model checking, to extract further information to quantify the degree of satisfaction, rather than just checking whether properties were satisfied or not, giving a finer-grained fitness measure.

In contrast to the work of Johnson, Katz and Peled, our goal is not to synthesise provably correct programs from scratch. Rather, we seek to modify existing programs that are already known to be provably correct, to trade-off non-functional properties by optimising a vector of probabilities. In addition to discrete true/false values returned by correctness properties, PRISM may return continuous values for non-functional properties, such as the expected number of steps to reach a goal state. The return values from such non-functional properties can be used as input to a fitness function, resulting in a much smoother fitness landscape than one based on correctness properties.

3   Distributed Systems

We consider a *distributed system* to be modelled by a connected directed graph $G = (V, E)$, where V denotes the set of nodes representing the processors, and E denotes the set of edges representing communication links between pairs of processors. In general, the topology of a distributed system is unrestricted. However, in this paper we focus on systems with unidirectional ring topologies, which consist of $|V| = N$ connected processors $P_1, P_2, \ldots, P_N$. Adopting the notation of Dolev [9], the *predecessor* of each processor P_i, $1 < i \leq N$, is P_{i-1}, and the predecessor of P_1 is P_N. Similarly, the *successor* of each processor P_i, $1 \leq i < N$, is P_{i+1}, and the successor of P_N is P_1. In such a system, each processor may only receive information from its predecessor, and send information to its successor.

3.1 Protocols and Configurations

Each processor in the ring executes a *protocol*, which comprises a set of variables and guarded actions. The guard of each action is a boolean expression involving the state of processor P_i and its predecessor. The corresponding statement of a guarded action atomically updates the state of the processor that executed it. A *configuration* of a distributed system is given by a vector containing the state of every processor at a particular time. A processor is said to be *enabled* in some system configuration, if any of the processor's action guards are satisfied by that configuration [5]. In the protocol considered here, no configuration of the distributed system will satisfy multiple guards within a single processor.

3.2 Daemons

The *daemon* is a scheduler that selects a subset of the enabled processors to be activated at each computation step, which then perform the actions corresponding to their satisfied guards. A *central daemon* is a scheduler that may only select a single processor to schedule from the set of enabled processors. In contrast, a *distributed daemon* may choose one or more enabled processors at each computation step [4]. A daemon is said to be *fair* if it will eventually schedule a continuously enabled processor. On the other hand, the only restriction on an *unfair* daemon is that it must schedule processors that can perform an action [4]. For reasons of computational tractability, we consider a protocol that assumes a *synchronous daemon* [11]. This special case of a fair distributed daemon schedules *all* enabled processors at each computation step, thus removing any choice and therefore non-determinism from the scheduling.

4   Self-stabilising Token Circulation

A fundamental problem in distributed systems is that of ensuring mutually exclusive access to shared resources. For systems consisting of a networked ring of processors, one solution to this problem is to implement a token circulation scheme. When a processor gains possession of the token it is considered privileged, and is granted access to the shared resource. However, mutual exclusion is only ensured if there exists just a single token in the network at any one time.

The concept of a *self-stabilising* system was first introduced by Dijkstra [8] in 1974, to describe a system that converges to a *stable* state within a finite number of steps, regardless of its initial state. This is a desirable trait, since it allows the system to automatically recover from the occurrence of transient faults [9]. In the context of token circulation, the system is considered to be in a stable state when exactly one processor holds a token. Dijkstra [8] gave three deterministic protocols that solve the problem of self-stabilising token circulation in a ring, which each require a distinguished processor to control the system. However, for distributed systems comprising identical anonymous processors, it is known that there exists no deterministic algorithm for self-stabilising token circulation, due to the inability to break symmetry [2]. Randomisation is a powerful tool in algorithm design, and is often used to break symmetry in such systems [9].

The protocol presented by Herman [11] is a classic example of randomised self-stabilising token circulation in unidirectional rings of anonymous identical processors, which assumes the presence of a synchronous daemon. Several other randomised self-stabilising protocols exist [9], but we have chosen the protocol presented by Beauquier et al. [3] as a case study due to its simplicity, and because the probability of each processor passing a token to its successor is directly controlled by the probability distribution it uses.

4.1 Case Study

Beauquier et al. [3] present a randomised self-stabilising token circulation protocol for unidirectional rings of anonymous processors. Each processor is identical, and executes the same protocol. The state of processor P_i is determined by the value of a single binary variable t_i. The configuration of the system at a particular time is therefore given by the vector $(t_1, t_2, \ldots, t_N)$. The locations of tokens in the ring are implicitly defined by the states of individual processors. A processor P_i holds a token if its state is equal to that of its predecessor. It is instructive to consider the example system configuration shown in Figure 1(a). Here, $t_1 = t_5$, therefore processor P_1 holds a token. Similarly, processors P_3 and P_5 each possess a token, because $t_3 = t_2$ and $t_5 = t_4$, respectively.

The system may begin in any state, and may therefore start with up to N tokens. In order to reach a stable state, the number of tokens in the ring must be reduced to one. For token rings comprising an odd number of processors,

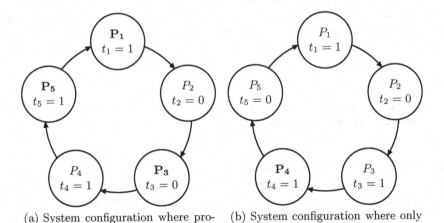

(a) System configuration where processors P_1, P_3 and P_5 hold tokens

(b) System configuration where only processor P_4 holds a token

Fig. 1. Example token rings of size $N = 5$

the protocol guarantees convergence to a stable state under a synchronous dae-mon. This is achieved through the random delay of tokens — a processor with the token randomly decides whether or not to pass it on to its successor. In-evitably, some tokens will temporarily remain stationary while others continue to circulate, eventually catching up with the others and eliminating them [13].

The synchronous daemon will schedule any processor in possession of a to-ken at each computation step. Each scheduled processor tosses a fair coin, to decide whether or not it should pass the token it holds to its successor. This is achieved by negating the value of its binary state variable t_i. Considering the system configuration in Figure 1(a) again, P_1, P_3, and P_5 will all be scheduled synchronously by the daemon, because they each hold a token. Supposing that P_3 and P_5 decide to pass on their tokens, but P_1 does not, the system configu-ration will change to the stable state shown in Figure 1(b), where only P_4 holds a token. Once the system has reached a stable state, the token will continue to be passed around the ring in a fair manner, until the system is perturbed due to the occurrence of a fault, and must self-stabilise again.

5   PRISM Model

We now consider the PRISM model for a token ring of size $N = 3$ as an example, since it is the simplest system we can construct that comprises multiple proces-sors. The concepts covered here generalise to larger token rings, which will be examined later. A system is constructed in the PRISM language by defining a number of interacting modules. These modules contain variables, that describe the state of the module, and a set of guarded commands of the form:

```
[] <guard> -> <prob_1> : <action_1> + ... + <prob_n> : <action_n>;
```

The guard is a predicate over the variables in the system. Each action describes a transition that will update the system if the guard is true, and is selected for execution with the corresponding probability [21]. The code snippet below, taken from our model for $N = 3$, shows the module definition for processor P_1.

```
module P1
    t1 : bool;
    [update]  t1=t3 -> p1 : (t1'=t1) + 1-p1 : (t1'=!t1);
    [update]  !t1=t3 -> true;
endmodule
```

This code first declares a boolean variable which represents the state variable t_1 belonging to processor P_1. Following this is a guarded command which models the action of probabilistically deciding whether to pass on the token. Our PRISM model of the algorithm differs slightly from the original Beauquier et al. [3] protocol, in that it is parameterised by the probability p_1 of processor P_1 holding on to a token (by leaving its state unchanged), which leaves a probability of $1-p_1$ of passing it on (by negating its state variable). In the original algorithm, the value of both these probabilities is 0.5, simulating the toss of a fair coin. The module definitions for processors P_2 and P_3 are defined in a similar way.

We model the algorithm as a Discrete Time Markov Chain (DTMC), and use the action label update to implement a synchronous daemon. This forces each module to make transitions simultaneously, resulting in a deterministic choice over which of the enabled processors should be scheduled at each time step. The second guarded command in the code snippet above causes the processor to do nothing when it does not hold a token, and is included simply to prevent deadlocks from occurring, as PRISM requires every module containing the action label update to take some action at each time step.

PRISM allows rewards (which can also be thought of as costs) to be assigned to specific states or transitions of the model. So that we may check non-functional properties of the system, we use a PRISM reward structure to assign every state in the model a reward value of 1. The expected value of these rewards can be checked using PRISM properties, to count the expected number of steps taken to reach a particular state. We also define a PRISM formula called num_tokens, which calculates the number of tokens present in the system at the current time step, given by the sum of the number of processors which have the same state as their predecessor. This then allows us to define the boolean label stable, for use in PRISM properties, which evaluates to true when num_tokens=1.

6   Objective Measures

The PRISM property specification language allows us to check PRISM models against specifications written in probabilistic temporal logics such as PCTL [18]. We will now discuss the PRISM properties we have written in this language, and how they are used to create objective functions.

6.1   Self-stabilisation Time

A non-functional objective of particular interest is the expected self-stabilisation time of the system, which we would like to minimise, since the system must be able to converge to a stable state faster than its expected failure rate if it is to make progress [12]. The following two PRISM properties check the average and maximum expected number of steps the system takes to reach a stable state, the values of which can be minimised directly, giving us the objective functions f_1 and f_2, respectively.

```
filter(avg, R=? [F "stable"], "init")
filter(max, R=? [F "stable"], "init")
```

The R=? operator can be used to analyse properties that relate to the expected value of rewards accumulated along an execution path until a certain state is reached. The path property F (meaning "future") is used here to check the probability of the system eventually reaching a state where the label "stable" evaluates to true, from a given initial state. Since we assign a reward of 1 to every state in the model, the property R=? [F "stable"] calculates the expected number of steps the system takes to reach a stable state, from a specified initial state. We instruct PRISM to calculate the expected number of steps to self-stabilise from every initial state by using a filter, which is specified using the syntax filter(operator, property, states). The avg and max operators are used to calculate the average and maximum value of property over states satisfying states, respectively. The set of states specified here is the set of all possible initial states, defined in the PRISM model using the label "init".

6.2   Fairness

Another interesting non-functional property of the system is the *fairness* of token circulation. In this paper, we consider a completely "fair" system to be one where each processor receives an equal time share of token possession. The following properties check steady-state *token residency* — the proportion of time each processor holds a token in the long-run — using the S=? operator.

```
filter(max, S=? [t1=t3], "init")
filter(max, S=? [t2=t1], "init")
filter(max, S=? [t3=t2], "init")
```

$$f_3 = 1 - \frac{\min\{R_1, R_2, \ldots, R_N\}}{\frac{1}{N}\sum_{i=1}^{N} R_i} \quad (1)$$

PRISM returns the minimum and maximum of a range of values over initial states, which are identical for these kinds of properties. We simply use a max filter operator to ensure a single value is returned, to save parsing the output. Let R_i be the steady-state token residency for processor P_i. Then, the above properties return the values R_1, R_2 and R_3, respectively. The objective measure for fairness can then be formally defined as shown in Equation 1. The second term of the equation evaluates to 1 when the token residency for each individual processor equals $1/N$, which only occurs when every processor P_i in the system shares the same value of p_i. Note that we attempt to maximise fairness, by minimising the value of f_3. This is simply because the evolutionary toolkit we are using attempts to minimise the value of objective measures by default.

6.3   Service Time

The last non-functional property we consider, is *service time* — how long it takes the token to circulate the ring of processors once a stable configuration is reached [13]. Minimising service time is desirable, as this reduces the time a processor must wait to regain privileged status. Each of the following properties checks, from every state where a processor P_i does not hold the token, the expected average number of steps until it obtains the token. Let A_i be the average expected service time for processor P_i, then the properties below return the values A_1, A_2 and A_3, respectively.

```
filter(avg, R=? [F "stable" & t1=t3], "stable" & t1!=t3)
filter(avg, R=? [F "stable" & t2=t1], "stable" & t2!=t1)
filter(avg, R=? [F "stable" & t3=t2], "stable" & t3!=t2)
```

Similarly, let M_i be the maximum expected service time for processor P_i. Then, the values of M_1, M_2 and M_3 are given by properties identical to those above, except with each filter's avg operator replaced with a max operator. The service time value is different for each processor when non-uniform probabilities are used, because the property does not take into account the behaviour of the processor being checked. We therefore find the average and maximum service times of every processor in the system, giving two more objective measures:

$$f_4 = \frac{1}{N} \sum_{i=1}^{N} A_i \qquad (2) \qquad\qquad f_5 = \max\{M_1, M_2, \ldots, M_N\} \qquad (3)$$

The objective measures f_4 and f_5 represent the average-case and worst-case service time of the system — the average and maximum time a processor must wait to receive the token again, respectively. These will both be minimised in system configurations where every processor P_i uses a value of $p_i = 0$, causing it to pass on any token it receives deterministically. If such a system is in a stable state, it will take $N - 1$ steps for each processor to receive the token again.

7   Multi-Objective Optimisation

Instead of having each processor in the system use an identically biased coin, we consider the case where each processor uses a coin with a different bias. In order to parameterise the system with a different probability distribution for each processor, we define a vector $(p_1, p_2, \ldots, p_N)$ of probabilities assigned to processors $(P_1, P_2, \ldots, P_N)$, respectively. Note that this assignment implicitly determines the values of $(1-p_1, 1-p_2, \ldots, 1-p_N)$, which control the probability of each processor passing on a token. The variables p1, p2, and p3 in the PRISM model for $N = 3$ are defined as uninitialised constants, the values for which are passed to PRISM via the command line. Due to the symmetrical ring network topology, the exact assignment of probabilities to processors is unimportant, so long as the order of their assignment is preserved. There exist N duplicate solutions for each probability vector, in which the probabilities are simply shifted along and assigned to different processors.

The PRISM model is parameterised by the vector $(p_1, p_2, \ldots, p_N)$ of probabilities, which may be searched over, to find solutions that trade-off non-functional objectives. In order to search for Pareto fronts we use a Genetic Algorithm to evolve individuals comprising a genome of N real numbers, each of which corresponds to a probability p_i in the parameter vector. The evolutionary computation toolkit ECJ [19] was used to evolve Pareto-optimal solutions using the Non-dominated Sorting Genetic Algorithm II (NSGA-II) [7], the main parameter settings for which are given in Table 1.

Table 1. Genetic Algorithm parameters

Parameter	Value
Population size	100
Generations	250
Crossover method	Simulated Binary Crossover
Mutation method	Polynomial
Crossover probability	0.9
Mutation probability	$1/N$
Crossover distribution index	20
Mutation distribution index	20
Selection method	Tournament selection, size 2

These parameters are the same as those originally used by Deb et al. [7], when NSGA-II was first presented. Any parameters not listed here are defaults inherited from the parameter files ec.params and simple.params which come packaged with ECJ. We have not attempted any parameter tuning, since the aim of these experiments was simply to confirm the existence of non-functional trade-offs for randomised algorithms in general, rather than to accurately approximate the underlying Pareto fronts for this particular case study.

7.1   Fitness Evaluation

The fitness evaluation process for a single individual is as follows: ECJ invokes the PRISM model checker via a command-line interface, specifying as input a PRISM model containing undefined constants corresponding to the probabilities used by each processor, along with the probability vector $(p_1, p_2, \ldots, p_N)$ representing the individual. PRISM uses these values to instantiate the model, assigning the probabilities to the corresponding constants. PRISM also takes as input a properties file, which contains a list of properties that we wish to check the model against. Model checking is computationally expensive, so instead of checking every PRISM property listed in Section 6, we only select the subset corresponding to the pair of non-functional properties we wish to optimise. The results from the model checking are output to a text file, which is then parsed by ECJ. The objective functions $f_1, f_2 \ldots, f_5$ are calculated from the return values of the individual PRISM properties, as described in Section 6, and their values are fed into the multi-objective fitness function for NSGA-II.

Although the original Beauquier et al. [3] protocol is provably correct, in our PRISM model the probability p_i of a processor P_i holding on to a token may potentially be set to 1, which would cause the processor to deterministically retain possession of any token it receives. While this would still guarantee convergence to a stable state, the single remaining token would not continue to circulate the ring of processors, and would instead be held forever by the same processor, causing others in the ring to wait indefinitely for privileged status. We automatically identify such invalid solutions during evolution, by checking values in the probability vector within ECJ, and discard them immediately.

8   Results

In summary, the problem to be solved was to determine which pairs of non-functional objectives were conflicting, and could therefore be traded-off, allowing us to find a Pareto-optimal set of solutions. The five objective functions $f_1, f_2, \ldots, f_5$ give rise to ten possible unique pairs of objectives to optimise.

8.1   Initial Exploration

Since there was no *a priori* indication of whether trade-offs would exist for any given pair of objectives, we enumerated the space of solutions for $N = 3$ with a fine granularity. Models were generated for every possible vector (p_1, p_2, p_3), where each probability was limited to increments of 0.01 in the closed interval $[0.05, 0.95]$ — essentially a 90-level factorial experiment with 3 factors. For each model, the values of $f_1, f_2, \ldots, f_5$ were checked using PRISM, and plotted against each other in pairs. Out of the ten possibilities, only six pairs of objective measures were found to be conflicting. Average and worst-case expected time to self-stabilise (f_1 vs f_2) were found to be non-conflicting, as were fairness and average/worst-case service time (f_3 vs f_4 and f_3 vs f_5), and average-case and worst-case service time (f_4 vs f_5). For each of the remaining six conflicting objective pairs, Pareto fronts were discovered. However, due to their similarity, we only show the results for worst-case properties here, for the sake of brevity.

Figure 2(a) shows the values of f_2 and f_3 measured for every individual in the factorial experiment, where each point corresponds to an individual's non-functional properties in objective space. The original solution, which uses uniform probability distributions, is circled in this and every subsequent plot. This graph clearly demonstrates that trade-offs between non-functional properties exist for this protocol in rings of size $N = 3$. The individuals on the Pareto front with the lowest values of f_3 (the most fair — including the original solution) are the slowest to converge to a stable state, and we can see that self-stabilisation

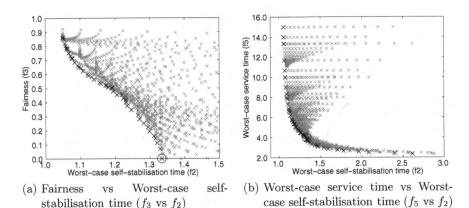

(a) Fairness   vs   Worst-case   self-stabilisation time (f_3 vs f_2)

(b) Worst-case service time vs Worst-case self-stabilisation time (f_5 vs f_2)

Fig. 2. Results of enumeration for $N = 3$

time can be improved at the expense of fairness. Note that unfairness does not directly result in faster self-stabilisation, as there exist many unfair solutions with far worse stabilisation times than those that are completely fair.

Figure 2(b) shows the Pareto front found for the conflicting objectives of worst-case expected self-stabilisation time and worst-case service time (f_2 vs f_5). Here, we can see that in order to improve the speed of self-stabilisation, we must sacrifice service time. These Pareto fronts allow us to visualise the relationship between pairs of non-functional properties, and we can see that the relationship for (f_2, f_3) compared to (f_2, f_5) is qualitatively very different for $N = 3$.

8.2   Larger Token Rings

After confirming the existence of trade-offs for token rings of size $N = 3$, we sought to investigate whether these trade-offs were restricted to this particular problem size. However, enumeration of the search space at any useful granularity is infeasible for problem sizes larger than $N = 3$, due to the computational effort required by PRISM to verify properties of larger systems. This is where the application of evolutionary search becomes useful, since it allows us to approximate the underlying Pareto fronts with far fewer fitness evaluations.

In order to provide some assurance that our evolutionary framework would be able to reasonably approximate any Pareto fronts that may have existed for larger problem sizes, we attempted to rediscover the trade-offs found for $N = 3$, but this time using evolutionary search. The rediscovered Pareto fronts were found to be almost identical to those originally discovered through enumeration, so we proceeded to apply our evolutionary method to search for trade-offs in token rings of sizes $N = 5, 7, 9$. As shown in Figures 3(a) and 3(b), Pareto fronts were found for these larger rings, demonstrating that trade-offs of non-functional properties are not limited to rings of size $N = 3$. In addition, we checked whether the non-conflicting objectives became conflicting at these higher values of N, but they did not.

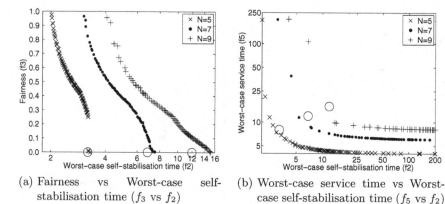

(a) Fairness   vs   Worst-case   self-stabilisation time (f_3 vs f_2)

(b) Worst-case service time vs Worst-case self-stabilisation time (f_5 vs f_2)

Fig. 3. Pareto fronts for larger token rings, found through evolution

The shape of the Pareto fronts for $N = 5, 7$ in Figure 3(a) are similar to that of $N = 3$ in Figure 2(a), except that the points of inflection occur at lower values of f_3, indicating that characteristics of the trade-off change with problem size. There does not even appear to be a point of inflection in the Pareto front found for $N = 9$, however this may be because 250 generations were not sufficient for NSGA-II to converge completely on the underlying Pareto front, as indicated by the position of the original solution in objective space.

The Pareto fronts shown in Figure 3(b) are also qualitatively similar to those found for $N = 3$ in Figure 2(b), illustrating that the relationship between f_2 and f_5 does not change radically as the size of the problem increases. However, the sparsity of solutions in the evolved Pareto fronts does increase with problem size. We conjecture that this is due to the parameters used for NSGA-II, rather than a feature of the underlying solution space, but we cannot be certain.

8.3   Example Individuals

We now present graphically the probability vectors of selected individuals, to illustrate how the probability distributions used by the processors in the system impact each non-functional objective. Figure 4(a) shows example individuals from the $N = 7$ Pareto front in Figure 3(a). Each shaded segment represents the relative magnitude of the probability p_i for each processor P_i. The sum of the probability vector $(p_1, p_2, \ldots, p_N)$ for each individual is also given, beneath the values of the objective functions, to demonstrate the influence of the absolute probability values used. It can be seen that individuals with the worst self-stabilisation times are very fair, but as stabilisation time improves, individuals begin to use probabilities that result in unfair token circulation. Those individuals with the fastest self-stabilisation times contain two selfish processors, while the rest are almost completely selfless. Tokens will stop at the selfish processors with high probability, allowing those circulating behind them to catch up and eliminate them, resulting in fast convergence to a configuration with a single token. Notice that individuals with nearly identical relative distributions

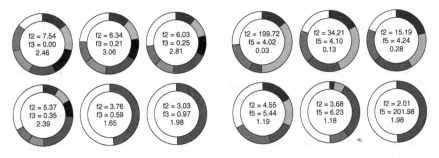

(a) Individuals from the Pareto front for $N = 7$ in Figure 3(a) (f_3 vs f_2)

(b) Individuals from the Pareto front for $N = 5$ in Figure 3(b) (f_5 vs f_2)

Fig. 4. Example individuals from the Pareto fronts found through evolution

of probabilities, but a different sum of probabilities, can have quite different non-functional properties, indicating that both relative values and absolute probabilities have a significant influence. Figure 4(b) shows example individuals from the $N = 5$ Pareto front in Figure 3(b). Again, we see here that the absolute probability values are important — fair rings of very selfless processors (indicated by a small sum of probabilities) result in fast token circulation. As service time is traded-off for stabilisation time, we begin see individuals where pairs of selfish processors dominate.

9   Conclusions and Future Work

To our knowledge, this paper represents the first application of Search-Based Software Engineering to randomised algorithms. Although this work constitutes only a proof of principle, and the objective measures examined for this case study algorithm are simply a means to an end, it is conjectured that similar trade-offs between non-functional properties will exist for other randomised algorithms.

Unfortunately, since our approach is based on model checking, verifying properties of systems larger than those considered here is computationally very expensive. This is because the number of states in the model grows exponentially with problem size, due to the well known state explosion problem [23]. To combat this, future work may investigate the potential of using approximate model checking or sampling methods to guide evolutionary search, by providing imperfect information about the non-functional properties of individual solutions at a reduced computational cost.

Acknowledgements. This work is part of the Software Engineering by Automated SEarch (SEBASE) project and is funded by an EPSRC grant (EP/D050618/1).

References

1. Afzal, W., Torkar, R., Feldt, R.: A systematic review of search-based testing for non-functional system properties. Information and Software Technology 51, 957–976 (2009)
2. Angluin, D.: Local and global properties in networks of processors. In: Proceedings of the twelfth annual ACM Symposium on Theory of Computing, pp. 82–93 (1980)
3. Beauquier, J., Cordier, S., Delaët, S.: Optimum probabilistic self-stabilization on uniform rings. In: Proceedings of the Second Workshop on Self-Stabilizing Systems, pp. 15.1–15.15 (1995)
4. Beauquier, J., Gradinariu, M., Johnen, C.: Memory space requirements for self-stabilizing leader election protocols. In: Proceedings of the Eighteenth Annual ACM Symposium on Principles of Distributed Computing, pp. 199–207. ACM (1999)
5. Beauquier, J., Gradinariu, M., Johnen, C.: Randomized self-stabilizing and space optimal leader election under arbitrary scheduler on rings. Distributed Computing 20, 75–93 (2007)

6. Clarke, E.M.: Model Checking. In: Ramesh, S., Sivakumar, G. (eds.) FST TCS 1997. LNCS, vol. 1346, pp. 54–56. Springer, Heidelberg (1997)
7. Deb, K., Pratap, A., Agarwal, S., Meyarivan, T.: A fast and elitist multiobjective genetic algorithm: NSGA-II. IEEE Transactions on Evolutionary Computation 6(2), 182–197 (2002)
8. Dijkstra, E.W.: Self-stabilizing systems in spite of distributed control. Communications of the ACM 17, 643–644 (1974)
9. Dolev, S.: Self-stabilization. The MIT Press (2000)
10. Harman, M., Mansouri, S., Zhang, Y.: Search Based Software Engineering: A Comprehensive Analysis and Review of Trends Techniques and Applications. Department of Computer Science, Kings College London, Tech. Rep. TR-09-03 (2009)
11. Herman, T.: Probabilistic Self-stabilization. Information Processing Letters 35(2), 63–67 (1990)
12. Higham, L., Myers, S.: Self-stabilizing token circulation on anonymous message passing rings. In: OPODIS 1998 Second International Conference on Principles of Distributed Systems (1999)
13. Johnen, C.: Service Time Optimal Self-stabilizing Token Circulation Protocol on Anonymous Undirectional Rings. In: Proceedings of the 21st IEEE Symposium on Reliable Distributed Systems, pp. 80–89. IEEE (2002)
14. Johnson, C.G.: Genetic Programming with Fitness Based on Model Checking. In: Ebner, M., O'Neill, M., Ekárt, A., Vanneschi, L., Esparcia-Alcázar, A.I. (eds.) EuroGP 2007. LNCS, vol. 4445, pp. 114–124. Springer, Heidelberg (2007)
15. Katz, G., Peled, D.: Genetic Programming and Model Checking: Synthesizing New Mutual Exclusion Algorithms. In: Cha, S(S.), Choi, J.-Y., Kim, M., Lee, I., Viswanathan, M. (eds.) ATVA 2008. LNCS, vol. 5311, pp. 33–47. Springer, Heidelberg (2008)
16. Katz, G., Peled, D.: Model Checking-Based Genetic Programming with an Application to Mutual Exclusion. In: Ramakrishnan, C.R., Rehof, J. (eds.) TACAS 2008. LNCS, vol. 4963, pp. 141–156. Springer, Heidelberg (2008)
17. Katz, G., Peled, D.: Synthesizing Solutions to the Leader Election Problem Using Model Checking and Genetic Programming. In: Namjoshi, K., Zeller, A., Ziv, A. (eds.) HVC 2009. LNCS, vol. 6405, pp. 117–132. Springer, Heidelberg (2011)
18. Kwiatkowska, M., Norman, G., Parker, D.: PRISM: Probabilistic Symbolic Model Checker. In: Field, T., Harrison, P.G., Bradley, J., Harder, U. (eds.) TOOLS 2002. LNCS, vol. 2324, pp. 200–204. Springer, Heidelberg (2002)
19. Luke, S.: ECJ, http://cs.gmu.edu/~eclab/projects/ecj/
20. Motwani, R.: Randomized Algorithms. Cambridge University Press (1995)
21. Norman, G.: Analysing Randomized Distributed Algorithms. Validation of Stochastic Systems, pp. 384–418 (2004)
22. Rabin, M.: Probabilistic algorithms. Algorithms and Complexity 21 (1976)
23. Valmari, A.: The State Explosion Problem. In: Reisig, W., Rozenberg, G. (eds.) APN 1998. LNCS, vol. 1491, pp. 429–528. Springer, Heidelberg (1998)

Automatically RELAXing a Goal Model to Cope with Uncertainty

Andres J. Ramirez, Erik M. Fredericks,
Adam C. Jensen, and Betty H.C. Cheng

Michigan State University,
East Lansing, MI, 48823, USA
{ramir105,freder99,acj,chengb}@cse.msu.edu

Abstract. Dynamically adaptive systems (DAS) must cope with changing system and environmental conditions that may not have been fully
understood or anticipated during development time. RELAX is a fuzzy
logic-based specification language for making DAS requirements more
tolerable to unanticipated environmental conditions. This paper presents
AutoRELAX, an approach that generates RELAXed goal models that
address environmental uncertainty by identifying which goals to RE-
LAX, which RELAX operators to apply, and the shape of the fuzzy logic
function that defines the goal satisfaction criteria. AutoRELAX searches
for RELAXed goal models that enable a DAS to satisfy its functional
requirements while balancing tradeoffs between minimizing the number
of RELAXed goals and minimizing the number of adaptations triggered
by minor and adverse environmental conditions. We apply AutoRELAX
to an industry-provided network application that self-reconfigures in response to adverse environmental conditions, such as link failures.

1   Introduction

A dynamically adaptive system (DAS) must identify and respond to changing
system and environmental conditions that may not have been fully understood
or anticipated during requirements analysis and design time. Within a DAS,
this contextual *uncertainty* arises from a combination of unpredictable environmental conditions [1,2,5,24] that can limit the adaptation capabilities of a DAS.
RELAX [2,24] is a specification language that can be used in goal-oriented modeling approaches for specifying and mitigating sources of uncertainty in a DAS.
This paper presents an approach for automatically RELAXing a goal-oriented
model to account for environmental uncertainty, thus potentially decreasing the
number of dynamic reconfigurations needed at run time.

It is unlikely for a DAS to always satisfy its requirements since it is often infeasible for a requirements engineer or a developer to identify all possible environmental conditions that the DAS may encounter throughout its lifetime [2,24]. In
light of this implication, RELAX extends goal-oriented requirements modeling approaches, such as KAOS [4,14], with fuzzy logic-based operators that specify the
extent to which a goal can become temporarily unsatisfied and yet deliver acceptable behavior. For instance, the "AS EARLY AS POSSIBLE" RELAX operator

G. Fraser (Ed.): SSBSE 2012, LNCS 7515, pp. 198–212, 2012.

enables a DAS to satisfy a goal over a longer period of time [24]. Nevertheless, it is difficult for a requirements engineer to assess, at design time, which goals to RELAX, what RELAX operators to apply, and how a goal's RELAXation will affect the overall behavior of the DAS at run time.

This paper introduces AutoRELAX, an approach that extends and automates an approach previously presented by Cheng *et al.* [2] for modeling sources of uncertainty in a DAS with RELAX. AutoRELAX explicitly handles environmental uncertainty in a DAS by automatically RELAXing goals in a KAOS goal model. In particular, AutoRELAX specifies whether a goal should be RELAXed, and if so, which RELAX operator to apply, and to what degree to lessen the constraints or bounds that define a goal's satisfaction criteria. AutoRELAX can be applied to automatically generate one or more RELAXed goal models, each of which enables a DAS to cope with specific manifestations of system and environmental uncertainty while reducing the number of adaptations performed.

AutoRELAX leverages a genetic algorithm [9] as a search heuristic to efficiently explore parts of the solution space comprising all possible RELAXed goal models. Throughout the search process, AutoRELAX uses an executable specification of the DAS to measure how candidate RELAXed goal models handle the effects of system and environmental uncertainty. AutoRELAX applies a set of fitness sub-functions that use this information to reward candidate RELAXed goal models that enable a DAS to satisfy its functional requirements while also reducing the number of adaptations the DAS performs and, consequently, the impact of a dynamic reconfiguration at run time.

We demonstrate AutoRELAX by applying it to an industry-provided application that handles the dynamic reconfiguration of a remote data mirroring (RDM) network [11,12] that improves data availability and protection by replicating and storing data at physically isolated locations. In particular, the RDM network must distribute data even under adverse system and environmental conditions, such as faulty network links and dropped messages. Experimental results show that AutoRELAX can automatically generate RELAXed goal models that are as good, if not better, than those manually created by a requirements engineer. Moreover, experimental results also demonstrate that RELAXing the satisfaction criteria of goals affected by uncertainty can reduce both the number of adaptations and the level of disruption of an adaptation.

The remainder of this paper is organized as follows. Section 2 provides background material on remote data mirroring, goal-oriented requirements modeling, RELAX, and genetic algorithms. Next, Section 3 presents the AutoRELAX process, followed by an experimental evaluation in Section 4. Section 5 overviews related work. Lastly, Section 6 summarizes findings and presents future directions.

2   Background

This section presents background material on remote data mirroring, goal-oriented modeling, the RELAX specification language, and genetic algorithms.

2.1   Remote Data Mirroring

Remote data mirroring (RDM) [11,12] is a data protection technique that prevents data loss and unavailability by storing replicates at physically remote locations. An RDM can be configured in terms of its network topology, such as a minimum spanning tree or a redundant topology, as well as the method and timing of data distribution among nodes. In particular, synchronous propagation automatically distributes every data modification to all other nodes. In contrast, asynchronous propagation batches data modifications to coalesce edits to the same data. While asynchronous propagation provides better network performance than synchronous propagation, it also provides a weaker form of data protection because batched data could be lost in the event of a site failure.

2.2   Goal-oriented Requirements Modeling

A goal declaratively specifies the objectives and constraints that a system must provide and satisfy, respectively. A functional goal specifies a service that the system must provide to its stakeholders. A goal can also be classified either as an invariant or a non-invariant. While a system must always satisfy invariant goals, a system may tolerate the temporary violation of a non-invariant goal.

A goal-oriented requirements model visually captures relationships between goals by using a directed acyclic graph where a node represents a goal and an edge represents a type of goal refinement [14]. For instance, KAOS [4,14] provides a framework for systematically refining high-level *functional* goals into finer-grained goals that are more amenable to analysis. Within KAOS, a goal can be refined via an AND or OR refinement. A goal that has been AND-refined can only be satisfied if every subgoal is also satisfied. Conversely, a goal that has been OR-refined is satisfied if at least one subgoal has been satisfied. Goal refinement continues until each leaf-level goal is assigned to an agent responsible for satisfying that goal, which then defines a requirement.

The KAOS goal model in Figure 1 captures functional requirements of the RDM application. RDMs must (A) maintain remotely stored copies of data. To this end, RDMs must (B) maintain operational costs within the allocated budget while (C) achieving acceptable levels of risk and (D) distributing data to all other nodes. Before distributing data, the RDM must (E) measure network properties and (F) construct a network by (P) activating and (Q) deactivating network links. The system must then (I) send and (J) receive data, using either (G) synchronous or (H) asynchronous propagation methods.

2.3   RELAX Specification Language

RELAX [24] is a language for specifying how sources of uncertainty that arise at the shared boundary [10] between the system and its environment affects requirements. RELAX organizes information about these sources of uncertainty into ENV, MON and REL elements: ENV specifies environmental properties that

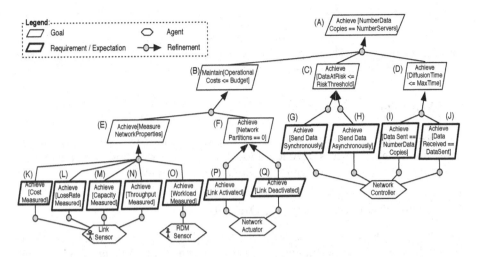

Fig. 1. KAOS goal model for the remote data mirroring application

can be observed by the DAS; MON indicates sensors in the monitoring infrastructure of the DAS; and REL establishes mathematical relationships for computing the values of ENV properties by aggregating values from MON elements.

Table 1 presents the fuzzy logic operators that RELAX provides for capturing uncertainty in requirements. These ordinal and temporal operators add flexibility in terms of how and when a requirement must be satisfied, respectively. For example, goal (F) in Figure 1 specifies that the RDM shall "achieve zero network partitions". Nevertheless, unpredictable network link failures can temporarily obstruct this goal. This goal can be RELAXed to state that the RDM shall "achieve AS FEW network partitions AS POSSIBLE", thus providing flexibility to account for unanticipated events while distributing data. In this manner, RELAX facilitates designing more flexible systems that might potentially require fewer dynamic reconfigurations.

Table 1. RELAX operators

Temporal Operators	Ordinal Operators
AS EARLY AS POSSIBLE	AS FEW AS POSSIBLE
AS CLOSE AS POSSIBLE TO [frequency]	AS CLOSE AS POSSIBLE TO [quantity]
AS LATE AS POSSIBLE	AS MANY AS POSSIBLE

2.4 Genetic Algorithms

A genetic algorithm [9] is a stochastic search-based heuristic for efficiently solving complex optimization problems. In a genetic algorithm, a population comprises a set of individuals, each encoding a candidate solution. A fitness function evaluates the quality of each individual, thereby guiding the search process towards promising areas in the solution space. New solutions can be generated with crossover and mutation operators. The crossover operator exchanges parts of existing solutions to form new individuals with, ideally, higher fitness values, and the mutation operator randomly modifies an individual to maintain diverse solution elements in the population. These operations are executed until the maximum number of generations, or iterations, are exhausted.

3 Approach

This section introduces the AutoRELAX approach. First, we state the assumptions, inputs, and outputs of AutoRELAX. We then describe how AutoRELAX can be applied to automatically generate RELAXed goal models.

3.1 Assumptions, Inputs, and Outputs

AutoRELAX needs three key input elements: a goal model, a set of utility functions for requirements monitoring, and an executable specification or prototype of the DAS. Next, we briefly describe the contents and purpose of each element.

Goal Model. AutoRELAX requires a goal model of the functional requirements that the DAS must satisfy. Currently, we target KAOS goal models [4,14]. Each goal must be designated as invariant or non-invariant.

Utility Function. A requirements engineer must derive utility functions that can monitor the satisfaction of requirements in a DAS [8,17,22]. Each utility function comprises mathematical relationships that map monitoring data to a scalar value between zero and one. This value is proportional to how well a given goal or requirement is satisfied at run time. For example, satisfaction of goal (B) in the RDM application (see Figure 1) can be evaluated with a utility function that returns one if operational costs have always been less than or equal to the allocated budget, and zero otherwise. AutoRELAX uses these utility functions to evaluate how goal RELAXations can affect DAS behavior.

Executable Specification. AutoRELAX requires an executable specification of the DAS, such as a simulation or a prototype, that applies the set of utility functions to measure how well the DAS satisfies its requirements in response to adverse conditions. In addition, a requirements engineer must also specify possible sources of uncertainty to which the DAS will be exposed. Ideally, these sources of uncertainty will exercise the adaptation logic of the DAS by subjecting it to unpredictable and adverse environmental conditions that can lead to a requirements violation. For example, in our remote data mirroring application, we configure the probability and frequency that a network link can fail or a message can be dropped. Changing either the sources of uncertainty, their likelihood, or frequency can lead to different types of RELAXed goal models.

3.2 AutoRELAX Process

Figure 2 presents a data flow diagram (DFD) that overviews the AutoRELAX process. We now present each step in the AutoRELAX process in detail:

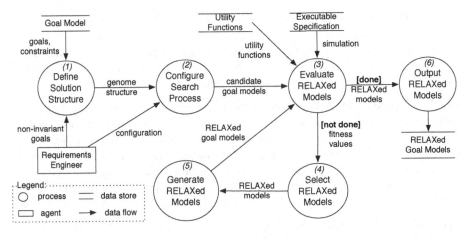

Fig. 2. DFD diagram of AutoRELAX process

(1) Define Solution Structure. Each candidate solution in AutoRELAX comprises a vector of n elements or *genes*, where n is equal to the total number of non-invariant goals in the KAOS goal model of the DAS. Figure 3(A), in turn, shows the structure of each gene. As this figure illustrates, each gene comprises a boolean variable that specifies whether a non-invariant goal will be RELAXed, a corresponding RELAX operator (see Table 1), and two floating point values that define the left and right boundaries of the fuzzy logic function, respectively.

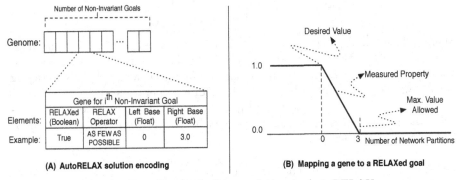

Fig. 3. Encoding a candidate solution in AutoRELAX

Figure 3(B) illustrates how each gene is mapped to a fuzzy logic function that can be used to evaluate the satisfaction of a goal. In this example, the unRELAXed satisfaction criteria (i.e., utility function) for goal (F) in Figure 1 returns 1.0 if the network is connected and 0.0 otherwise. Nevertheless, if the network partition is transient, then it may be possible to continue the data replication process amongst those nodes that are connected while the network is reconfigured. Thus, as the bolded lines show in Figure 3(B), the satisfaction criteria of this goal can be RELAXed by applying the "AS FEW AS POSSIBLE" ordinal operator that maps to a *left shoulder* fuzzy logic function shape [24].

For this RELAXed goal, the apex is still centered upon the ideal value of a system or environmental property. In this case, zero network partitions and the downward slope from the apex to the right endpoint reflects values that, while not ideal, might be tolerated at run time. Note that the *left endpoint* value encoded in the gene is not used for this particular fuzzy logic shape.

(2) Configure Search Process. A requirements engineer must configure AutoRELAX by specifying a population size, crossover and mutation rates, and a termination criterion. The population size determines how many candidate RELAXed goal models AutoRELAX can explore in parallel during each generation; the crossover and mutation rates specify how AutoRELAX will generate new RELAXed goal models; and the termination criteria specifies when AutoRELAX will stop searching for new solutions and output the resulting RELAXed goal models.

(3) Evaluate RELAXed Models. To evaluate the quality of a RELAXed goal model, AutoRELAX first maps the RELAX operators encoded in an individual to their corresponding utility functions for requirements monitoring in the executable specification (see Step 1). Next, AutoRELAX simulates the executable specification and records the satisfaction of each goal as well as the number of adaptations performed by the DAS. Two fitness sub-functions use this information to reward candidate RELAXed goal models for minimizing both the number of RELAXed goals as well as how many adaptations are triggered by minor environmental conditions.

The first fitness sub-function, FF_{nrg}, rewards candidate solutions that minimize the number of RELAXed goals:

$$FF_{nrg} = 1.0 - \left(\frac{|relaxed|}{|Goals_{non\text{-}invariant}|}\right)$$

where $|relaxed|$ and $|Goals_{non\text{-}invariant}|$ are the number of RELAXed and non-invariant goals in the goal model, respectively. The intent of this fitness sub-function is to preserve the intent of the original goal model by discouraging AutoRELAX from unnecessarily introducing RELAX operators.

The second fitness sub-function, FF_{na}, rewards candidate solutions that minimize the number of adaptations performed by the DAS in response to minor and transient environmental conditions:

$$FF_{na} = 1.0 - \left(\frac{|adaptations|}{|faults|}\right)$$

where $|adaptations|$ represents the total number of adaptations performed by the DAS, and $|faults|$ measures the total number of adverse environmental conditions introduced throughout a simulation. This fitness sub-function rewards RELAXed goal models that tolerate unanticipated environmental conditions and reduce the number of adaptations a DAS performs, thereby reducing the number of passive and quiescent components at run time [13]. While a *passive* component may service transactions from other components, it may not initiate transactions. In contrast, a *quiescent* component cannot initiate new transactions nor service transactions from other components. Reducing the number of passive and quiescent components minimizes the impact of adaptation upon the DAS behavior.

These two fitness sub-functions can be combined into a linear weighted sum:

$$\text{Fitness Value} = \begin{cases} \alpha_{\text{nrg}} * \text{FF}_{\text{nrg}} + \alpha_{\text{na}} * \text{FF}_{\text{na}} & \text{iff invariants true} \\ 0.0 & \text{otherwise} \end{cases}$$

where α_{nrg} and α_{na} coefficients reflect the relative importance of each fitness sub-function, the sum of which must equal 1.0.[1] The fitness value of a RELAXed goal model depends upon the satisfaction of all invariant goals. For example, if a RELAXed goal model in our RDM application does not replicate every data item, then its fitness value is 0.0. This penalty ensures only *viable* RELAXed goal models, where all invariant goals are satisfied, are output as solutions.

(4) Select RELAXed Models. Using the fitness value associated with each evaluated RELAXed goal model, AutoRELAX *selects* the most promising individuals from the population to guide the search process towards that area of the solution space. To this end, AutoRELAX applies tournament selection [9], a technique that randomly selects k individuals from the population and *competes* them against one another. The RELAXed goal model with the highest fitness value amongst these k solutions survives onto the next generation.

(5) Generate RELAXed Models. AutoRELAX uses two-point crossover and single-point mutation to generate new RELAXed goal models, which were set to 50% and 40% for this work, respectively. As Figure 4(A) shows, two-point crossover takes two individuals from the population as *parents* and produces two new RELAXed goal models as *offspring*. As this figure illustrates with different shading, two-point crossover exchanges the genes between two randomly chosen indices. In contrast, Figure 4(B) shows how single-point mutation takes an individual from the population and randomly modifies the values of a single gene. In this particular example, the effect of the mutation operator is to change a gene such that its corresponding non-invariant goal is now RELAXed with the "AS MANY AS POSSIBLE" RELAX operator. In this manner, while crossover attempts to construct better solutions by combining good elements from existing RELAXed goal models, mutation introduces diverse sets of goal RELAXations that might not be obtainable otherwise.

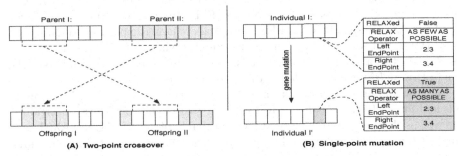

Fig. 4. Generating new RELAXed goal models with crossover and mutation operators

---

[1] Although fitness sub-functions can be combined in different ways, we find that a linear-weighed sum facilitates the balancing of competing concerns.

(6) Output RELAXed Models. AutoRELAX iteratively applies steps (3) through (5) until it reaches its generational limit. Then, AutoRELAX outputs one or more RELAXed goal models with the highest fitness values in the population.

4   Experimental Results

This section describes our experimental setup and discusses the experimental results where we apply AutoRELAX to a RDM application.

4.1   Experimental Setup

For this work, we modeled the RDM network as a completely connected graph where a node represents an RDM and an edge represents a network link. In particular, the network consists of 25 RDMs and 300 network links that can be activated and used to transfer data between RDMs. We leverage an RDM operational model previously presented by Keeton *et al.* [12] to generate performance attributes of each RDM and network link. Each network simulation executes for 150 time steps. Throughout each simulation, 20 data items are randomly inserted at different RDMs that are then responsible for distributing those data items to all other RDMs.

The RDM network is subject to environmental uncertainty in the form of unpredictable network link failures and dropped messages. As such, the RDM network might need to self-adapt in response to these adverse system and environmental conditions. To this end, each RDM implements the dynamic change management (DCM) protocol previously introduced by Kramer and Magee [13]. Furthermore, we implemented a rule-based adaptation engine that monitors the satisfaction of each goal to determine if the network structure and propagation parameters need to be reconfigured. If an adaptation is warranted, then Plato [18] and the DCM protocol are executed to generate a target system configuration and a series of reconfiguration steps that safely transitions the executing system from its current configuration to its target configuration, respectively.

We compare and evaluate the resulting RELAXed goal models produced by AutoRELAX with two different goal models of the same RDM application, the unRELAXed goal model shown in Figure 1 and a goal model manually RELAXed by a requirements engineer. The manually RELAXed goal model consists of five goal RELAXations: goal (C) is RELAXed to allow larger exposures to data loss; goal (D) is RELAXed to add temporal flexibility when diffusing data; goal (F) is RELAXed to allow up to three simultaneous network partitions; and goals (I) and (J) were RELAXed to tolerate dropped messages.

We use the fitness functions presented in Section 3 to compare these models and illustrate the benefits of RELAXing a goal model to address uncertainty as well as demonstrate that AutoRELAX is capable of generating RELAXed goal models that are as good, if not better, than those manually created by a requirements engineer. We set α_{nrg} to 0.3 and α_{na} to 0.7, thereby emphasizing the reduction in the number of adaptations performed. For statistical purposes, we conducted 50 trials of each experiment and, where applicable, plot or report the mean values with corresponding error bars or deviations.

4.2   Uncertain Environment

For this experiment, we define the first null hypothesis, $H1_0$, to state that there is no difference between a RELAXed and an unRELAXed goal model. In addition, we also define a second null hypothesis, $H2_0$, to state that there is no difference between RELAXed goal models generated by AutoRELAX and those manually created by a requirements engineer.

Figure 5 presents three box plots with the fitness values obtained by generated AutoRELAX models, a manually created RELAXed goal model, and an unRELAXed goal model, respectively, with the latter two being conducted only once. As these box plots illustrate, despite the fitness boost unRELAXed goal models obtain by not introducing any goal RELAXations (see Section 3, FF_{nrg}), RELAXed goal models achieved statistically significant higher fitness values than unRELAXed goal models ($p < 0.001$, Welch Two Sample t-test). These results enable us to reject our first null hypothesis, $H1_0$, as well as conclude that RELAX does indeed reduce the number of adaptations when addressing system and environmental uncertainty.

The box plots in Figure 5 also demonstrate that AutoRELAX generated RELAXed goal models achieved statistically significant higher fitness values than those manually created by a requirements engineer ($p < 0.001$, Welch Two Sample t-test). As a result, we also reject our second null hypothesis, $H2_0$ and conclude that AutoRELAX is capable of generating RELAXed goal models that better address specific sources of uncertainty than manually RELAXed goal models.

Figure 6 presents three sets of box plots that capture the adaptation costs incurred by generated AutoRELAX models, a manually created RELAXed goal model, and an unRELAXed goal model, respectively. Specifically, each set of box plots measures the amount of time components in the RDM network were in active, passive, and quiescent modes *during* a reconfiguration (these plots do not include time outside of a reconfiguration). As this figure illustrates, a key

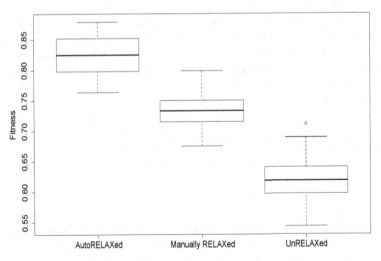

Fig. 5. Fitness values comparison between RELAXed and unRELAXed goal models.

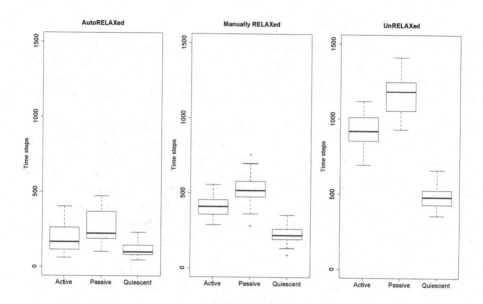

Fig. 6. Adaptation costs comparison between RELAXed and unRELAXed goal models

reason for why RELAXed goal models outperform unRELAXed goal models is that by carefully lessening the satisfaction criteria of non-invariant goals, the number of adaptations decrease and so does the amount of time components spend in passive and quiescent modes during a reconfiguration.

Both Figures 5 and 6 show that AutoRELAX is able to generate RELAXed goal models that perform better than manually RELAXed goal models. Examining the automatically generated RELAXed goal models suggests two primary reasons for this difference in fitness values and, consequently, in adaptation costs. First, while the manually RELAXed goal model introduced RELAXations to goals (C), (D), (F), (I), and (J), AutoRELAX mostly introduced RELAX operators to goals (F), (I), and (J), thereby slightly boosting its fitness value in comparison. Second, the manually RELAXed goal model contained some goal RELAXations that were too constrained. For instance, AutoRELAX was able to extend the goal satisfaction boundary of goal (F) beyond the bounds applied in the manually RELAXed goal model. As a result, the goal model produced by AutoRELAX was able to tolerate a greater number of temporary network partitions while allowing components to remain actively distributing data throughout the network.

Threat to Validity. This research was a proof of concept study to determine if it is *possible* to automatically RELAX goal models to produce viable goal models. We applied the technique on a problem provided to us from industrial collaborators. As a point of reference, we compared the AutoRELAXed goal models to those developed by a requirements engineer. Threats to validity include whether this technique will achieve similar results with other goal models and other applications. While, we did not intend to study human effectiveness [20], the positive results motivate its study explicitly as part of future work.

5   Related Work

This section presents related work on obstacle mitigation, expressing uncertainty in requirements, and requirements monitoring.

Obstacle Mitigation. van Lamsweerde *et al.* [14,15] proposed a set of strategies to facilitate the systematic identification, analysis, and resolution of obstacles, or conditions that prevent a system from satisfying its objectives. If an obstacle cannot be prevented, then one of their proposed mitigation strategies consists of tolerating the violation of a goal. This strategy, however, does not specify the extent to which a goal can become unsatisfied without adversely affecting other goals. From this perspective, AutoRELAX complements their proposed mitigation strategies by automatically determining if, and to what extent, a goal can become unsatisfied at run time.

Although the heuristics proposed by van Lamsweerde *et al.* [14,15] facilitate the systematic identification and analysis of obstacles, unpredictable environments may still prevent a DAS from satisfying its requirements. Letier and van Lamsweerde [16] also introduced a probabilistic framework for specifying the probability of a goal being satisfied. These probabilities, which can be obtained from a domain expert or derived from actual system usage data, can be used to identify previously unknown obstacles. In contrast to AutoRELAX, however, their probabilistic framework treats requirements as being either strictly satisfied or not. AutoRELAX could leverage the probability of a goal being satisfied when identifying goals that might benefit from RELAXation.

Expressing Uncertainty in Requirements. Fuzzy set theory has been recently applied to represent and analyze the effects of uncertainty in requirements. For instance, Whittle *et al.* [24] introduced RELAX to facilitate the identification and analysis of sources of uncertainty in a DAS. Cheng *et al.* [2] extended RELAX to support the modeling of RELAXed goals in a KAOS goal model [4,14]. Similarly, Baresi *et al.* [1] presented FLAGS, a KAOS goal modeling framework that introduces the concept of a fuzzy goal whose satisfaction is evaluated through fuzzy logic functions. Both RELAX and FLAGS depend on a requirements engineer to manually determine goals that may become unsatisfied, as well as how much flexibility to introduce for each goal's satisfaction criteria. AutoRELAX automates this process.

Welsh *et al.* [23] introduced Claims as markers of uncertainty that capture doubts about how a given goal realization strategy contributes to the satisficement [3], or satisfaction of a degree, of a soft goal. If a Claim is proven false at run time, then the DAS self-reconfigures towards a more desirable goal realization strategy. Although AutoRELAX focuses on RELAXing the satisfaction of functional non-invariant goals, it could be extended to also automatically RELAX the satisfaction criteria of soft goals.

Requirements Monitoring and Reflection. Feather *et al.* [6,7] developed requirements monitoring frameworks that can detect the occurrence of obstacles and reconfigure the system in response if necessary. More recently, Sawyer *et al.* [19]

suggested promoting requirements to live run-time entities whose satisfaction can be evaluated in support of adaptation decisions. Their concept is similar to the feedback-loop Awareness Requirements (AwReqs) construct proposed by Souza *et al.* [21] where meta-level requirements manage the satisfaction of other requirements. None of these approaches, however, support the management or run-time monitoring of RELAXed requirements. If these requirements monitoring and management frameworks were extended to support RELAXed requirements, then AutoRELAX could be applied to automatically specify the satisfaction criteria of RELAXed goals while these frameworks handle the run-time logistics of monitoring their satisfaction at run time.

6   Conclusions

In this paper we presented AutoRELAX, an approach that applies a genetic algorithm to automatically generate RELAXed goal models that can mitigate the effects of uncertainty upon the self-assessment capabilities of a DAS. By providing automated tool support, AutoRELAX relieves requirements engineers from the daunting task of considering a large number of strategies for dealing with uncertainty. We applied AutoRELAX to a RDM application that must distribute data to all nodes within the network while self-reconfiguring in response to network link failures and dropped messages. When compared with an unRELAXed goal model, experimental results show that RELAX is able to reduce the number of adaptations, and therefore adaptation costs, that would otherwise be incurred by minor and transient environmental conditions. Results also show that AutoRELAX can generate RELAXed goal models of equal or greater quality than those manually created by a requirements engineer.

AutoRELAX automatically provides feedback about sources of uncertainty. In these experiments, the goal RELAXations introduced by AutoRELAX concurred with the specific sources of uncertainty. Specifically, AutoRELAX introduced goal RELAXations to goals (F) and (J) depending on whether network links were more likely to fail than messages dropped, and vice-versa. Interestingly enough, AutoRELAX always RELAXed goal (J). By analyzing our goal models with this information we noted that, in contrast to goals (F) and (J), the satisfaction of goal (I) is affected by both network failures and dropped messages. Thus, the environmental uncertainty introduced in our experiments would frequently cause goal (I) to be violated and the network to self-reconfigure in response. As such, AutoRELAX automatically identified how sources of uncertainty affected this specific goal and then introduced RELAXed operators to lessen its impact.

Future directions include extending the search parameters to also optimize the underlying shape (i.e., triangle versus trapezoid) of the fuzzy logic function that defines the satisfaction criteria of a RELAXed goal, as well as extending AutoRELAX to support the automatic RELAXation of non-functional goals.

Acknowledgements. This work has been supported in part by NSF grants CCF-0541131, IIP-0700329, CCF-0750787, CCF-0820220, DBI-0939454, CNS-0854931, Army Research Office grant W911NF-08-1-0495, Ford Motor Company.

Any opinions, findings, and conclusions or recommendations expressed in this material are those of the author(s) and do not necessarily reflect the views of the National Science Foundation, Army, Ford, or other research sponsors.

References

1. Baresi, L., Pasquale, L., Spoletini, P.: Fuzzy goals for requirements-driven adaptation. In: Proceedings of the 18th IEEE International Requirements Engineering Conference, pp. 125–134. IEEE, Sydney (2010)
2. Cheng, B.H.C., Sawyer, P., Bencomo, N., Whittle, J.: A Goal-Based Modeling Approach to Develop Requirements of an Adaptive System with Environmental Uncertainty. In: Schürr, A., Selic, B. (eds.) MODELS 2009. LNCS, vol. 5795, pp. 468–483. Springer, Heidelberg (2009)
3. Chung, L., Nixon, B., Yu, E., Mylopoulos, J.: Non-Functional Requirements in Software Engineering. Kluwer Academic Publishers (2000)
4. Dardenne, A., van Lamsweerde, A., Fickas, S.: Goal-directed requirements acquisition. Science of Computer Programming 20(1-2), 3–50 (1993)
5. Esfahani, N., Kouroshfar, E., Malek, S.: Taming uncertainty in self-adaptive software. In: Proceedings of the 19th ACM SIGSOFT Symposium and the 13th European Conference on Foundations of Software Engineering, Szeged, Hungary, pp. 234–244 (2011)
6. Feather, M.S., Fickas, S., van Lamsweerde, A., Ponsard, C.: Reconciling system requirements and runtime behavior. In: Proceedings of the 8th International Workshop on Software Specification and Design, pp. 50–59. IEEE Computer Society, Washington, DC (1998)
7. Fickas, S., Feather, M.S.: Requirements monitoring in dynamic environments. In: Proceedings of the Second IEEE International Symposium on Requirements Engineering, pp. 140–147. IEEE Computer Society, Washington, DC (1995)
8. de Grandis, P., Valetto, G.: Elicitation and utilization of application-level utility functions. In: The Proceedings of the Sixth International Conference on Autonomic Computing (ICAC 2009), June 2009, pp. 107–116. ACM, Barcelona (2009)
9. Holland, J.H.: Adaptation in Natural and Artificial Systems. MIT Press, Cambridge (1992)
10. Jackson, M., Zave, P.: Deriving specifications from requirements: an example. In: Proceedings of the 17th International Conference on Software Engineering, pp. 15–24. ACM, Seattle (1995)
11. Ji, M., Veitch, A., Wilkes, J.: Seneca: Remote mirroring done write. In: USENIX, 2003 Annual Technical Conference, pp. 253–268. USENIX Association, Berkeley (2003)
12. Keeton, K., Santos, C., Beyer, D., Chase, J., Wilkes, J.: Designing for disasters. In: Proceedings of the 3rd USENIX Conference on File and Storage Technologies, pp. 59–62. USENIX Association, Berkeley (2004)
13. Kramer, J., Magee, J.: The evolving philosophers problem: Dynamic change management. IEEE Trans. on Soft. Eng. 16(11), 1293–1306 (1990)
14. van Lamsweerde, A.: Requirements Engineering: From System Goals to UML Models to Software Specifications. Wiley (March 2009)
15. van Lamsweerde, A., Letier, E.: Handling obstacles in goal-oriented requirements engineering. IEEE Transactions on Software Engineering 26(10), 978–1005 (2000)

16. Letier, E., van Lamsweerde, A.: Reasoning about partial goal satisfaction for requirements and design engineering. In: Proceedings of the 12th ACM SIGSOFT International Symposium on Foundations of Software Engineering, pp. 53–62. ACM, Newport Beach (2004)

17. Ramirez, A.J., Cheng, B.H.C.: Automatically deriving utility functions for monitoring software requirements. In: Proceedings of the 2011 International Conference on Model Driven Engineering Languages and Systems Conference, Wellington, New Zealand, pp. 501–516 (2011)

18. Ramirez, A.J., Knoester, D.B., Cheng, B.H.C., McKinley, P.K.: Applying genetic algorithms to decision making in autonomic computing systems. In: Proceedings of the Sixth International Conference on Autonomic Computing, Barcelona, Spain, pp. 97–106 (June 2009)

19. Sawyer, P., Bencomo, N., Letier, E., Finkelstein, A.: Requirements-aware systems: A research agenda for re self-adaptive systems. In: Proceedings of the 18th IEEE International Requirements Engineering Conference, Sydney, Australia, September 2010, pp. 95–103 (2010)

20. de Souza, J.T., Maia, C.L.B., de Freitas, F.G., Coutinho, D.P.: The human competitiveness of search based software engineering. In: Proceedings of the 2nd International Symposium on Search Based Software Engineering (SSBSE 2010), September 7-9, pp. 143–152. IEEE, Benevento (2010)

21. Souza, V.E.S., Mylopoulos, J.: From awareness requirements to adaptive systems: A control-theoretic approach. In: Proceedings of the Second International Workshop on Requirements at Run Time, pp. 9–15. IEEE Computer Society Press, Trento (2011)

22. Walsh, W.E., Tesauro, G., Kephart, J.O., Das, R.: Utility functions in autonomic systems. In: Proceedings of the First IEEE International Conference on Autonomic Computing, pp. 70–77. IEEE Computer Society, New York (2004)

23. Welsh, K., Sawyer, P.: Understanding the Scope of Uncertainty in Dynamically Adaptive Systems. In: Wieringa, R., Persson, A. (eds.) REFSQ 2010. LNCS, vol. 6182, pp. 2–16. Springer, Heidelberg (2010)

24. Whittle, J., Sawyer, P., Bencomo, N., Cheng, B.H.C., Bruel, J.M.: RELAX: Incorporating uncertainty into the specification of self-adaptive systems. In: Proceedings of the 17th International Requirements Engineering Conference (RE 2009), pp. 79–88. IEEE Computer Society, Atlanta (2009)

Boosting Search Based Testing
by Using Constraint Based Testing

Abdelilah Sakti, Yann-Gaël Guéhéneuc, and Gilles Pesant

Department of Computer and Software Engineering
École Polytechnique de Montréal, Québec, Canada
{abdelilah.sakti,yann-gael.gueheneuc,gilles.pesant}@polymtl.ca

Abstract. Search-Based Testing (SBT) uses an evolutionary algorithm to generate test cases. Traditionally, a random selection is used to generate an initial population and also, less often, during the evolution process. Such selection is likely to achieve lower coverage than a guided selection. We define two novel concepts: (1) a constrained population generator (CPG) that generates a diversified initial population that satisfies some test target constraints; and (2) a constrained evolution operator (CEO) that evolves test candidates according to some constraints of the test target. Either the CPG or CEO may substantially increase the chance of reaching adequate coverage with less effort. In this paper, we propose an approach that models a relaxed version of the unit under test as a constraint satisfaction problem. Based on this model and the test target, a CPG generates an initial population. Then, an evolutionary algorithm uses a CEO and this population to generate test input leading to the test target being covered. Our approach combines constraint-based testing (CBT) and SBT and overcomes the limitations associated with each of them. Using eToc, an open-source SBT tool, we implement a prototype of this approach. We present the empirical results of applying both CPG or CEO on three open-source programs and show that CPG or CEO improve SBT performance in terms of branch coverage by 11% while reducing computation time.

Keywords: Search Based Testing, Constraint Based Testing, Initial Population Generator, Evolution Operator.

1 Introduction

Proving that some software system corresponds to its specification or exposing hidden errors in its implementation is a time consuming and tedious process, accounting for 50% of the total software cost [13]. Test case generation is one of the most expensive parts of the software testing phase. Therefore, automating testing can significantly reduce software cost, development time, and time to market [6]. Constraint Based Testing (CBT) and Search Based Testing (SBT) have become the dominant approaches to test case generation, because they achieve high code coverage. Over the last decade, theses two approaches have been extensively explored [3,5,9,10,14,18,19].

G. Fraser (Ed.): SSBSE 2012, LNCS 7515, pp. 213–227, 2012.

The main advantages of the CBT approach are its precision in test data generation and its ability to prove that some paths are unreachable. The main disadvantage of CBT is its inability to manage the dynamic aspects of a unit under test (dynamic structures, native function calls, and communication with the external environment) [19]. Using CBT, it is not always possible to generate an exact test input due to source code complexity or unavailability.

In contrast, SBT approaches handle any sort of programs but these approaches depend on the search space size, the diversity of their initial populations, and the effectiveness of their fitness functions. It is inefficient to use evolutionary testing in a large search space without a diversified population or without sufficient guidance: for example, in a program that contains conditional statements on boolean variables (flags), evolutionary testing may not have the necessary information to guide its search [11]. Even though SBT is largely used in industry, it still suffers from many problems [11]. Those problems can be dealt with by using CBT. Recently, a strong combination of constraint programming and of an evolutionary algorithm has shown great promise to solve optimization problems [8]. Such combination leads us to believe that combining SBT and CBT is interesting for an efficient test data generation.

In this paper, we propose a hybrid approach, *CSBT*, that combines *CBT* and *SBT* to generate test data. We define a novel CBT framework to replace any random generation in the SBT approach. To generate test input candidates for SBT, the CBT models and solves a relaxed version of the unit under test (UUT). Then, the SBT framework takes these input candidates and generates the actual test input. We implemented a prototype of this hybrid approach and applied it to generate test input data leading to branch coverage on a set of programs. We report the comparison of CBT, SBT, and our novel approach CSBT and show that the CSBT technique outperforms the others. These empirical results will show that CSBT reduces the effort needed to reach a given test target and that it achieves higher branch coverage than the test input generated by each approach alone.

The contributions of this paper are: a novel approach to combine both *search based* and *constraint based* techniques to generate test input data; a framework to model a relaxation of a UUT as a constraint satisfaction problem; an empirical comparison of CBT, SBT, and CSBT on some programs.

The remainder of the paper is organized as follows: Section 2 presents a brief overview of CBT and SBT. Section 3 introduces the problem and the approach by using a motivating example. Section 4 describes our testing approach. Section 5 describes how we implemented a prototype of our approach by extending the SBT tool eToc [20]. Section 6 presents the empirical study used to evaluate the CSBT and the analysis of the results. Section 7 summarizes related work. Section 8 concludes with some future work.

2   Background

Constraint-based testing is applied in two different ways: static [2] (symbolic execution) and dynamic [4] (Dynamic Symbolic Execution DSE).

Static CBT. We distinguish between two static CBT approaches: *path-oriented* [2] and *goal-oriented* [5]. The first finds test inputs of a given execution path. *Symbolic execution* (SE) is a well known path-oriented technique that was introduced by Clarke [2] in the 1970s. SE consists of statically selecting an execution path from the control-flow graph (CFG) and then executing it symbolically and creating a path constraint over the program's input variables. A solution to this path constraint is a test data that will drive the execution down the selected path. The second approach finds test input that reaches a given statement (test target). Generally, this approach translates a whole program into a constraint programming problem. The given test target is translated into a constraint. Solving the conjunction of this constraint and the generated constraint programming problem yields a test data that will reach the test target.

Dynamic CBT. In the literature, the dynamic CBT technique most used is *dynamic symbolic execution* (DSE) [4,17]. DSE combines symbolic and concrete execution. It consists of exploring execution paths at runtime: First, it executes an instrumented version of a UUT; second it gets the executed path condition and some concrete values; and then it derives a new path condition; it uses concrete values to simplify complex (unsupported) expressions. Solving this new path condition generates test inputs to explore a new path. This procedure is repeated until the test target is covered (e.g., all-paths, all-branches, all-statements) or some condition limit is met.

Search-Based Testing. SBT was introduced by Miller and Spooner [12] in the 1970s. To generate test inputs SBT uses an evolutionary algorithm that is guided by an objective function or fitness function. The fitness function is defined according to a desirable test target. The commonly used fitness functions are branch-distance and approach-level [11]. To generate test input, SBT starts by generating a random set of test input candidates (initial population). For each test input candidate, an instrumented version of the UUT is executed and its fitness is computed. Based on the fitness ranking, the evolutionary algorithm evolves the current population to generate a new one. It continues evolving populations until the test target is achieved or a stopping criterion is reached.

3   Motivating Example

We use the program in Fig. 1 as a running example. It considers the problem of generating a test input to reach Targets 1, 2, and 3. The three targets reflect different problems of test input generation.

- Target 1 is easy to reach;
- Target 2 is unreachable because $!(x \leq 0 || y \leq 0) \& !(x < y/2) \& (y > 3 \times x)$ is unsatisfiable;
- Target 3 is hard to reach because it contains nested predicates and involves the native function call, fun, that returns x^2/y.

CBT can generate test inputs for Target 1 and prove that Target 2 is unreachable. However, if we suppose that the function *fun* cannot be handled by a particular constraint solver, a static CBT approach cannot generate test inputs for Target 3. A dynamic CBT can also fail to derive test inputs for Target 3 in a reasonable amount of time.

```
int foo(int X, int Y){
  if(X<=0 || Y<=0)
    return 0;
  int Z;
  if ((X < Y/2)|| (Y==0))
    Z= 1; //Target 1
  else if (Y>3*X)
    Z=2; //Target 2
  else{
    Z = fun(X,Y);
    if ((Z >8) && (Y==10))
      if(Z==Y)
        Z=3;//Target 3
  }
  return Z;
}
```

Fig. 1. Running example

SBT can derive test inputs for Target 1 but it takes more time than CBT. Therefore, if the search space is large, SBT may take a very long time before generating a test input for Target 3 (it needs $x = 10$ and $y = 10$). For Target 2, SBT may search forever without proving its unreachability.

Target 3 is problematic for both approaches: it is a nested branch predicate [11] for SBT and it contains an unsupported function for CBT. However a hybrid approach can take advantage of both to generate test input for Targets 1 and 3, and it may prove the unreachability of Target 2. We confirmed this fact by generating a test input to reach Target 3 using all three approaches: SBT (eToc [20] and a hill climbing implementation), CBT (CP-SST [15], a goal-oriented static CBT approach), and a combination of these approaches. We concluded that eToc and Hill Climbing were unable to generate test inputs to reach Target 3 after 45000 fitness calculations in domain $[-20000, 20000]$. As well, CP-SST was unable to reach Target 3. However, a hybridization eToc+CP-SST (resp. HC+CP-SST) was able to generate test input just after 200 (resp. 600) fitness calculations.

The key idea of our hybridization is to proceed in two phases. First, generating a relaxed version of *foo* by replacing the function *fun* with an uninitialized variable typed as *foo*'s return value. Then CBT uses the relaxed version to generate pseudo test inputs. In a second phase, SBT uses those pseudo test inputs as candidates to generate the actual test inputs.

Now, consider for example the program in Fig.1. In order to reach Target 3, the conjunction of the negation of the first three conditional statements and the last two conditional statements must be fulfilled. For the last two conditions, z depends on $fun(x, y)$'s return value. Using CP-SST [15] the path condition can be written as follow:

$$not((x_0 \leq 0)||(y_0 \leq 0)) \wedge not((x_0 < y_0/2)||(y_0 = 0)) \wedge not(y_0 > 3 \times x_0) \wedge z_3 = fun(x_0, y_0) \wedge ((z_3 > 8) \wedge (y_0 = 10)) \wedge (z_3 = y_0).$$

The relaxed version is obtained by ignoring the constraint $z_3 = fun(x_0, y_0)$. Then, the path condition becomes $not((x_0 \leq 0)||(y_0 \leq 0)) \wedge not((x_0 < y_0/2)||(y_0 = 0)) \wedge not(y_0 > 3 \times x_0) \wedge ((z_3 > 8) \wedge (y_0 = 10)) \wedge (z_3 = y)$. As the relaxed path condition is less restrictive than the actual one, generated pseudo test inputs won't necessarily trigger Target 3, but they satisfy a big part of the path condition. By solving the relaxed constraint, we obtain solutions of the form $(x_0 \geq 5, y_0 = 10)$. Therefore, the search space size has been reduced. Using those pseudo test inputs as an initial population for an evolutionary algorithm can reduce significantly the effort needed to generate test data.

In this example, only one path led to Target 3. If there are many paths that can reach the test target, generating the test data that allows covering different paths or different branches may assure a diversified population for SBT.

4   CSBT: Constrained Search Based Testing

The assumption underlying the research work presented in this paper supposes that using a diversified initial population that partly satisfies the predicates leading to the test target can reduce significantly the effort required to reach this test target. Our approach is called Constrained Search Based Testing (CSBT) because it starts, in a first phase, by using CBT to generate an initial population and then, in a second phase, it uses SBT to generate test inputs. We propose to replace the random generation of an initial population used by SBT with a set of pseudo test input generated using a CBT approach.

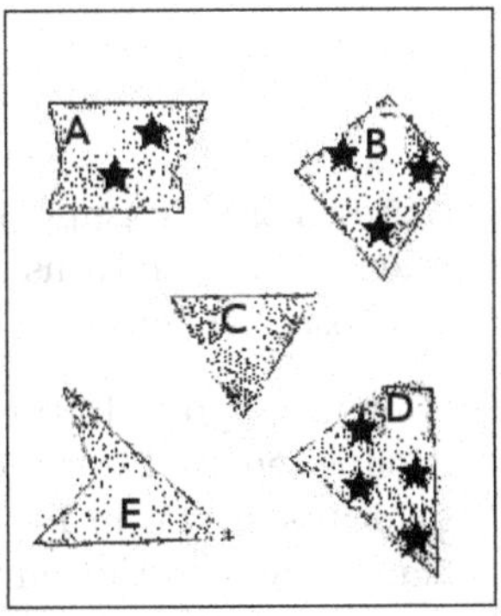

Fig. 2. Inputs search space

Fig. 2 presents the whole inputs search space of a program P: the parts A, B, C, D, and E are the CBT solutions space of a relaxed version of P which are called *pseudo test inputs*, while stars are actual test inputs. This example shows that a random test input candidate is likely to be too far from an actual test inputs compared to some pseudo test inputs. The parts C and E don't contain any test input, so a pseudo input from these two parts may also be far from an actual test input. A population that takes its candidates from different parts is likely to be near of a test input, we call it a *diversified population*. We can offer an acceptable level of diversity by generating a pseudo test input for every consistent branch with the test target (All-branch), and a high level of diversity by generating a pseudo test input for every path leads to the test target (All-path).

4.1   Unit Under Test Relaxation

To avoid traditional CBT problems, we introduce a preliminary phase called program relaxation. We propose to apply CBT on a relaxed version of the UUT. A relaxed UUT version is a simplified version, of the original one, that contains only expressions supported by the constraint solver: the expressions that generate constraints whose consistency can be checked by the constraint solvers are kept in the relaxed version, while all the other expression are relaxed or ignored, e.g., for an integer solver, expressions that can generate constraints over string or

```
intStr(int X,int Y,
  String S1, String S2){
  int y= X<<Y;
  int x=y+X/Y;
  String s=S1+S2;
  if((s.equals("OK")
    && x>0)
    && s.length()>x)
    return 1; //Target
  return 0;
}
```

```
intStr(int X,int Y){
  int R1,R2;
  int y= R1;
  int x=y+X/Y;
  if(
    x>0)
    && R2>x)
    return 1; //Tar
  return 0;
}
```

```
intStr(String S1,
       String S2){

  String s=S1+S2;
  if((s.equals("OK")
    )
    )
    return 1; //Tar
  return 0;
}
```

Fig. 3. intStr function   **Fig. 4.** Relaxed version for an integer solver   **Fig. 5.** Relaxed version for a string solver

float are ignored, unsupported operators (expressions) are relaxed, and a native function call that returns an integer is relaxed.

A relaxed version is obtained by applying the following rules:

- Any unsupported expression (function call, operator) is relaxed: the expression is replaced by a new variable. Fig. 4 shows a relaxed version of *intStr*, which is shown in Fig.3, for an integer solver. We suppose that the solver cannot handle the shift operator, so the expression that uses this operator has been replaced by a new variable $R1$. The variable $R1$ is not initialized. Therefore, when we will translate the relaxed version into a CSP, the CSP variable $R1$ can take any integer value.
- Any statement over unsupported data type is ignored. In Fig. 4, The relaxed version of *intStr* ignores the statement $String\ s = S1 + S2$; and the condition $s.equals("OK")$ because those two expressions are over strings.
- For each data type that needs a different solver a new relaxed version is created. The function *intStr* needs integer and string type as inputs. If we have an available solver for string, then we generate another relaxed version over string. Fig. 5 shows a relaxed version of *intStr* for a string solver.
- For loops, CBT cannot model an unlimited number of iterations. In general a constant k-path (equal 1, 2, or 3) is used to limit the number of loop iterations. We can model a loop in two different ways: first, we force a loop to stop at most after k-path iterations, in this case some feasible paths may become infeasible; second, we don't force a loop to stop and we model just k-path iterations, after which the value of a variable assigned inside a loop is unknown. We use the second case and we relax any variable assigned inside a loop just after this loop. The variable is assigned a new uninitialized variable.

4.2   Collaboration between CBT and SBT

Every test input generated is a result of a collaborative task between CBT and SBT: CBT generates a pseudo test input, and then SBT generates an actual

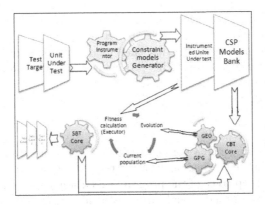

Fig. 6. Implementation overview

test input. Our main contribution here is to define the information exchanged and connection points that can make CBT useful for SBT. SBT needs new test input candidates at three different points:

1. During the generation of the initial population;
2. When it restarts, i.e., when it reaches the attempt limit of evolving;
3. During its evolving procedure.

For the first and the second points, a CBT can generate the whole or a part of the population. If CBT is unable to generate the required number (population size) of candidates the usual generation procedure (random) can be called to complete the population. We called this technique *Constrained Population Generator*(CPG). For the third point, CBT can participate to guide the population evolution by discarding candidates that break the relaxed model and only allowing candidates that satisfy the relaxed model, we called this technique *Constrained Evolution Operator* (CEO). But frequent calls to CBT seem to be too expensive in practice, this can weaken the main advantage of evolutionary algorithms which is their speed. Therefore, we propose to limit the number of CBT calls during the population evolving procedure.

5   Implementation

We have implemented an integer version (with an integer solver) of our approach CSBT by using a new implementation of our CBT [15] and by extending eToc [20]. Fig. 6 shows the overview of the implementation, built over six components: Program instrumentor, Constraint models Generator, CBT Core, a CEO, a CPG, and SBT Core. The main components are CBT Core and SBT Core. These two components communicate via shared target and population. They identify sub-targets (branches) in the same way by using the Program Instrumentor component as a common preprocessing phase.

In the architecture, the SBT Core acts as the master and CBT Core plays the role of slave. When SBT Core needs test input candidates, it requests CBT Core by sending its current test target. To answer the request, CBT Core uses CEO or CPG and replies by sending a pseudo test input, proving the inaccessibility of the target, or by simply saying that the execution time limit is reached. Then SBT evolves its new population. This process is repeated until the test target is covered or some condition limit is met.

5.1   SBT Core

We use eToc that implements a genetic algorithm to generate test cases for object oriented testing. eToc begins with instrumenting the UUT to identify the test target and to keep trace of the program execution. eToc generates a random initial population whose individuals are a sequence of methods. eToc follows the GA principle: to evolve its population it uses a crossover operator and four mutation operators. The main mutation operator that interests us mutates method arguments. To adapt eToc to our requirements, we modified the argument values generator from a random generator to a CPG. Thus, we modified one mutation operator to make it a CEO.

5.2   CBT Core

Constraint Models Generator. This component takes a UUT as input optionally instrumented with eToc program instrumentor. For each Java class method a specific structure of control flow graph is generated. In addition, using the Choco[1] language a constraint model for a relaxed version of this method is generated. All generated models constitute a CSP Models Bank that is used by CBT Core during test input generation.

CEO. CEO can be implemented as crossover operator, mutation operator, or neighbourhood generator. In this work, we implemented CEO as a mutation operator. With a predefined likelihood the genetic algorithm calls CEO by sending the methods under test, the current test target, current parameters values, and the parameter to change. CEO uses this information to choose the adequate CSP model and to fix the test target and parameters except those required to change. After that the model is solved and a new value is assigned to the requested parameter. If the solver does not return a solution then all parameters are assigned arbitrary values.

CPG. When the eToc genetic algorithm starts or restarts, randomly it generates its chromosomes (methods sequences), and then it calls CPG to generate parameters values. For each function in the population, this generator check a queue of pseudo test inputs, if there is a pseudo test input for this method, then the method's parameters are assigned and this pseudo input test is deleted. Otherwise the CPG generates a pseudo test input for each test target not yet

---

[1] Choco is an open source java constraint programming library.
  http://www.emn.fr/z-info/choco-solver/pdf/choco-presentation.pdf

covered in this method. One of these pseudo test inputs is immediately assigned to this method's parameters, the rest are pushed into the queue. During solving test target if a target is proved infeasible, then the generator drops it from the set of target to cover.

6   Empirical Study

The *goal* of our empirical study is to compare our proposed approach against previous work and identify the best variant among the two proposed techniques in Section 4.2 and a combination thereof. The *quality focus* is the performance of the CPG technique, the CEO technique, CPG+CEO, CBT, and SBT to generate test inputs set that covers all-branches. The *context* of our research includes three case studies: Integer, BitSet, and ArithmeticUtils were taken from the *Java* standard library and *Apache Commons project*[2]. BitSet was previously used in evaluating white-box test generation tools [20,7]. These classes have around 1000 lines of java code. The SBT used (eToc) was not able to manage array structures and long type. We fixed the long type limitation by using *int* type instead, but we had to avoid testing methods over array structures. Therefore, we tested the whole *BitSet* and *ArithmeticUtils*, but only a part from *Integer* because it contains array structure an input data for some methods. The number of all branches is 285: 38 in Integer, 145 in BitSet, and 102 in ArithmeticUtils. These classes were chosen specifically because they contain function calls, loops, nested predicates, and complex arithmetic operations over integers.

6.1   Research Questions

This case study aims at answering the following three research questions:

RQ1: Can our approach CBST boost the SBT performance in terms of branch coverage and runtime? This question shows the applicability and the usefulness of our approach.

RQ2: When and at what order of magnitude is using our approach useful for SBT? This question shows the effectiveness of our approach.

RQ3: Which of the three proposed techniques is best suited to generate test inputs in an efficient way?

6.2   Parameters

As shown in the running example, during the empirical study we observed that the difference, in terms of fitness calculations, between CSBT and SBT is very large. We think that comparing approaches using this metric is unfair because CSBT uses CBT to reduce the number of fitness calculations. So to provide a fair comparison across the five approaches (CBT, SBT, CPG, CEO, CPG+CEO), we measure the cumulative branch coverage achieved by these approaches every

---

[2] The Apache Software Foundation. http://commons.apache.org

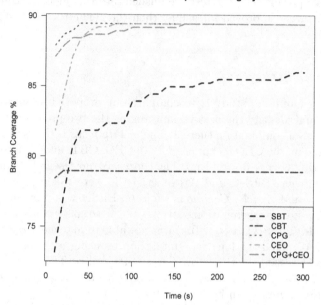

Fig. 7. Comparing all techniques on Integer.java

10 seconds for a sufficient period of *runtime* (300 s). This period was empirically determined as sufficient for all approaches. To reduce the random aspect in the observed values, we repeated each computation 10 times. We think that the default integer domain in eToc $[-100, 100]$ is unrealistically small so to get a meaningful empirical study, we chose to fix the domain for all the input variables to a larger domain $[-2 \times 10^4, 2 \times 10^4]$. We kept the rest of eToc's default parameters as is. We used identical parameters values for all techniques.

For CBT, the solver uses the minDom variable selection heuristic on the CSP variables that represent CFG nodes as a first goal and on CSP variables that represent parameters as a second goal. The variable value is selected randomly from the reduced domain. To make the study scalable, we restrict the solver runtime per test target to 500 ms. This avoids the solver hanging or consuming a large amount of time.

6.3 Analysis

Fig. 7 summarizes the branch coverage percentage in terms of execution time for the BitSet class. Overall, the CPG technique is the most effective, achieving 89.5% branch coverage in less than 40 s. Also, CEO and CPG+CEO reach 89.5% but after 70 s. eToc was unable to go beyond 85.78%, which is attained after 120 s. CBT performs badly, its best attained coverage is 78.94%. This figure shows that the proposed three techniques can improve the efficiency of eToc in terms of percentage coverage and execution time.

Comparing all techniques on all classes

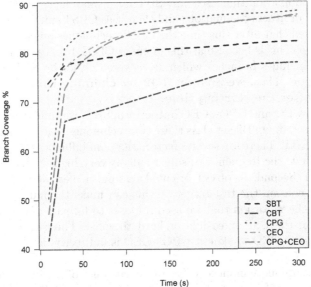

Fig. 8. Comparing all techniques on all the three java classes

We analysed branch coverage percentage in terms of execution time on BitSet and on ArithmeticUtils. We got two graphics that resemble Fig. 7 with a slight difference: eToc starts better than the three proposed techniques for the first twenty seconds, but CPG rapidly makes up for lost time outperforming eToc just after 20 s. Also CEO and CPG+CEO outperformed eToc, but they needed a little more time especially on ArithmeticUtils. CBT performs always worse than even the weakest proposed techniques.

Finally, Fig. 8 reflects the achieved results for all classes Integer, BitSet, and ArithmeticUtils. It confirms that the CPG technique is the most effective, and eToc starts better than the proposed techniques during the first twenty seconds.

6.4 Study Discussions

For the given classes, it is clear that CPG outperforms all techniques in terms of execution time and branch coverage. Also CEO and CPG+CEO perform better than eToc. Therefore, the proposed techniques boost SBT implemented in eToc. This result may be due to the kind of UUTs tried, which essentially use integer data types. More evidence is needed to verify whether the advantage of the proposed techniques represents a general trend. Yet, on the selected UUTs and the tool eToc **we answer RQ1 by claiming that the CBST techniques can boost the SBT.**

Over almost all graphics, we observed that eToc starts better than the proposed techniques. Therefore the latter are not useful for the first twenty seconds.

This behaviour is due to the frequency of solver calls: at the start time all targets are not yet covered even the easiest one which can be covered randomly. Therefore, a combination that uses only SBT for a small lapse of time or until a certain number of fitness calculations and then uses CBST, may perform better. Also, we observed that after this time the CBST techniques quickly reach a high level of coverage. This is because at this moment CBST takes advantage of both approaches: it includes branches which are covered either by SBT, CBT or by their combination. Thus, **we answer RQ2 by claiming that the CBST is more useful after the starting time**.

Even though CEO and CPG+CEO outperform SBT, we think that they don't perform as expected. On BitSet class these two techniques take a significant time before beating SBT. There are several factors that can influence the performance of the CEO. First, the frequency of solver calls is very high; it makes a call for every mutation. Second, in object oriented testing, a method under test does not necessarily contain the test target — this can make the mutation operator useless. Third, the mutation that we used imposes to fix part of the parameters which can make the CSP infeasible or hard to solve. These three factors are the main sources of CEO weakness. CPG+CEO is indirectly influenced by these factors by using CEO.

The CPG technique enhanced eToc by an average of 6.88% on all classes: 3.60% on Integer, 5.65% on BitSet, and 11.37% on ArithmeticUtils. These values did not take into account the proved infeasible branches: just on ArithmeticUtils CPG has proved 4 infeasible branches. We confirmed manually that these branches were infeasible because they use in their predicates some values out of the domain used. According to [7] the branches not covered by eToc are very difficult to cover. Therefore, a percentage ranging between 3.6% and 11.37% is a good performance for CPG. Thus, **we answer RQ3 by claiming that CPG is more useful to generate test inputs**.

6.5   Threats to Validity

The results showed the importance of using CSBT to generate test data, especially to improve the SBT performance in terms of runtime and coverage.

Yet, several threats potentially impact the validity of the results of our empirical study. A potential source of bias comes from the natural behaviour of any search based approach: the random aspect in the observed values. This can influence the internal validity of the experiments. In general, to overcome this problem, the approach should be applied multiple times on samples with a reasonable size. In our empirical study, each experiment took 300 seconds and was repeated 10 times. The coverage was traced every 10 s. The observed values become stable after 120 s. Each tested class contains around 1,000 LOCs and more than 100 branches. Therefore, experiments provided a reasonable size of data from which we can draw some conclusions, but more experiments are strongly recommended to confirm or refute such conclusions.

Another source of bias comes from the eToc genetic algorithm parameters: we didn't try different combinations of parameters values to show empirically that the approach is robust to eToc parameters. This can affect the internal validity of our empirical study.

Another potential source of bias includes the selection of the classes used in the empirical study, which could potentially affect its external validity. The BitSet class has been used to evaluate different structural testing approaches before [20,7], thus it is a good candidate for comparing the different proposed techniques. The two other classes were chosen because they feature integers and because they contain common problems in CBT and SBT (e.g, path that contains nested predicates and native function calls) and they represent widely used classes with non-trivial sizes.

7 Related Work

Several approaches have been proposed to use CBT or combine it with SBT in order to solve some input test generation problems, but they apply CBT on a complete version of the UUT [7,9,10]. All these approaches are limited by the size and the complexity of the UUT, and the fact that the input test generation problem is undecidable (reachability problem).

EVACON [7] was the first tool proposed that combines a CBT approach and a SBT approach. It bridges eToc [20] and jCute [16] to generate test data for classes. eToc is used to generate method sequences and jCute is used to generate test inputs. In this approach CBT and SBT work in a cooperative way: each of them has a separate task. CBT is dedicated to test input generation. In this way both approaches can complete each other but cannot solve common problems. In contrast, in our approach CBT and SBT work in a collaborative way: CBT starts the task by generating a pseudo test input and SBT complete this task by evolving pseudo test input and then generating the actual test input. A common problem is solved partially by using CBT, and then is completed by using SBT.

To solve constraints over floating point, the CBT tool PEX [19] has been extended by using FloPSy [9] which is a SBT approach that solves floating point constraints. FloPSy deals with a specific issue by using SBT to help CBT in solving constraints over floating point. In contrast our approach works in the opposite direction: it uses CBT to help SBT to improve its performance in terms of time and coverage.

Lakhotia et al. [1] propose a fitness function based on symbolic execution. The proposed fitness function analyses and approximates symbolic execution paths: some variables are approximated and any branches involving these variables are ignored. Then, the fitness value is computed based on the number of ignored conditions and the path distance (branch distance). This fitness function uses constraint programming to enhance SBT. Our approach is different because we propose an initial population generator and some evolution operators. Their fitness and our approach enhance SBT by using CBT, but in different ways. This fitness can be used in the same framework with our approach.

Recently, J. Malburg and G. Fraser [10] proposed a hybrid approach that combines GA and DSE on Java PathFinder. This approach uses GA to generate the test inputs; during the GA evolution the approach calls CBT to explore a new part in the program. The CBT is used as a mutation operator which uses SE to derive a constraint path, and then it uses DSE to solve this constraint. Therefore, the test data that are generated based on the combination are actually generated by using DSE. As explained in Section 2 DSE uses concrete values to simplify complex expressions. It is well known that these concrete values may make the path constraint infeasible. In general, these concrete values are randomly chosen, in which case DSE falls in a random search. In contrast our CEO uses SBT to find the values of complex expressions. In addition in our approach all actual test inputs are generated by using SBT.

Our approach differs from previous work in that it provides a general framework for combining a SBT and a CBT to generate test inputs. It uses CBT to improve SBT, i.e., CBT is used to reduce the SBT search space. The initial population of SBT is generated using CBT solutions domain; and SBT evolution procedure is constrained to evolve the population in the CBT solutions space. Test inputs are generated using an incremental combination: CBT proposes pseudo test inputs, SBT materializes test inputs.

8   Conclusion

In this paper, we presented a novel combination of SBT and CBT to generate test inputs, called CBST. The novelties of our approach lie in its use of a relaxed version of a UUT with CBT and of CBT solutions (pseudo test inputs) as test input candidates in service of SBT. We identified three main points where CBT can be useful for SBT. For each point we proposed a technique of combination: a CPG that uses CBT to generate test input candidates for SBT; and a CEO that uses CBT to evolve test input candidates. We implemented a prototype of this approach. Then we compared three variants of CBST with CBT and SBT. Results of this comparison showed that CPG outperforms all techniques in terms of runtime and branch coverage. It is able to reach 89.5% branch coverage in less than 40 s. Also CEO and CPG+CEO perform better than SBT in terms of branch coverage. The obtained results are promising but more experiments must be performed using different sort of solvers (String, floating point) to confirm if the absolute advantage of the proposed techniques represents a general trend.

In the future, we will focus on extending our approach by exploring new combination techniques. In particular, we would like to enhance our technique CEO and to use several solvers at the same time.

References

1. Baars, A., Harman, M., Hassoun, Y., Lakhotia, K., McMinn, P., Tonella, P., Vos, T.: Symbolic search-based testing. In: 26th IEEE/ACM International Conference on ASE, pp. 53 –62 (November 2011)
2. Clarke, L.: A system to generate test data and symbolically execute programs. IEEE Transactions on Software Engineering, SE 2(3), 215–222 (1976)
3. Collavizza, H., Rueher, M., Hentenryck, P.V.: Cpbpv: a constraint-programming framework for bounded program verification. Constraints 15, 238–264 (2010)

4. Godefroid, P., Klarlund, N., Sen, K.: Dart: directed automated random testing. SIGPLAN Not. 40, 213–223 (2005)
5. Gotlieb, A.: Euclide: A constraint-based testing framework for critical c programs. In: ICST, pp. 151–160 (2009)
6. Ince, D.C.: The automatic generation of test data. The Computer Journal 30(1), 63–69 (1987)
7. Inkumsah, K., Xie, T.: Evacon: A framework for integrating evolutionary and concolic testing for object-oriented programs. In: Proc. 22nd IEEE/ACM ASE, pp. 425–428 (November 2007)
8. Khichane, M., Albert, P., Solnon, C.: Strong Combination of Ant Colony Optimization with Constraint Programming Optimization. In: Lodi, A., Milano, M., Toth, P. (eds.) CPAIOR 2010. LNCS, vol. 6140, pp. 232–245. Springer, Heidelberg (2010)
9. Lakhotia, K., Tillmann, N., Harman, M., de Halleux, J.: FloPSy - Search-Based Floating Point Constraint Solving for Symbolic Execution. In: Petrenko, A., Simão, A., Maldonado, J.C. (eds.) ICTSS 2010. LNCS, vol. 6435, pp. 142–157. Springer, Heidelberg (2010)
10. Malburg, J., Fraser, G.: Combining search-based and constraint-based testing. In: Proceedings of the 2011 International Symposium on Software Testing and Analysis, ISSTA 2011. ACM, New York (2011)
11. McMinn, P.: Search-based software test data generation: a survey. Software Testing Verification & Reliability 14, 105–156 (2004)
12. Miller, W., Spooner, D.: Automatic generation of floating-point test data. IEEE Transactions on Software Engineering, SE 2(3), 223–226 (1976)
13. Myers, G.J.: The art of software testing. John Wiley and Sons (1979)
14. Păsăreanu, C.S., Rungta, N.: Symbolic pathfinder: symbolic execution of java bytecode. In: Proceedings of the IEEE/ACM International Conference on Automated Software Engineering, ASE 2010, pp. 179–180. ACM, New York (2010)
15. Sakti, A., Guéhéneuc, Y.G., Pesant, G.: Cp-sst: approche basée sur la programmation par contraintes pour le test structurel du logiciel. In: Septitièmes Journées Francophones de Programmation par Contraintes (JFPC), pp. 289–298 (June 2011)
16. Sen, K., Agha, G.: CUTE and jCUTE: Concolic Unit Testing and Explicit Path Model-Checking Tools. In: Ball, T., Jones, R.B. (eds.) CAV 2006. LNCS, vol. 4144, pp. 419–423. Springer, Heidelberg (2006)
17. Sen, K., Marinov, D., Agha, G.: Cute: a concolic unit testing engine for c. SIGSOFT Softw. Eng. Notes 30, 263–272 (2005)
18. Staats, M., Păsăreanu, C.: Parallel symbolic execution for structural test generation. In: Proceedings of the 19th international Symposium on Software Testing and Analysis, ISSTA 2010, pp. 183–194. ACM, New York (2010)
19. Tillmann, N., de Halleux, J.: Pex–White Box Test Generation for.NET. In: Beckert, B., Hähnle, R. (eds.) TAP 2008. LNCS, vol. 4966, pp. 134–153. Springer, Heidelberg (2008)
20. Tonella, P.: Evolutionary testing of classes. SIGSOFT Softw. Eng. Notes 29(4), 119–128 (2004)

Domain-Driven Reduction Optimization
of Recovered Business Processes

Alex Tomasi, Alessandro Marchetto, and Chiara Di Francescomarino

FBK-CIT, Trento, Italy
{aletomasi,marchetto,dfmchiara}@fbk.eu

Abstract. Process models play a key role in taking decisions when existing procedures and systems need to be changed and improved. However, these models are often not available or not aligned with the actual process implementation. In these cases, process model recovery techniques can be applied to analyze the existing system implementation and capture the underlying business process models. Several techniques have been proposed in the literature to recover business processes, although the resulting processes are often complex, intricate and thus difficult to understand for business analysts.

In this paper, we propose a process reduction technique based on multi-objective optimization, which minimizes at the same time process complexity, non-conformances, and loss of business content. This allows us to improve the process model understandability by decreasing its structural complexity, while preserving the completeness of the described business and domain-specific information. We conducted a case study based on a real-life e-commerce system. Results indicate that by balancing complexity, conformance and business content our technique produces understandable and meaningful reduced process models.

Keywords: Business Process Recovery, Multi-Objective Optimization, and Ontology.

1   Introduction

Business process models are often used to drive the evolution of the associated informative software system. Evolving a system without an accurate and faithful documentation (e.g., a model) of the underlying process, in fact, could lead to a quite difficult and time consuming activity. However, in most of the cases, an adequate and accurate documentation of the running software system does not exist or it is not aligned with the actual system implementation. In these cases, one option is trying to reconstruct the missing or inaccurate documentation through process model recovery and mining.

Several works [1,6] in the literature propose techniques to recover process models starting from the analysis of different artifacts, such as the source code or execution traces collected in log files. In our previous work [6], we used dynamic analysis to recover processes realized through Web systems. The typical problems to face when process models are recovered from logs include: (i) models' size and complexity; (ii) their under-generalization: models may describe only a subset of the actual system behaviors; and (iii) their over-generalization: recovery algorithms may generalize the actual observations beyond the possible behaviors. The use of a large set of traces and of an

G. Fraser (Ed.): SSBSE 2012, LNCS 7515, pp. 228–243, 2012.

appropriate algorithm can limit the impact of problems (ii) and (iii), but state of the art techniques have still a hard time with problem (i). In fact, process recovery and mining tools tend to produce process models that are quite difficult to understand, because they are overly complex and intricate (they are also called "spaghetti" processes [15]).

Existing approaches dealing with the process model size and complexity problem belong to two groups: clustering and frequency based filtering. Clustering takes advantage of the possibility of modularization offered by sub-processes [6]. Frequency-based process filtering removes process elements that are rarely executed [14]. Empirical evidence [10] shows that modularity positively affects process understandability, but its effect may be negligible due to factors such as the process domain, structure and complexity. Pruning processes according to the execution frequency [14] is useful to remove noise, but may lead to overly simplified processes, that do not capture the business domain and application-specific properties and activities.

In this work we reformulate the process reduction technique as an intrinsically multi-objective optimization problem. The reduced process models (having lower size and complexity) are expected to be simpler to understand and manage than unreduced ones, though at the same time they should maximize their capability to represent possible process flows, in a way that is meaningful and expressive for business analysts. This work extends our previous work [9] by explicitly considering the business content of the recovered process models (see Sec. 6). We, hence, propose an approach to reduce the recovered business process model by balancing: (1) the model complexity; (2) its non-conformances (i.e., inability to represent some execution traces); and (3) its loss of business content (i.e., its inability to preserve business domain content). The underlying assumption is that limiting, at the same time, non-conformance and loss of business content we can reduce (i.e., cut) the process model preserving its capability of representing business relevant flows and information, thus improving readability and effectiveness of reduced business process models. The proposed technique has been implemented in a tool and applied to a real-life e-commerce Web application. Results indicate that balancing complexity, conformance and business content leads to an improvement of the recovered business processes.

2 Business Process Reduction

Recovered process models are reduced by removing some process elements (e.g., sequence flows and activities), with the aim of limiting process size and complexity, while preserving the completeness of the described business and domain-specific information. In our previous work [9], we formally defined the notion of process reduction. For sake of completeness, we summarize here the main intuition about this notion.

By *business process reduction* we mean a sequence of *atomic reduction operations*, each basically consisting of the removal of a direct connection between a pair of process activities, possibly followed by a cascade of further process element removals, occurring whenever an entire sub-process becomes unreachable from the start due to the initial removal. To perform business process reduction, we rely upon an intermediate process representation, called *activity graph*. The mapping between the original process model P and the activity graph AG_P must be invertible, since we want to be able to convert the reduced activity graph back to the original process.

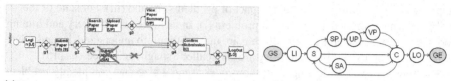

(a) Process models: unreduced (with) and reduced (without crossed   (b) Activity graph for the unreduced process
elements) models

Fig. 1. Paper submission system: BPMN process models and activity graph

The process model (obtained by ignoring the three crosses on the process elements) in Figure 1(a) represents the process for an on-line submission system. Such a process, *unreduced*, is described in BPMN[1], the process modeling notation used in this work for representing models. The corresponding activity graph (Figure 1(b)) contains a node for each activity/gateway of the process and two nodes (named GS and GE) representing the start and end event of the process. In the graph, the removal of the edge (S, SA), makes node SA unreachable from the start node. By reachability analysis, we can easily determine that node SA as well as its outgoing edge (SA, C) must be also removed, since unreachable from GS. The corresponding reduced BPMN process, P_r, is the one reported in Figure 1(a), in which the three crossed BPMN elements (i.e., the sequence flow $(g2, SA)$, the process activity *Submit Abstract* and its outgoing sequence flow $(SA, g4)$) have been removed.

3 Process Metrics

To evaluate the quality of recovered process models, three main factors have been identified in the literature [4,11]: *complexity*, *conformance* and *business content*.

- **Process Complexity.** The existing literature [4] proposes to measure the process complexity by resorting to the cyclomatic complexity, commonly used with software, and adapting it for processes. Hence, given a process model P, we measure its control-flow complexity $CFC(P)$ as: $CFC(P) = \sum_{g \in G(P) \wedge FOUT(g) > 1} FOUT(g)$; where $G(P)$ is the set of all process control-flow elements (gateways, in BPMN) and $FOUT(g)$ is the number of the sequence flows outgoing from g (fanout). We take into account only decision points of the process, i.e., elements with fanout greater than one. A high value of CFC indicates a high number of alternative execution flows, thus denoting a process potentially difficult to read and understand for the analyst.

For example, the online submission process *unreduced* in Figure 1 contains five gateways, one with fanout 3, two with fanout 2 and two with fanout 1. The resulting process $CFC(unreduced)$ is hence 7. For the reduced process P_r in Figure 1, i.e., the one without the crossed elements, $CFC(P_r)$ is 6.

- **Process Conformance.** The conformance of a process model is its ability to reproduce the execution traces [11]. A trace is a sequence of events executed during the run of a process and stored in a log file. Table 1 shows some examples of execution traces for the submission process in Figure 1. For instance, trace t_3 represents an execution in which

---
[1] http://www.omg.org/spec/BPMN/1.2

the user logs into the submission system (activity *LI*), submits the information about the paper (*S*), confirms the submission (*C*) and logs out (*LO*). Such a trace is admitted by the model in Figure 1 and it hence positively contributes to the process model conformance to the traces. Conversely, it negatively contributes to the model non-conformance, which we would like to minimize together with the model complexity. To compute the non-conformance $NConf(P)$ for a process P, we interpret the model as a parser that can either accept or refuse a trace, regarded as its input string. A process model parses a trace if there is an execution flow in the process model that complies with the process model semantics and consists of the sequence of events in the trace. We assume that the parser associated with a process model is a *robust* parser, i.e., it has a parsing resume strategy which allows the parser to skip an input subsequence that cannot be accepted and to resume parsing from a successive input symbol. Specifically, if the trace consists of three parts $t =< p, m, s >$, such that the prefix p can be parsed and brings the process model execution to state S while the subsequence m cannot be accepted from state S, a resume strategy is a rule that transforms the process state S into S', such that the suffix s can be parsed from S'. Usually, the resume strategies implemented by parsers are heuristic rules to modify the state S by performing minimal changes to it, until an accepting state S' is reached.

Table 1. Example of execution traces

Trace	Event sequence	Trace	Event sequence
t_1	< LI, S, SP, UP, C, LO >	t_2	< LI, S, SA, C, LO >
t_3	< LI, S, C, LO >	t_4	< LI, S, SP, UP, VP, C, S, SP, UP, C, LO >
t_5	< LI, S, SP, UP, C, S, SA, C, LO >		

Given a process model P and a trace $t \in T$ (the set of traces), we start parsing t with P in the start state. When event $e \in t$ is provided as input to process P (in the execution state S), if there does not exist any transition $S \longmapsto_e S'$ in P and there does not exist any resume rule R such that $S \longmapsto_R S' \longmapsto_e S''$, we recognize a non conformance $\langle S, e \rangle$, which is added to the non-conformance set. The size of the non-conformance set after trying to reproduce all the traces in T is our non-conformance metrics $NConf$. It should be noticed that if the same non-conformance $\langle S, e \rangle$ occurs multiple times (it appears in many traces) it contributes only as a +1 to $NConf$.

Under the assumption that the process model being considered contains only exclusive gateways (which is the case of the process recovery tool used in our study [6]), the state of the process model execution when event e is provided as input consists of the last accepted activity. Hence, when the input activity b cannot be accepted by P, the preceding input activity a in the trace is considered. If the process model P contains the edge (a, b), parsing is resumed from state a. $NConf$ is computed as follows:

$$NConf(P) = \left| \bigcup_{t \in T} \{(a,b) | (a,b) \in t \land dc(a,b) \notin P\} \right|$$

where $dc(a, b)$ indicates the existence of a *direct connection* (i.e., an edge or a path containing only gateways as intermediate elements) between a and b in the process model P. Hence, we measure the number of unique transitions in the traces that are not reproducible in the process model. Let us consider the trace $t_2 = \langle LI, S, SA, C, LO \rangle$ from Table 1. The reduced process model P_r in Figure 1 accepts the first two events, but then SA is not even present in the reduced process model. Thus, $(S, SA), (SA, C)$ are non-conformances. However, when LO is considered as input, parsing can restart from C according to the resume rule given above, such that (C, LO) is accepted. The final value of $NConf$ is 2 for this reduced process. $NConf(P)$ is 0 if the process model reproduces all traces, while a high value of $NConf(P)$ indicates that the process model is not able to reproduce many transitions in the traces.

If the considered process model P contains other types of gateways, in addition to the exclusive ones, we need to resort to a more general definition of $NConf(P)$ that requires more complex resume rules such as the one proposed by Rozinat et al. [11].

- Process Business Content. The *business content* of a process model represents the amount of information, related to a specific business domain, which it conveys. We assume that such a business domain can be described by means of an ontology. Indeed, *domain ontologies* are often used to model specific domain of interest thanks to their capability of providing semantics to the entities of a specific domain. For example, RosettaNet[2] is used in the purchase-order domain. The domain ontology, moreover, can be also used to elicit the semantics of the process activities in terms of the business domain and hence to evaluate the business domain information contained in the process model, i.e., its business content. To estimate the amount of business content of the recovered processes two preliminary steps are required:

(1) *Domain ontology construction/retrieval*: A business domain ontology, which models the business domain of the process is created or retrieved, if it already exists. By means of concepts, their hierarchical organization according to the "is-a" relationship, and properties, the domain ontology provides semantics to the domain entities and allows to formally describe the specific business domain of interest (e.g., the e-commerce domain).

(2) *Process Semantic Annotation*: The business domain semantics of the process activities is made explicit by associating to each of them the corresponding concept in the ontology, i.e., the domain concept that provides the semantics of the activity with respect to the business domain. The process activities are hence enriched/annotated with semantic information (e.g., [12], [7]). This operation is called *semantic annotation* of process models and it is conducted, for example, to support business analysts in documenting, sharing and reasoning on processes and models. Indeed, by providing a precise semantics to process elements, it overcomes the pure syntactical level allowing to reach the semantic one.

For instance, Figure 2 (left) shows a fragment of a possible ontology for the paper submission domain, that could be used to semantically annotate the process in Figure 1. Figure 2 (right) shows the resulting annotated process: we exploited the BPMN textual annotations by prefixing the annotation concept with a "@" symbol, to represent the association between activities and concepts.

Since the process semantic annotation can be consuming both in terms of time and effort, approaches have been proposed in the literature to support business analysts in

---

[2] http://www.w3.org/2002/ws/sawsdl/spec/ontology/rosetta.owl

this task. For example, in [8], the automatic semantic annotation of process activities is investigated. In detail, by exploiting a similarity metric between activity label short sentences and ontology concept names, the proposed technique allows to automatically establish associations between activity labels and ontology concepts. Whenever no concept in the ontology satisfies the similarity threshold required to annotate an activity, new concepts can be suggested by the technique and added to the ontology, thus allowing to annotate the considered activity; if instead the short label of the activity is not recognized as meaningful, the activity remain unannotated[3].

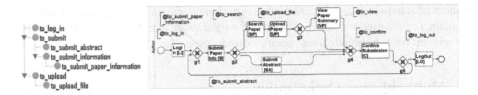

Fig. 2. Paper submission system: ontology (left) and semantically annotated process (right)

Given the domain ontology, which describes the business domain of the process and the association between ontology concepts and process model activities, the business content BC of a process model P can be computed as the sum of the values of the business content associated to the process model activities: $BC(P) = \sum_{a \in Activities(P)} bcv(a)$. In detail, for each activity $a \in Activities(P)$, the value of its business content $bcv(a)$ is approximated with the depth of the corresponding concept in the domain ontology; if the activity has no ontology concept associated (i.e., the semantics of activity does not coincide with any business domain entity), $bcv(a)$ is 0. The reason behind the choice to consider the depth of a concept (i.e., its distance from the root concept in the ontology) as an indicator of its business value is that concepts placed deep in the hierarchy of concepts characterizing ontologies (e.g., the leaves), are more specific and hence contain more business information, than generic ones. In case more than one parent exists for a concept (i.e., multiple inheritance) the maximal depth of the parent concepts is used to compute the business content for the corresponding activity.

As done for the conformance, we consider the complement of the business content as metric to be minimized, that is the loss of business content (with respect to the unreduced process model) LBC. LBC is computed as follows: $LBC(P) = BC(unreduced) - BC(P)$; where $BC(unreduced)$ is the business information conveyed by the unreduced process model (indeed, the process ideally containing the whole information also contains the business relevant knowledge); and $BC(P)$ is the business content of the considered reduced model.

For example, considering the *unreduced* process semantically annotated in Figure 2 (right) and assuming that all the concepts used for annotating the activities that do not appear in the ontology fragment in Figure 2 (left) have depth equals to 1, the business content of the process is 12. One activity (*Submit Paper Info*) has in fact business value 3, two activities business value 2 and the remaining five have business value 1. Since this process is the unreduced one, $LBC = 0$, i.e. there is no loss of business content.

---

[3] A detailed introduction about the process semantic annotation technique used is in [8].

The business content of the reduced process P_r in Figure 1, instead is $LBC(P_r) = 10$ (the only activity in the unreduced process that does not appear in *unreduced* has an associated concept with depth equals to 1).

The idea underlying the metric is that the business content of a process can be computed by measuring how well the process "covers" the domain-specific concepts associated to its activities and hence how much business content it conveys. The loss of business content, on the contrary, measures how much business information has been lost by reducing the model. This metric is considered, when reducing a process model, with the aim of preserving as much as possible the business domain activities, i.e., the annotated activities, while discarding thoses without a semantics related to the business domain, i.e., the non-annotated ones, though we are aware that these latter ones can still be relevant for some application-specific business aspects.

4 Multi-objective Optimization

We use multi-objective optimization to produce a reduced process model that minimizes process complexity, non-conformance, and loss of business content. Specifically, we rely on the Non-dominated Sorting Genetic Algorithm II (NSGA-II, Deb et al. [5]).

NSGA-II uses a set of genetic operators (crossover, mutation, selection) to iteratively evolve an initial population of candidate solutions (i.e., processes). The evolution is guided by an objective function (fitness function) that evaluates the quality of each candidate solution along the considered dimensions. In each iteration, the Pareto front of the best alternative solutions is generated from the evolved population. The front contains the set of non-dominated solutions, i.e., those solutions that are not inferior to any other solution in *all* considered dimensions. Population evolution is iterated until a maximum number of iterations is reached.

In our case, the obtained Pareto front represents the optimal balance among complexity, non-conformance and loss of business content determined by the algorithm. The business analyst can inspect the Pareto front to find the best compromise among the three dimensions, e.g., having a model that conforms to the observed traces or with the maximum business content, but quite complex; or having a simpler model, but less adherent to reality or loosing relevant business information.

Solution Encoding: In our instantiation of the algorithm, a candidate solution is an activity graph (representing a process) in which some edges are kept and some are removed. We represent such a solution by means of a standard binary encoding, i.e., a binary vector. The length of the vector is the number of edges in the activity graph extracted from the unreduced process model. While the binary vector for the unreduced process is entirely set to 1, for a reduced process each binary vector element is set to 1 when the associated activity graph edge is kept; it is set to 0 otherwise. For instance, the encoding of the unreduced submission process (Figure 1) consists of a vector of 13 elements (i.e., all the edges of the activity graph), all set to 1. In the vector encoding the reduced process in Figure 1, three bits (representing the edges *(S, C)*, *(S, SA)* and *(SA, C)*) are 0.

Initialization: We initialize the starting population in two ways, either randomly or resorting to the frequency-based edge-filtering heuristics. Random initialization consists of the generation of random binary vectors for the individuals in the initial population. The frequency based edge-filtering heuristics consists of removing (i.e., flipping from 1 to 0) the edges that have a frequency of occurrence in the traces below a given, user-defined threshold [14]. An initial set population has been obtained by assigning to the frequency threshold (used to decide which edges to filter) all possible values between minimum and maximum frequencies among those of the edges in the initial process.

Genetic Operators: NSGA-II resorts to three genetic operators for the evolution of the population: mutation, crossover and selection. As mutation operator, we used the bit-flip mutation: one randomly chosen bit of the solution is swapped. The adopted crossover operator is the one-point crossover: a pair of solutions is recombined by cutting the two binary vectors at a randomly chosen (intermediate) point and swapping the tails of the two cut vectors. The selection operator we used is binary tournament: two solutions are randomly chosen and the fitter of the two is the one that survives in the next population.

Fitness Functions: Our objective functions are the three metrics CFC, $NConf$, and LBC measuring respectively the process complexity, non-conformance, and loss of business content.

5   Case Study

We implemented our process recovery algorithm in a set of tools[4]: *JWebTracer* [6] traces the run of a Web application; *JBPRecovery* [6] infers a BPMN model; *BPOntoManger* [7] supports the ontology construction and process semantic annotation; *JBPFreqReducer* [9] implements the frequency-based heuristics Fbr [14]; and *JBPEvo2* (extension of *JBPEvo* [9]) implements the presented multi-objective optimization (called $MGA2$).

We applied our tools to a real-life e-commerce application: Softslate Commerce[5]. Softslate is a Java shopping cart application that allows companies to manage their on-line stores. It consists of more than 200k lines of Java/JSP code, it uses several frameworks (e.g., Struts, Wsdl4j), and it can be interfaced with database managers (e.g., MySql)

The aim of the case study was answering the following research questions investigating effectiveness and viability of the process reduction techniques in making the recovered processes more understandable and manageable.

- *RQ1: Does the shape of the Pareto fronts offer a set of solutions which includes a wide range of tunable trade-offs among complexity, conformance and business-domain content?* It addresses the variety of different, alternative solutions produced by $MGA2$. In particular, the shape of the Pareto front and the density of the solutions in the front determine the possibility to choose among a wide range of interesting alternatives vs. the availability of a restricted number of choices. The Pareto front, in fact, might consist of points spread uniformly in the interesting regions, or it may be concentrated in limited, possibly uninteresting regions of the plots (e.g., near the unreduced process).

- *RQ2: Are the reduced processes in the Pareto front understandable and meaningful for business analysts?* It deals with the quality of the solutions produced by $MGA2$,

---

[4] http://selab.fbk.eu/marchetto/tools/rebpmn
[5] http://www.softslate.com/

as perceived by business analysts. We want to understand the quality of the processes reduced by $MGA2$ in terms of their business meaningfulness.

- *RQ3: How is the domain-specific ontology used to annotate recovered processes?* It deals with the impact on the produced processes of the quality, mainly in terms of completeness from a business perspective, of the domain ontology and of the process activities annotation.

The procedure followed in the experimentation consists of: (step1) *JWebTracer* has been used to trace 30 executions of Softslate Commerce (involving, on average, 15.6 user actions per execution). These executions exercise the main application functionality at least once. (step2) *JBPRecovery* has been used to build the *unreduced* process model. We obtained a process having 376 sequence flows, 140 activities and 77 gateways. (step3) *JBPFreqReducer* has been used to reduce the initial process model by applying the Fbr heuristic. A set of solutions has been obtained by varying the frequency threshold. (step4) *BPOntoManger* has been used to annotate the process activities with domain concepts. To this aim we used a domain ontology for e-commerce applications previously used by one of the authors in a different work [7]. This ontology has been expanded according to the suggestions automatically produced by *BPOntoManger* and checked by one of the authors. (step5) *JBPEvo2* has been ran with two initial populations: generated randomly ($MGA2r$) and by *JBPFreqReducer* ($MGA2f$). We ran $MGA2$ with the following setup: population size=100; we considered several iterations ranging from 100 to 1000000; crossover probability=0.9; mutation probability=1/376, where 376 is the number of sequence flows of the unreduced process considered in the case study.

5.1 Results

- **RQ1.** We plotted the process models in the Pareto fronts generated by the different iterations of $MGA2r$ and $MGA2f$. Figures 3 and 4 show some Pareto fronts we obtained and Table 2 summarizes data related to Fbr and to the processes in each front. We can observe that with a relatively small number of iterations $MGA2r$ generates Pareto fronts with a limited number of unique solutions (e.g., with 100 iterations it generates only 31 different solutions); while the increase of the number of iterations also enlarges the set and the distribution of unique solutions in the fronts. $MGA2r$, in fact, required at least 10000 iterations to generate a front containing a non-trivial number of solutions (in the range 70-90) distributed across the three considered dimensions. Conversely, with $MGA2f$, the number of unique solutions in the Pareto fronts is limited (in the range 20-25) for all the considered iterations (up to 1000000), though higher is the number of iterations, more distributed the solutions in the front are.

By inspecting the solutions, in fact, we found that, after 1000000 iterations: (i) $MGA2r$ generated a Pareto front that contains many solutions widely spread across the region with $50 < CFC < 130$, $80 < NConf < 160$, and $20 < LBC < 90$; while (ii) the Pareto front generated by $MGA2f$ contains less solutions than $MGA2r$, mainly spread across the region with $50 < CFC < 314$, $20 < NConf < 150$, and $10 < LBC < 130$. Both $MGA2r$ and $MGA2f$ improved the initial populations. In particular, $MGA2r$ showed substantial improvements even after a limited number of iterations (10000), though major improvements have been observed only after 1000000 iterations. $MGA2f$ instead required a large number of iterations, i.e., at least 300000, to substantially improve the initial population generated by Fbr.

Moreover, we can observe that: (i) processes generated by Fbr have higher CFC and lower $NConf$ and LBC than those reduced by $MGA2$; (ii) the processes generated by $MGA2r$ and $MGA2f$ are, on average, comparable in terms of CFC and LBC, while their $NConf$ differs ($MGA2f$ has $NConf$ lower than the one of $MGA2r$).

We can conclude that $MGA2$ potentially provides the analyst with a Pareto front which includes a wide range of tunable and effective solutions. However, to this aim, a high enough number of iterations is required to avoid potentially limited and sub-optimal solutions.

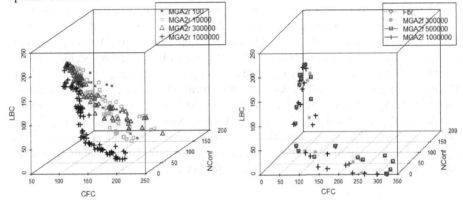

Fig. 3. $MGA2r$ Pareto fronts   **Fig. 4.** $MGA2f$ Pareto fronts

Table 2. Process and times in Pareto fronts. The execution time of each algorithm has been computed on a desktop PC with an Intel(R) Core(TM) 2 Duo CPU working at 2.66GHz and with 3GB of RAM memory.

Algo	Iter	Proc.	Unique Proc.	Time (min.)	Algo	Iter	Proc.	Unique Proc.	Time (min.)
$MGA2r$	100	100	31	0.08	Fbr	0	20	12	0.01
$MGA2r$	10000	100	72	2.86	$MGA2f$	300000	20	19	71.1
$MGA2r$	300000	98	80	70.9	$MGA2f$	500000	21	20	363.7
$MGA2r$	1000000	99	95	576.1	$MGA2f$	1000000	21	20	558.1

- **RQ2.** We analyzed the unreduced process model, the processes reduced by Fbr, and all the processes in the best Pareto fronts generated by $MGA2r$ and $MGA2f$. For each process we measured: (i) structural metrics: the number of process activities and gateways ($Size$) and the number of sequence flows (SF); (ii) CFC, $NConf$ and LBC; and (iii) the quality of the process model in terms of its *application-specific business content*, i.e., the quality, from the point of view of the analyst, of a process with respect to the application implementation.

In this study to measure the application-specific business content, we exploited the 18 core business activities ($asBacts$) and 23 core business properties ($asBprops$), defined by one of the authors of our previous work [9] (not involved in the case study and unaware of $MGA2$ and of the recovered processes) and that ideally should pertain to the business process of the application being modeled, i.e., Softslate. We only conducted a preliminary analysis on these sets of activities and processes by filtering out those related to application behaviors non-exercised when collecting the 30 execution traces (step1 of the case study procedure), e.g., activity $ac16$ *Update the password* has never been exercised this time. For instance, *Login* and *Checkout* (activities $ac11$ and $ac8$) are core application-specific business activities, i.e., activities that any analyst would expect to be modeled explicitly, while an example of a core application property is: the user must be logged-in in order to be allowed to complete the Checkout operation (property $pr21$). We assume that a good process model should represent such application-specific business activities and be compliant with the properties.

Table 3 (top) reports the metrics computed for the Softslate unreduced process; their median values for the process models reduced by Fbr, as well as for those in the best Pareto fronts of $MGA2r$ and $MGA2f$. It shows that the unreduced process captures all the business activities ($asBacts = 18$, that is 100% of application-specific business acitivities) while it fails in capturing some core properties ($asBprops = 17$, i.e., 73%). By inspecting the unreduced process we can confirm [9] that the reason for this is that the recovery algorithm we used (implemented by $JBPRecovery$) generalized the behaviors observed in execution traces, which contain noise and irrelevant details. This allows $JBPRecovery$ to capture more behaviors than the observed ones but it could result in an over-generalization of the possible application executions, making it potentially incompatible with some business properties (see [9] for details).

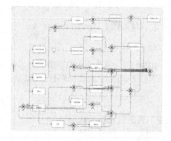 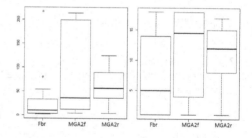

Fig. 5. *P5* generated by $MGA2f$ **Fig. 6.** Distribution of *Size* (left) and *asBacts* (rigth) of processes reduced by Fbr, $MGA2f$, $MGA2r$

In terms of complexity a big difference can be observed between unreduced and reduced processes: reduced models have, on average, substantially lower values of structural and complexity metrics than the unreduced one (e.g., Figure 5 shows process *P5*). However, this has a cost in terms of non-conformance and loss of business content, as shown in Table 3. Table 3 shows that, for example, the reduced models generated by Fbr have very low complexity and size, thus missing several application-specific business activities ($asBacts_{median} = 5$, i.e., 27%) and properties ($asBprops_{median} = 4$, i.e., 17%). $MGA2$ instead reduces the process models less than Fbr and preserves more application-specific business activities and properties (e.g., for $MGA2f$, the median value of core activities and properties is 14.5 and 13, i.e., 80% and 56%, respectively).

Nevertheless, differences exist also between the process models generated by $MGA2r$ and $MGA2f$. Figure 6 shows that, on average, the process models reduced by $MGA2f$ are less similar to one another than those reduced by $MGA2r$. The process models reduced by $MGA2f$, in fact, have lower and more variable average values of structural metrics (i.e., $Size, SF, CFC$) than the ones recovered by $MGA2r$. For instance, $Size(MGA2f)$ mainly ranges between 10 and 200 instead $Size(MGA2r)$ ranges between 40 and 90. This variability in the structural characteristics of process models is also reflected in a high variability in the number of business activities contained in the process and of enforced business properties. Most of the processes of $MGA2f$, in fact, contains from 5 to 18 business activities versus the range 9-14 of $MGA2r$ (Figure 6). Besides a lower complexity, models generated with $MGA2f$ also have, on average, non-conformance and loss of business content lower than those reduced by $MGA2r$, thus allowing $MGA2f$ to recover processes containing, on average, more application-specific business activities and enforcing a higher number of properties.

Table 3. Results for the unreduced process and (median value) for reduced process models

Algo	Unique Proc.	#Size	#SF	#CFC	NConf	LBC	asBacts	asBprops
unreduced	1	215	376	323	0	0	18	17
Median values for reduced process models								
Fbr	20	10	19	5	185.5	131.5	5	4
MGA2f	20	36	71	84	68	20	14.5	13
MGA2r	95	57	84	71	121	54	12	11
Examples of reduced process models								
MGA2f - P6	1	40	78	50	73	30	15	19
MGA2f - P12	1	177	318	270	30	15	17	16
MGA2r - P44	1	77	122	69	125	72	15	15
MGA2f - P85	1	96	160	71	124	29	17	20

Despite the good results obtained by $MGA2$ in terms of enforced business properties (Table 3, bottom, details the metrics computed for specific process models generated with $MGA2r$ and $MGA2f$), by looking at the generated processes we noticed that none of the considered reduced processes enforces all the 23 application-specific business properties. We confirmed the three main problems observed in our previous work [9]: (1) one or more activities involved in the property have been removed by the reduction algorithm (e.g., removal of activity *DeleteItem* $ac9$ from $P5$ causes the failure of $pr3$ and $pr5$); (2) a sequence flow that would allow to verify the business property has been removed by the reduction algorithm (e.g., removal of the sequence flow between the activities *Add To Cart* and *Save This Cart* causes the failure of $pr28$ in $P5$); (3) over-generalization, which as explained above limits the recovered processes in enforcing application-specific properties. Cases 1 and 2 are instances of under-generalization (i.e., the process does not represent some application behaviors), while case 3 is an over-generalization (i.e., the process represents some spurious application behaviors).

We analyzed the trend of under and over-generalization problems for the reduced processes, at decreasing complexity and increasing non-conformance and loss of business content (Table 3). Under-generalization problems grow at decreasing of process size and complexity; they, in fact, are strictly related with the level of non-conformance. In such cases, however, limiting the loss of business content can allow processes to preserve domain and application-specific business activities as well as to enforce business properties. Indeed, we observed that a decrease of the loss of business content for processes having comparable size, complexity and non-conformance could guarantee an increase of the capability of processes to preserve business activities and enforce business properties (e.g., *P44* and *P85* in Table 3). Over-generalization problems, instead, grow at increasing size and complexity, due to the excessive number of paths in the model. In such cases, a limited loss of business content does not seem to prevent the processes from cutting application-specific business activities and violating business properties. The reason is that business activities can be related to domain-specific business aspects (those captured in the domain ontology) and/or to aspects that are specific for the considered application (those hard-coded in the application implementation). For instance, the activity *Add To Cart* is a business activity for both the domain and the application. Its preservation, driven by LCA, will hence guarantee also the enforcement of application-specific properties (e.g., *pr2*). *Edit*, instead, is an application-specific business activity (it allows the user to configure her cart, e.g., the number of items to buy) but is not a business domain activity; it could hence be cut in the reduced models, thus preventing them to enforce 5 application-specific properties we defined that involves *Edit* (e.g., *pr3*). An enhancement in the quality and completeness of the domain ontology and process activity annotations could hence be beneficial for $MGA2$.

Summarizing, few application-specific business activities and properties result, on average, missing in the reduced processes, which makes them meaningful to business analysts, especially in comparison to the unreduced process and in case of strongly reduced processes. $MGA2$ is hence overall effective in optimizing the trade-off, i.e., in reducing the complexity of the process while preserving the business-relevant information captured by the process model.

- **RQ3**. The domain ontology we adopted in the study has been previously used by one of the authors for a different purpose [7]. It contains 149 concepts hierarchically organized in three layers (i.e., the maximal depth is 3): 76 concepts in layer 1, 49 in layer 2 and 24 in layer 3. In the ontology, 51 out of 149 (34%) concepts have been suggested by *BPOntoManager* and added (39 of them in layer 1, 9 in layer 2 and 3 in layer 3) after a confirmatory check by one of the authors. In detail, out of these new 51 concepts, 12 are completely new terms, while the remaining 39 are the composition of existing or new terms. *BPOntoManger* annotated 122 out of 150 (81%) activities of the unreduced process with different labels, using 34 (22%) unique concepts of this ontology (24 of layer 1 and 10 of layer 2); 16 out of the 34 concepts are among the 51 new concepts added to the initial ontology.

These numbers make evident that the domain ontology has been only partially used. The possible reasons for this result can be traced back, on one side to the level of abstraction of the domain ontology (too high for activity labels extracted from logs), and, on the other, to potential inaccuracies of the automatic semantic annotation of processes. Often, in fact, recovered labels are composed of non-meaningful terms that are difficult to analyze syntactically or with a semantics only known at application level (e.g., *Edit*). Further investigation will be devoted to support analysts in the ontology customization and to improve the algorithm used to semantically enrich recovered process models.

Threats to validity It is always hard to generalize the results obtained on a single case study. The selected application, however, seems to be representative of the e-commerce domain, where Web applications are used to implement the trading processes. Three threats to validity impact the procedure of the case study. The first concerns the used executions traces. Since different models can be recovered considering different sets of traces, we applied the functional coverage criterion for the selection of application executions to be used. The second threat concerns the stochastic nature of the $MGA2$ algorithm. We considered different runs of the algorithm and, obtaining comparable results, we detailed only one of them. Third threat concerns the used business domain ontology which can bring to achieve different results. To limit such a threat we resorted to an *On-line shop* ontology already used in other research works (e.g., [7]).

6 Related Works

Specifications mined from existing software systems are largely proposed in the literature to support analysis and re-documentation of software systems. A huge literature exists on mining and recovering of finite state models from observed execution traces. For instance, Tonella et al. [13] proposed the use of a multi-objective optimization algorithm to recover specification models that balance the amount of over- and under-approximation of an application's behavior observed in traces. Process recovery and mining deals with the analysis of system implementation to extract models of the underlying processes. Process mining techniques (e.g.,[14], [3]) infer processes by analyzing the workflow logs, containing information about organizations and process executions. Most of the effort in this area has been devoted to control flow mining (e.g., α algorithm [14]).

Mining techniques often produce large and intricate processes. Assuming that this is due to spurious information contained in the traces, some works prune the model by removing process elements having low execution frequencies (e.g., Fbr [14]). Other works cluster segments of traces and mine a number of smaller process models, one for each different cluster [3]. We used clustering [6] to modularize the processes into sub-processes, thus improving their understandability. The effectiveness of modularity could be however limited by the process structure and complexity [10]. Process reduction, hence, is applied after process recovery and after the use of these existing complexity-reduction techniques.

Some works propose the use of evolutionary algorithms to balance under and over-generalization when mining a process. For instance, Alves de Medeiro et al. [2] proposed a genetic algorithm that optimizes a single-objective which combines under and over-generalization; it balances the ability of models in reproducing the traced and extra-behaviors. On the contrary, we aim at three objectives at the same time: complexity, non-conformance and loss of business content. The former and the latter, absent in [2], derive from the analysts' needs to understand models that are meaningful within a business domain.

In our previous work [9], we presented a multi-objective technique to reduce process models by balancing their complexity and conformance. However, this approach, reducing processes by only looking at low-level information (extracted from logs), could be not enough in case of heavily reduced models. Indeed, on one side, low-level information extracted from logs can contain a huge amount of noise which can persist in the reduced processes. On the other side, business analysts are not enabled to drive the process reduction. In this work, we extended our previous work explicitly considering the business domain content of recovered processes, thus limiting their domain-irrelevant activities and allowing analysts to drive, by customizing the description of the domain, the process reduction according to their needs.

7   Conclusion

We have presented a multi-objective process reduction technique for reducing large processes by balancing complexity, conformance and business content. We conducted a case study using a real-life Web application. Results indicate that: (1) $MGA2$ produces a rich, fine grained, evenly distributed set of alternatives; (2) convergence of $MGA2$ could depend on the choice of the initial population; (3) though reduced, processes produced by $MGA2$ include relevant business activities and properties (i.e., we deem them as meaningful for analysts). Future works will be devoted to perform further experiments, involving additional case studies and including other process mining tools.

References

1. van der Aalst, W., Weijter, A., Maruster, L.: Workflow mining: Discovering process models from event logs. IEEE Transactions on Knowledge and Data Engineering 16, 2004 (2003)
2. Alves de Medeiros, A., Weijters, A., van der Aalst, W.: Genetic process mining: An experimental evaluation. Journal of Data Mining and Knowledge Discovery 14(2), 245–304 (2006)
3. Bose, R., van der Aalst, W.: Context aware trace clustering: Towards improving process mining results. In: Proc. of Symp. on Discrete Algorithms (SDM-SIAM), pp. 401–412 (2009)
4. Cardoso, J., Mendling, J., Neumann, G., Reijers, H.: A discourse on complexity of process models. In: Proc. of Workshop on Business Process Intelligence (BPI), pp. 115–126 (2006)
5. Deb, K., Pratap, A., Agarwal, S., Meyarivan, T.: A fast and elitist multiobjective genetic algorithm: NSGA-II. IEEE Transactions on Evolutionary Computation 6(2), 182–197 (2002)
6. Di Francescomarino, C., Marchetto, A., Tonella, P.: Cluster-based modularization of processes recovered from web applications. Journal of Software Maintenance and Evolution: Research and Practice (2010), doi: 10.1002/smr.518
7. Di Francescomarino, C., Ghidini, C., Rospocher, M., Serafini, L., Tonella, P.: Reasoning on Semantically Annotated Processes. In: Bouguettaya, A., Krueger, I., Margaria, T. (eds.) ICSOC 2008. LNCS, vol. 5364, pp. 132–146. Springer, Heidelberg (2008)
8. Di Francescomarino, C., Tonella, P.: Supporting Ontology-Based Semantic Annotation of Business Processes with Automated Suggestions. In: Halpin, T., Krogstie, J., Nurcan, S., Proper, E., Schmidt, R., Soffer, P., Ukor, R. (eds.) BPMDS 2009 and EMMSAD 2009. LNBIP, vol. 29, pp. 211–223. Springer, Heidelberg (2009)
9. Marchetto, A., Di Francescomarino, C., Tonella, P.: Optimizing the Trade-Off between Complexity and Conformance in Process Reduction. In: Cohen, M.B., Ó Cinnéide, M. (eds.) SS-BSE 2011. LNCS, vol. 6956, pp. 158–172. Springer, Heidelberg (2011)
10. Reijers, H., Mendling, J.: Modularity in Process Models: Review and Effects. In: Dumas, M., Reichert, M., Shan, M.-C. (eds.) BPM 2008. LNCS, vol. 5240, pp. 20–35. Springer, Heidelberg (2008)

11. Rozinat, A., van der Aalst, W.: Conformance checking of processes based on monitoring real behavior. Information Systems 33(1), 64–95 (2008)
12. Thomas, O., Fellmann, M.: Semantic epc: Enhancing process modeling using ontology languages. In: SBPM. CEUR Workshop Proceedings, vol. 251. CEUR-WS.org (2007)
13. Tonella, P., Marchetto, A., Nguyen, C., Jia, Y., Lakhotia, K., Harman, M.: Finding the optimal balance between over and under approximation of models inferred from execution logs. In: Int. Conference on Software Testing, Verification and Validation (ICST), pp. 21–30 (2012)
14. van der Aalst, W., van Dongen, B., Herbst, J., Maruster, L.G., Schimm, W.A.: Workflow mining: A survey of issues and approaches. Journal of Data and Knowledge Engineering 47(2), 237–267 (2003)
15. Veiga, G.M., Ferreira, D.R.: Understanding spaghetti models with sequence clustering for prom. In: Proc. of Workshop on Business Process Intelligence (BPI), Ulm, Germany (2009)

Evolving Human Competitive Spectra-Based Fault Localisation Techniques

Shin Yoo

University College London

Abstract. Spectra-Based Fault Localisation (SBFL) aims to assist debugging by applying risk evaluation formulæ (sometimes called suspiciousness metrics) to program spectra and ranking statements according to the predicted risk. Designing a risk evaluation formula is often an intuitive process done by human software engineer. This paper presents a Genetic Programming (GP) approach for evolving risk assessment formulæ. The empirical evaluation using 92 faults from four `Unix` utilities produces promising results[1]. Equations evolved by Genetic Programming can consistently outperform many of the human-designed formulæ, such as Tarantula, Ochiai, Jaccard, Ample, and Wong1/2, up to 6 times. More importantly, they can perform equally as well as Op2, which was recently proved to be optimal against `If-Then-Else-2` (`ITE2`) structure, or even outperform it against other program structures.

1 Introduction

Despite the advances in software testing techniques, faults still prevail in many software systems and debugging remains a hard task. Fault localisation aims to guide the programmer towards the program statement that contains the fault, using the information observed during test execution.

Spectra-Based Fault Localisation (SBFL) is a class of fault localisation techniques that uses program spectra (i.e., a summary of program's execution trace) to predict the likelihood of each program statement containing the fault [1,13,14]. The key element is a risk evaluation formula, or sometimes a suspiciousness metric, that converts the program spectra to relative risk value for each statement. SBFL subsequently ranks program statements according to the relative risk: the programmer can investigate the source code following the rank order. The intuition is that the faulty statement will be high in the ranking, reducing the number of statements the programmer has to check.

The performance of a SBFL technique depends mostly on the quality of the risk evaluation formula. The majority of the existing, widely studied formulæ are either inherited from other fields [12, 18] or designed by human intuition [5, 14, 15, 17, 24]: there is no guarantee that one formula is optimal for all classes of faults. Designing a risk evaluation formula that performs universally

---

[1] The program spectra data used in the paper, as well as the complete empirical results, are available from: http://www.cs.ucl.ac.uk/staff/s.yoo/evolving-sbfl.html

G. Fraser (Ed.): SSBSE 2012, LNCS 7515, pp. 244–258, 2012.

well against all possible combination of various program structures, test suites, and potential locations of faults remains a difficult task for a human. The only available methodology is that of trial and error: to design intuitively and evaluate empirically. Recent work includes efforts to design a risk evaluation formula that can be proven to be *optimal*, but only with respect to the case that the fault is contained within a specific program structure [17].

We presents an alternative approach: to evolve risk evaluation formulæ from program spectra directly. Using program spectra from test executions and known fault locations, we use Genetic Programming (GP) to evolve risk evaluation formulæ. By using a non-biased sample of known faults as the training data for GP, we try to obtain formulæ that are effective against various program structures. It is true that the evolved formulæ will be only as good as the input data for the GP. However, compared to proving optimality of risk evaluation formulæ against all possible program structures, providing common program structures that contain fault is a significantly easier task. In fact, this bears a strong resonance to the mantra of Search Based Software Engineering (SBSE) [10], namely:

> It is easier to compare solutions and choose the better one than to design a perfect solution from the scratch.

This paper introduces an evolutionary approach to designing risk evaluation formulæ for SBFL. GP uses program spectra from four Unix utilities from Software Infrastructure Repository [6] and the location information of 92 injected faults. The contributions of this paper are as follows:

- The paper presents the first evolutionary approach to generating risk evaluation formulæ for SBFL. All existing formulæ have been manually designed, often relying only on intuition. The introduced approach is evaluated with empirical studies, using test spectra data from real world Unix utilities.
- The empirical evaluation shows that GP-generated risk evaluation formulæ can outperform those designed by human. GP-generated formulæ can outperform some of the widely studied formulæ. Moreover, GP-generated formulæ can perform equally well or even better than an existing formula that has been proven to be optimal against a specific program structure. The equal performance provides evidence that GP can match the human design efforts; the outperformance provides evidence that GP can produce formulæ that are very effective for structures against no proof of optimality is currently available.
- All data used for the empirical study in the paper have been made available online to encourage replication and further research.

The rest of the paper is structured as follows. Section 2 introduces the concept of Spectra-Based Fault Localisation and the role of risk evaluation formulæ. Section 3 explains how we formulate the design of risk evaluation formulæ using Genetic Programming. Section 4 describes the experimental setup. Section 5 presents and analyses the results from the empirical evaluation. Section 6 presents the related work. Section 7 concludes and discusses future work.

2   Spectra-Based Fault Localisation

2.1   Basic Concept

Fault location aims to reduce the cost of debugging by guiding the process of searching for the location of the fault in the program. Various techniques rely on different software artefact to aid the developer: delta debugging [27,28] uses the cause-effect chain between the test input and the failure to guide the developer to the specific part of test input that causes the failure. Program Dependence Graph (PDG) has been used to construct a causal inference model for the location of fault [4].

One branch of fault localisation techniques that have attracted a significant amount of interest is Spectra-Based Fault Localisation (SBFL). Program spectra is a summary of a set of program executions [11]. For many of the SBFL techniques, we observe the execution of the test suite for System Under Test (SUT). Suppose SUT has n statements, and the test suite contains m test cases: the program spectrum for SBFL can be described as a matrix of n rows and 4 columns. Each row corresponds to individual statement of SUT, and contains four counters: (e_p, e_f, n_p, n_f). Counter e_p and e_f represent the number of times the corresponding program statement has been executed by tests, with pass and fail as a result respectively. Similarly, n_p and n_f represent the number of times the corresponding program statement has *not* been executed by tests, with pass and fail as a result respectively[2]. SBFL techniques subsequently use a risk evaluation formula, which is a formula based on the four counters, to predict the relative risk of each statement containing the fault. Compared to the case in which the developer investigates the structural elements in the order from s_1 to s_9, the ranking according to Tarantula produces 66.66% reduction in debugging effort (i.e. the developer will encounter s_7 6 elements earlier.

$$\text{Tarantula} = \frac{\frac{e_f}{e_f+n_f}}{\frac{e_p}{e_p+n_p} + \frac{e_f}{e_f+n_f}} \tag{1}$$

For example, Table 1 illustrates how the Tarantula metric [13], defined in Equation 1, can be applied to a small exemplar program spectrum. Suppose the structural element s_7 contains the fault. The coverage relationship between structural elements and the given test suite $T = \{t_1, t_2, t_3\}$ is given in the second column, with the corresponding test results. The Spectrum column contains the program spectrum data for T; the column Tarantula contains the resulting risk evaluation metric values. Finally, the column Rank contains the ranking of structural elements according to the Tarantula metric values. The faulty statement, s_7, is assigned with the highest Tarantula metric value, and therefore ends up in the first place.

2.2   Risk Evaluation formulæ

The effectiveness of a SBFL technique is determined by the risk evaluation formula, such as Equation 1. All existing formulæ are generated by human [17].

---

[2] The sum of $e_p, e_f, n_p,$ and n_f should be m.

Table 1. Motivating Example: the faulty statement s_7 achieves the 1st place when ranked according to the Tarantula risk evaluation formula in Eq 1.

Structural Elements	Test t_1	Test t_2	Test t_3	Spectrum e_p e_f n_p n_f	Tarantula	Rank
s_1	•			1 0 0 2	0.00	9
s_2	•			1 0 0 2	0.00	9
s_3	•			1 0 0 2	0.00	9
s_4	•			1 0 0 2	0.00	9
s_5	•			1 0 0 2	0.00	9
s_6	•		•	1 1 0 1	0.33	4
s_7 (faulty)		•	•	0 2 1 0	1.00	1
s_8	•	•		1 1 0 1	0.33	4
s_9	•	•	•	1 2 0 0	0.50	2
Result	P	F	F			

Table 2 contains several of the most widely studied formulæ. Interestingly, Jaccard [12] and Ochiai [18] were first studied in Botany and Zoology respectively but have been subsequently studied in the context of fault localisation [1, 17]. Tarantula was originally developed as a visualisation method [14, 15] but also increasingly considered as an SBFL risk evaluation formula independent from visualisation [13, 19]. AMPLE [5] and three different versions of Wong metric [24] have been introduced specifically for fault localisation.

Op1 and Op2 metrics are recent additions to SBFL techniques that showed an interesting research direction: these metrics are proven to produce optimal ranking, as long as the fault is located in a specific program structure (two consecutive If-Then-Else blocks, called ITE2) [17]. Although the proof does not guarantee that Op1 and Op2 are optimal for all locations of faults (and not just limited to ITE2), the empirical evaluation showed that both Op1 and Op2 are very strong formulæ.

Table 2. Risk Evaluation formulæ

Name	Formula	Name	Formula
Jaccard [12]	$\frac{e_f}{e_f+n_f+e_p}$	Ochiai [18]	$\frac{e_f}{\sqrt{(e_f+n_f)\cdot(e_f+e_p)}}$
Tarantula [15]	$\frac{\frac{e_f}{e_f+n_f}}{\frac{e_p}{e_p+n_p}+\frac{e_f}{e_f+n_f}}$	AMPLE [5]	$\left\lvert\frac{e_f}{e_f+n_f} - \frac{e_p}{e_p+n_p}\right\rvert$
Wong1 [24]	e_f	Wong2 [24]	$e_f - e_p$
Wong3 [24]	$e_f - h$, where $h = \begin{cases} e_p & \text{if } e_p \leq 2 \\ 2+0.1(e_p-2) & \text{if } 2 < e_p \leq 10 \\ 2.8+0.001(e_p-10) & \text{if } e_p > 10 \end{cases}$		
Op1 [17]	$\begin{cases} -1 & \text{if } n_f > 0 \\ n_p & \text{otherwise} \end{cases}$	Op2 [17]	$e_f - \frac{e_p}{e_p+n_p+1}$

2.3   Designing Risk Evaluation formulæ

This subsection discusses why Genetic Programming can be an ideal tool for designing risk evaluation formulæ.

Difficulties in Formal Approaches: Although the optimality proof of Naish et al. [17] presents a complete approach towards designing a risk evaluation formula, it will require significant human efforts to provide optimality proofs for a wider range of program structures. Moreover, SBFL can be applied to other testing criteria such as the existing work in concurrency testing [19], for which the possibility of optimality proof remains unknown.

Data-driven Iteration: Barring the formal proof of optimality, the most intuitive process of designing a risk evaluation formula would be an iterative modification of a candidate formula, against as a wide range of spectra datasets as possible, until its performance reaches an acceptable level. Not only the amount of data will burden the human designer, but this process also is, in fact, how GP operates, i.e., a data-driven, systematic trial-and-error.

Providing Insights: The goal of using GP for designing risk evaluation formulæ does not have be to replace human designs completely. It can actually be a powerful tool that the human software engineer can use to explore the design space with, to identify building blocks of better formulæ, and to gain insights into the specific domain under consideration.

2.4   Research Questions

Based on the discussions in Section 2.3, this paper investigates the performance of GP-designed risk evaluation formulæ for structural SBFL.

- **RQ1. Effectiveness:** How much debugging effort can be reduced by the GP-generated risk evaluation formulæ compare to existing human-designs?
- **RQ2. Design Space:** How much diversity is observed among the GP-generated formulæ? Does GP re-discover human-designed formulæ? How much problem does GP-bloat cause?
- **RQ3. Insights:** Are there design insights we can obtain by analysing the GP-generated formulæ? Do more complex formulæ perform better? Are certain spectra elements more important than the others?

RQ1 directly concerns the performance of the GP-evolved risk evaluation formulæ. It will be answered by performing statistical hypothesis testing to the reduction of debugging effort produced by GP and human generated formulæ. **RQ2** aims to investigate how much diversity can be allowed in the design space. It will be answered by comparing the GP-generated formulæ, both the whole and its parts, to the existing ones. Finally, **RQ3** is about the design insights we can expect to learn by evolving risk evaluation formulæ using GP.

3   Genetic Programming for SBFL

3.1   Representation

We use a simple tree-based representation and a set of simple operators on the ground that they can sufficiently represent most of the existing risk evaluation formulæ. Table 3 presents the GP operators used in the paper. Addition (gp_add), subtraction (gp_sub), and multiplication (gp_mul) do not require any treatment, because these operations cannot result in numerical exceptions. The division operator gp_div will return 1 when division by zero error is expected. Similarly, the square root operator gp_sqrt uses the absolute value of the given input. For terminal symbols, we use the program spectra data $\{e_p, e_f, n_p, n_f\}$, as well as one constant, 1.

Table 3. List of GP operators

Operator Node	Definition
gp_add(a, b)	a + b
gp_sub(a, b)	a - b
gp_mul(a, b)	ab
gp_div(a, b)	1 if $b = 0$, $\frac{a}{b}$ otherwise
gp_sqrt(a)	$\sqrt{|a|}$

3.2 Fitness Function

The aim of risk evaluation formula is not only to assign high risk value to the faulty statement, but also to ensure that the assigned high risk value results in a high ranking of the faulty statement. That is, the performance of a risk evaluation formula is measured by the relative position of the faulty statement when ranked by the formula.

In literature, this relative measurement is often referred to as the Expense metric [21], which is a normalised ranking of the faulty statement. Given a risk evaluation formula τ, a program p, and a fault b in p, the Expense metric E is calculated as in Equation 2:

$$E(\tau, p, b) = \frac{\text{Ranking of } b \text{ according to } \tau}{\text{Number of statements in } p} * 100 \qquad (2)$$

Expense is an *a-posteriori*, evaluative metric: it can be calculated only when the faulty statement is known. Because we are evolving a risk evaluation formula from locations of the known faults, Expense can be used as a fitness function. To avoid over-fitting to the location of a specific fault, we calculate Expense metric for a candidate formula using multiple faults from different programs and take the average as the fitness function. For a set of n known faults $B = \{b_1, \ldots, b_n\}$ from corresponding n programs $P = \{p_1, \ldots, p_n\}$, the fitness value of a candidate risk evaluation formula τ is calculated as follows:

$$\text{fitness}(\tau, B, P) = \frac{1}{n} \sum_{i=1}^{n} E(\tau, p_i, b_i) \text{ (to be minimised)} \qquad (3)$$

Depending on the risk evaluation formula, multiple statements may get assigned the same risk evaluation value and, thereby, tie in the ranking. Because it is not immediately clear what will be the appropriate tie-breaker for a candidate formula, we do not break ties and assign the most conservative ranking to all tied statements, which is equal to the sum of the number of the tied statements and the number of statements ranked before them [21, 26]. In the context of the fault localisation, this means that we assume the developer has to check all of the tied statements to locate the fault.

4   Experimental Setup

4.1   Subjects

Table 4 lists the subject programs whose faults are studied in the paper: `flex` (a lexical analyzer), `grep` (a text-search utility), `gzip` (a compression utility), and `sed` (a stream text editor). All four programs are obtained from Software Infrastructure Repository (SIR) [6] along with their test suites. Statement coverage information was collected using the GNU profiler, `gcov` version 4.3.2 on Linux version 2.6.27. We use the test suites provided by SIR.

Table 4. Subject Programs from SIR

Subject	Number of Tests	Lines of Code	Executable Lines of Code	Number of Faults
flex	567	12,407–14,244	3,393–3,965	47
grep	199	12,653–13,363	3,078–3,314	11
gzip	214	6,576– 7,996	1,705–1,993	18
sed	360	8,082–11,990	1,923–2,172	16

SIR provides a total of 219 (both real and seeded) faults across the five versions of the four subject programs [6]. We exclude 35 of these faults because these faults were unreachable when compiled for the experimental environment, and additional 92 faults because these are not detected by the chosen test suites. This leaves 92 faults, the distribution of which are listed in Table 4.

4.2   Implementation and Configuration

We use `pyevolve` [20] version 0.6 to implement the Genetic Programming. The algorithm was executed using Python runtime version 2.7.3. The population size was iteratively configured to 40. The initialisation uses the ramping method with the maximum tree depth of 4: the maximum tree depth was chosen to be able to express the most of the existing formalæ. The stopping criterion is a fixed run of 100 generations. The GP is configured with a rank selection operator, a single point crossover operator with the rate of 1.0, and a subtree replacement mutation operator with the rate of 0.08.

4.3   Evaluation

The Genetic Programming algorithm was repeated 30 times to cater for its stochastic nature. Each individual run of the GP uses a random sample of 20 faults out of 92 to evolve a risk evaluation formula; the remaining 72 faults are reserved for evaluation purposes.

We use Vargha & Delaney's A-test [22] to compare the Expense metric values of GP-evolved formulæ to those of existing ones. Vargha & Delaney's A-test is a non-parametric statistical test for determining stochastic superiority/inferiority of one sample X over another sample Y: the value of A is the probability that a

Table 5. Comparison of mean Expense for 72 faults in evaluation sets. Rows in bold correspond to GP-results that perform as well as or better than any human-designed formulæ.

ID	GP	Op1	Op2	Ochiai	AMPLE	Jacc'd	Tarant.	Wong1	Wong2	Wong3
GP01	5.73	9.20	5.30	32.66	10.96	6.10	15.06	22.24	17.10	6.63
GP02	12.04	9.67	5.72	32.60	11.91	6.63	14.92	23.45	19.49	8.92
GP03	14.46	11.35	6.11	29.99	12.18	6.99	15.68	23.55	18.55	8.85
GP04	7.80	9.70	4.46	30.98	8.83	5.03	13.88	22.62	14.64	6.33
GP05	9.35	11.04	5.83	29.95	10.63	6.42	14.46	23.15	18.54	8.53
GP06	12.15	11.11	5.87	28.02	12.51	6.79	15.35	23.12	16.70	7.01
GP07	8.93	11.18	5.94	29.53	12.19	6.85	14.81	23.88	19.74	8.68
GP08	**6.32**	**10.23**	**6.34**	**30.91**	**11.67**	**7.04**	**16.21**	**23.54**	**19.94**	**9.05**
GP09	9.66	10.58	5.33	31.56	11.40	6.17	14.06	22.58	18.31	8.20
GP10	**6.31**	**11.55**	**6.31**	**29.83**	**12.51**	**7.16**	**15.79**	**22.99**	**19.74**	**8.56**
GP11	**5.83**	**11.07**	**5.83**	**33.52**	**12.12**	**6.69**	**16.77**	**22.05**	**18.16**	**6.96**
GP12	12.09	8.84	6.23	32.15	11.65	7.02	16.65	22.91	19.42	9.09
GP13	**5.11**	**9.05**	**5.11**	**31.67**	**10.27**	**5.90**	**15.92**	**22.03**	**17.00**	**6.69**
GP14	9.91	8.52	5.91	31.69	11.10	6.55	15.88	23.15	18.10	8.65
GP15	5.62	9.54	5.59	33.02	10.23	6.19	15.16	23.85	17.17	8.44
GP16	6.79	8.32	5.71	30.52	10.74	6.41	14.60	23.06	18.36	8.42
GP17	7.67	11.46	6.22	33.62	12.06	6.98	16.85	22.44	17.94	8.59
GP18	9.42	10.78	5.54	34.17	11.46	6.33	15.45	22.17	17.46	8.14
GP19	6.42	9.01	5.11	31.28	10.18	5.78	15.03	22.84	15.26	7.79
GP20	**5.69**	**10.93**	**5.69**	**29.34**	**10.88**	**6.38**	**15.23**	**23.41**	**19.30**	**8.42**
GP21	10.17	10.13	6.24	29.82	10.86	6.89	15.70	23.01	19.85	9.43
GP22	7.58	8.50	5.91	28.06	10.46	6.60	13.67	23.25	18.60	8.63
GP23	6.14	10.76	5.52	30.86	10.57	6.16	14.69	21.77	16.90	7.25
GP24	9.18	10.15	6.21	28.74	12.53	7.10	15.76	23.41	20.16	8.35
GP25	9.34	10.19	6.29	32.56	12.36	7.18	17.59	22.63	20.19	9.48
GP26	**6.38**	**11.62**	**6.38**	**32.83**	**12.27**	**7.25**	**18.28**	**23.77**	**16.18**	**7.69**
GP27	9.75	8.53	5.89	33.28	12.01	6.85	16.42	22.99	19.23	7.81
GP28	5.56	9.18	5.25	30.02	11.18	6.15	13.52	22.86	17.17	6.85
GP29	7.16	10.12	6.17	34.17	12.83	7.14	17.00	22.94	20.18	8.88
GP30	10.68	9.10	5.14	30.02	10.17	5.78	14.49	22.79	17.09	8.34

single subject taken randomly from group X has higher/lower value than another single case randomly taken from group Y. For $A(X > Y)$, the value of A closer to 1 represents a higher probability of $X > Y$, 0 a higher probability of $X < Y$, and 0.5 no effect (i.e., $X = Y$).

However, the statistical interpretation of the results should be treated with caution. There is no guarantee that the studied programs and faults are representative of all possible programs and faults and, therefore, it is not clear whether they are legitimate *samples* of the entire group. On the other hand, if the cost of designing risk evaluation formulæ is significantly reduced by the use of GP, the possibility of project-specific formulæ should not be entirely ruled out.

5   Results and Analysis

5.1   Effectiveness

Table 5 contains the mean Expense values for all 30 GP-evolved formulæ and human-designed formulæ in Table 7[3]. Each row reports the mean Expense values

---

[3] The complete results for individual faults are available from:
http://www.cs.ucl.ac.uk/staff/s.yoo/evolving-sbfl.html.

from 72 faults in corresponding evaluation set. Note that the evaluation set differs between GP runs, as the training set is sampled randomly to avoid bias.

Rows in bold typefaces represent the GP runs that produced formulæ that performed as well as or better than all of the human-designed formulæ: this was observed 6 times out of 30 runs. The human-designed formula that performs the best is Op2; its relative performance confirms the trend observed in the previous work [17]. In 5 runs out of the aforementioned 6 (GP10, GP11, GP13, GP20, and GP26), GP-evolved formulæ always produce the same ranking, and subsequently the same Expense value, as Op2 and outperforms all other human-designed formulæ. In GP8, the remaining one run, the GP-evolved formula does not completely agree with Op2, but the mean Expense value from GP-evolved formula is lower than that from Op2.

The biggest improvement over human-designed formula is found in GP13 between GP and Ochiai: the expense from GP-evolved formula is less than one sixth of that from Ochiai. In fact, Ochiai, Tarantula, Wong1, and Wong2 are outperformed by GP in all runs. Based on this observation, we focus our comparative statistical analysis to the better performing formulæ: Op1, Op2, Ample, Jaccard, and Wong3. Table 6 presents the statistical analysis of the comparison between GP-evolved formulæ and the five human-designed formulæthat can produce Expense value below 10. Column A contains Varghar & Delaney's A test results, with which we test whether GP-based Expenses are lower than those based on existing formulæ. Column Count contains a tuple $(x/y/z)$: x is the number of faults for which GP produces lower Expense than the corresponding human-designed formula, y is the number of faults for which the Expense values are equal, and finally z is the number of faults for which GP produces higher Expense[4]. Combined with the A-test, these numbers provide a summary of how GP-evolved formulæ compare to existing ones.

The overall trend in Table 6 is that the results from A-test are mostly close to 0.5, suggesting that there is no overall difference in Expense values produced by GP and other formulæ overall. This confirms the results in Table 5: GP-evolved formulæ perform as equally well as human-designed formulæ. However, observing the details in Column Count reveals that there exist faults for which GP outperforms existing formulæ and vice versa. Figure 1 provides a scatterplot with fault-by-fault comparison between some of GP-evolved formulæ and other metrics[5]. GP08 produces lower Expense values for only 3 faults and higher values for 10, but the mean Expense of GP08 is still lower(Table 5). GP11 performs exactly as well as Op2 (i.e., the rankings are identical). For GP15 and GP27, the story is mixed: GP15 comfortably outperforms Tarantula, but GP27 produces Expense values significantly higher than those from Jacard for a few faults.

Considering that the aim of our approach is to *design* a formula that will be repeatedly used, we argue that it is not unrealistic to apply GP to existing program spectra data repeatedly and choose the best performing outcome: the cost of multiple GP execution will be amortised over the saved effort in fault

---

[4] Therefore $x + y + z$ is equal to 72, i.e., the size of the evaluation set.

[5] Scatterplot comparisons for all GP-evolved formulæ are also available online.

Table 6. Vargha & Delaney's A-test between GP and the better performing formulæ. Rows in bold correspond to GP-results that perform as well as or better than any human-designed formulæ.

ID	Op1		Op2		AMPLE		Jaccard		Wong3	
	A	Count	A	Count	A	Count	A	Count	A	Count
GP01	0.51	3/63/6	0.50	2/64/6	0.53	25/46/1	0.51	22/47/3	0.50	7/60/5
GP02	0.38	9/16/47	0.35	8/16/48	0.39	22/8/42	0.36	19/10/43	0.39	13/15/44
GP03	0.45	4/52/16	0.42	0/56/16	0.45	21/33/18	0.42	20/33/19	0.44	5/54/13
GP04	0.37	11/9/52	0.34	7/9/56	0.37	16/9/47	0.34	10/9/53	0.37	9/9/54
GP05	0.49	6/53/13	0.47	4/53/15	0.49	19/42/11	0.47	15/41/16	0.50	10/51/11
GP06	0.49	4/48/20	0.47	3/48/21	0.50	6/56/10	0.47	5/48/19	0.48	6/46/20
GP07	0.46	6/38/28	0.44	2/42/28	0.47	19/30/23	0.44	14/31/27	0.46	7/38/27
GP08	**0.51**	**3/59/10**	**0.50**	**3/59/10**	**0.54**	**25/47/0**	**0.51**	**26/46/0**	**0.52**	**9/54/9**
GP09	0.50	6/51/15	0.48	2/55/15	0.50	17/43/12	0.48	17/42/13	0.50	4/53/15
GP10	**0.52**	**4/67/1**	**0.50**	**0/71/1**	**0.53**	**23/45/4**	**0.50**	**24/44/4**	**0.51**	**8/63/1**
GP11	**0.52**	**4/68/0**	**0.50**	**0/72/0**	**0.53**	**24/45/3**	**0.50**	**23/46/3**	**0.52**	**5/67/0**
GP12	0.48	2/53/17	0.47	2/53/17	0.50	19/46/7	0.48	19/45/8	0.49	2/55/15
GP13	**0.51**	**3/69/0**	**0.50**	**0/72/0**	**0.52**	**23/47/2**	**0.50**	**22/48/2**	**0.50**	**6/66/0**
GP14	0.50	2/59/11	0.49	2/59/11	0.52	20/49/3	0.49	18/49/5	0.50	5/56/11
GP15	0.51	3/63/6	0.50	3/63/6	0.51	21/48/3	0.50	21/48/3	0.52	10/56/6
GP16	0.50	2/58/12	0.49	2/58/12	0.53	22/47/3	0.50	17/50/5	0.52	10/53/9
GP17	0.48	5/50/17	0.45	1/53/18	0.49	22/33/17	0.46	18/35/19	0.48	8/49/15
GP18	0.50	4/61/7	0.48	0/65/7	0.50	21/42/9	0.48	20/43/9	0.50	2/64/6
GP19	0.50	4/49/19	0.49	3/49/20	0.52	20/46/6	0.50	16/46/10	0.51	8/49/15
GP20	**0.52**	**4/68/0**	**0.50**	**0/72/0**	**0.52**	**23/46/3**	**0.50**	**23/46/3**	**0.53**	**9/63/0**
GP21	0.50	3/61/8	0.49	3/61/8	0.51	22/46/4	0.49	20/46/6	0.51	9/55/8
GP22	0.50	2/67/3	0.49	0/69/3	0.52	22/47/3	0.50	20/49/3	0.52	5/65/2
GP23	0.52	4/63/5	0.50	0/67/5	0.52	23/45/4	0.50	19/47/6	0.52	5/64/3
GP24	0.51	3/56/13	0.50	3/56/13	0.52	20/50/2	0.50	19/49/4	0.51	6/54/12
GP25	0.48	11/46/15	0.47	8/47/17	0.50	17/37/18	0.48	18/36/18	0.50	12/43/17
GP26	**0.52**	**4/68/0**	**0.50**	**0/72/0**	**0.52**	**23/46/3**	**0.50**	**22/47/3**	**0.51**	**5/67/0**
GP27	0.51	2/58/12	0.50	2/58/12	0.52	21/51/0	0.50	11/51/10	0.51	6/54/12
GP28	0.52	3/60/9	0.51	3/60/9	0.53	22/50/0	0.51	21/49/2	0.52	8/57/7
GP29	0.51	6/45/21	0.49	5/45/22	0.52	19/41/12	0.50	18/39/15	0.52	11/42/19
GP30	0.50	3/60/9	0.49	1/62/9	0.50	18/46/8	0.49	17/46/9	0.51	4/59/9

localisation. Therefore, we answer **RQ1** positively: GP-evolved risk evaluation formulæ can reduce debugging effort more effectively than many of human-designed formulæ, sometimes over 6 times. For many faults, GP-evolved formulæ perform as equally well as the best known formula, Op2. Finally, for some faults, GP-evolved formulæ can outperform even Op2.

5.2 Design Space

Table 7 contains the GP-evolved formulæ in their refined forms. The original solutions were refined by removing syntactic bloats (such as $n_f - n_f$) and improving readability. Explicit bloats were only observed only twice among the 30 formulæ. Since we are evolving formulæ rather than programs, GP-trees do not contain non-reachable nodes. Therefore, it is not clear whether any subcomponents of evolved formulæ can be definitely labelled as bloats, apart from the explicit, syntactic ones.

The GP-evolved formulæ show strong diversity. There is only one formula that is evolved twice by the GP: both GP14 and GP24 evolved $e_f + \sqrt{n_p}$. The same subcomponent is found in GP02, GP22, and GP28. Finally, a similar pattern,

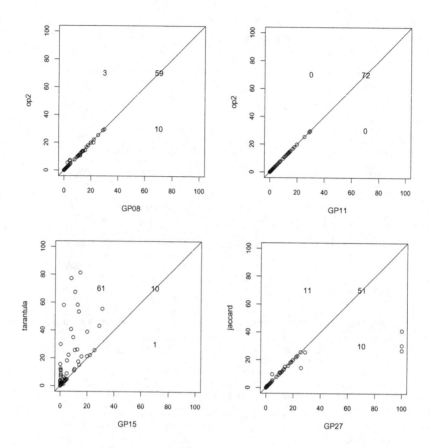

Fig. 1. Scatterplot comparisons of Expense for faults in evaluation set. Each dot represents a fault: the x-axis represents Expense produced by GP-evolved formula, and the y-axis by the specified formula. The solid line represents $y = x$: dots above the line correspond to faults which GP-evolved formulæ can rank higher. The upper two plots show that GP can perform equally or better than Op2. The lower left plot shows that GP can outperform Tarantula for most of the studied faults; the lower right plot shows a mixed results for GP against Jaccard.

$(ae_f^x + bn_p^y)$, where $a, b \in \mathbf{I}, x, y \in \{\frac{1}{2}, 1, \frac{3}{2}, 2, 3\}$, is also frequently observed as in GP01/09/12/21 (which contain $e_f + n_p$), **GP08** $(2e_f + 3n_p)$, **GP11/22/25/26** $(ae_f^2 + \sqrt{n_p})$, GP16 $(e_f^{\frac{3}{2}} + n_p)$, and GP18 $(e_f^3 + 2n_p)$. Interestingly, both $e_f + n_p$ and $\sqrt{e_f + n_p}$ are studied in existing literature [17]. However, GP did not rediscover these two metrics in their exact forms; rather, GP evolved variations of these formulæ as parts of larger formulæ. Apart from this, GP did not rediscover any of the existing formulæ.

To answer **RQ2**, the level of diversity observed in GP-evolved formulæ suggests the possibility that there may exist risk evaluation formulæ that are different from, but at least as effective as, the existing formulæ designed by the human. The observation made in Section 5.1, i.e., the fact that some GP-evolved formulæ can outperform existing ones for certain faults, provides further evidence that there may exist more effective formulæ for various program structures other than ITE2. However, the existence of common subcomponents suggest that a hybrid design approach may be even more successful: such an approach would introduce existing formulæ or partially-designed subcomponents into the GP population to assist the evolution.

5.3 Insights

Analysis of GP-evolved formulæ in Table 7 suggests that the most significant program spectra element, with respect to the faults we have studied, is e_f, i.e., the number of times a statement has been executed by failing tests. In all of the 8 GP-evolved formulæ that are equally as effective as Op2 in Table 5, e_p is the element that is either the only component proportional to the risk evaluation value, or the component that is the most dominant. The discussion of common subcomponent in Section 5.2 suggests that n_p is perhaps the second most significant element. Similarly, the least significant element appears to be n_f.

These observations do confirm our intuitions about the relationship between program spectra elements and fault localisation. A statement that contains fault will display a relatively higher e_f value (i.e., frequently covered by failing tests) and a relatively lower n_p value (i.e., less frequently covered by passing tests). In fact, human-designed formulæ such as Wong1/2/3 and Op2 are also designed to translate higher e_f and lower n_p values to higher rankings.

However, there are also some new design insights that can be gained by observing GP-evolved formulæ, which provide answers to **RQ3**. Most interestingly, it appears that ratio-type subcomponents (such as the ratio of a statement being covered by failing tests in Tarantula formula, $\frac{e_f}{e_f+n_f}$) are not necessarily required for a well performing formula: polynomials of spectra elements often seem to be sufficient. Similarly, the results achieved by polynomials of spectra elements suggests that specific constants, such as those found in Wong3, may not be necessary for designing a well performing formula.

6   Related Work

Various Spectra-Based Fault Localisation techniques have been developed to reduce the cost of debugging. One of the most widely studied risk evaluation formula, Tarantula, was initially developed as a visualisation aid for debugging process [14, 15]: subsequently, it has been studied independently from the visualisation [13, 19, 26]. Other notable formulæ include the family of Wong metrics [24], Statistical Bug Isolation (SBI) [16], and AMPLE [5]. Recently, Naish et al. provided an optimality proof against a specific program structure (ITE2: two consecutive If-Then-Else blocks) for their proposed metrics, Op1 and Op2 [17]. Naish et al. also provides an empirical evaluation of their metrics against a wide range of other formulæ, albeit using a set of relatively small subject programs. All existing metrics have been designed by human; this paper present the first

Table 7. GP-evolved risk evaluation formulæ. Trivial bloats, such as $n_f - n_f$, were removed.

ID	Refined Formula	ID	Refined Formula
GP01	$e_f(n_p + e_f(1 + \sqrt{e_f}))$	GP16	$\sqrt{e_f^{\frac{3}{2}}} + n_p$
GP02	$2(e_f + \sqrt{n_p}) + \sqrt{e_p}$	GP17	$\frac{2e_f + n_f}{e_f - n_p} + \frac{n_p}{\sqrt{e_f}} - e_f - e_f^2$
GP03	$\sqrt{|e_f^2 - \sqrt{e_p}|}$	GP18	$e_f^3 + 2n_p$
GP04	$\sqrt{|\frac{n_p}{e_p - n_p} - e_f|}$	GP19	$e_f \sqrt{|e_p - e_f + n_f - n_p|}$
GP05	$\frac{(e_f + n_p)\sqrt{e_f}}{(e_f + e_p)(n_p n_f + \sqrt{e_p})(e_p + n_p)\sqrt{|e_p - n_p|}}$	GP20	$2(e_f + \frac{n_p}{e_p + n_p})$
GP06	$e_f n_p$	GP21	$\sqrt{e_f + \sqrt{e_f + n_p}}$
GP07	$2e_f(1 + e_f + \frac{1}{2n_p}) + (1 + \sqrt{2})\sqrt{n_p}$	GP22	$e_f^2 + e_f + \sqrt{n_p}$
GP08	$e_f^2(2e_p + 2e_f + 3n_p)$	GP23	$\sqrt{e_f}(e_f^2 + \frac{n_p}{e_f} + \sqrt{n_p} + n_f + n_p)$
GP09	$\frac{e_f \sqrt{n_p}}{n_p + n_p} + n_p + e_f + e_f^3$	GP24	$e_f + \sqrt{n_p}$
GP10	$\sqrt{|e_f - \frac{1}{n_p}|}$	GP25	$e_f^2 + \sqrt{n_p} + \frac{\sqrt{e_f}}{\sqrt{|e_p - n_p|}} + \frac{n_p}{(e_f - n_p)}$
GP11	$e_f^2(e_f^2 + \sqrt{n_p})$	GP26	$2e_f^2 + \sqrt{n_p}$
GP12	$\sqrt{e_p + e_f + n_p} - \sqrt{e_p}$	GP27	$\frac{n_p \sqrt{(n_p n_f - e_f)}}{e_f + n_p n_f}$
GP13	$e_f(1 + \frac{1}{2e_p + e_f})$	GP28	$e_f(e_f + \sqrt{n_p} + 1)$
GP14	$e_f + \sqrt{n_p}$	GP29	$e_f(2e_f^2 + e_f + n_p) + \frac{(e_f - n_p)\sqrt{n_p e_f}}{e_p - n_p}$
GP15	$e_f + \sqrt{n_f + \sqrt{n_p}}$	GP30	$\sqrt{|e_f - \frac{n_f - n_p}{e_f + n_f}|}$

GP-based approach to the design of risk evaluation formulæ, reformulating it as a predictive modelling based on GP. Machine learning techniques have been also applied to fault localisation work, but the aim was to classify failing tests together rather than to identify the location of the fault directly [23].

Although SBFL originally started as a debugging aid for human developers, the technique is increasingly used to enable other automated Search-Based Software Engineering (SBSE) techniques. Goues et al. use SBFL to identify the parts of a program that needs to be automatically patched [7]. Yoo et al. use SBFL to measure the Shannon entropy of fault locality, so that the test suite can be prioritised for faster fault localisation [25]. GP may be able to help these techniques by evolving SBFL techniques with a specific set of characteristics, improving the synergy between predictive modelling and SBSE even further [9].

Other approaches towards fault localisation include slicing [2], consideration of test similarity [3,8], delta debugging [27,28], and causal inference [4]. While this paper only concerns the spectra-based approach, the positive results suggest that GP may be successfully employed to evolve a wider range of fault localisation techniques.

7   Conclusion

This paper reports the first application of Genetic Programming to evolving risk evaluation formulæ for Spectra-Based Fault Localisation. We use a simple tree-based GP to evolve risk evaluation formulæ that take program spectra elements as terminals. Empirical evaluation based on 92 different faults from four Unix utilities shows three important findings. First, GP-evolved formulæ can outperform widely studied human-designed formulæ by up to 5.9 times. Second, GP-evolved formulæ can perform optimally against the ITE2 program structure, for which existing formulæ, Op1 and Op2, have been proven to be optimal. Finally, GP-evolved formulæ can outperform Op1 and Op2 for certain studied faults.

Future work will include the use of more sophisticated GP representation (so that GP can evolve conditional formulæ as in Wong3), the inclusion of elements other than program spectra (e.g., code churn, dependency, or data-flow information), and the investigation of the possibility for the evolution of project-specific formalæ.

References

1. Abreu, R., Zoeteweij, P., van Gemund, A.J.C.: On the accuracy of spectrum-based fault localization. In: Proceedings of the Testing: Academic and Industrial Conference Practice and Research Techniques - MUTATION, pp. 89–98. IEEE Computer Society (2007)
2. Agrawal, H., Horgan, J., London, S., Wong, W.: Fault localization using execution slices and dataflow tests. In: Proceedings of IEEE Software Reliability Engineering, pp. 143–151 (1995)
3. Artzi, S., Dolby, J., Tip, F., Pistoia, M.: Directed test generation for effective fault localization. In: Proceedings of the 19th International Symposium on Software Testing and Analysis, ISSTA 2010, pp. 49–60. ACM, New York (2010)
4. Baah, G.K., Podgurski, A., Harrold, M.J.: Causal inference for statistical fault localization. In: Proceedings of the 19th International Symposium on Software Testing and Analysis (ISSTA 2010), pp. 73–84. ACM Press (July 2010)
5. Dallmeier, V., Lindig, C., Zeller, A.: Lightweight bug localization with ample. In: Proceedings of the Sixth International Symposium on Automated Analysis-driven Debugging, AADEBUG 2005, pp. 99–104. ACM, New York (2005)
6. Do, H., Elbaum, S.G., Rothermel, G.: Supporting controlled experimentation with testing techniques: An infrastructure and its potential impact. Empirical Software Engineering 10(4), 405–435 (2005)
7. Goues, C.L., Dewey-Vogt, M., Forrest, S., Weimer, W.: A systematic study of automated program repair: Fixing 55 out of 105 bugs for $8 each. In: Proceedings of the 34th International Conference on Software Engineering, pp. 3–13 (2012)
8. Hao, D., Zhang, L., Pan, Y., Mei, H., Sun, J.: On similarity-awareness in testing-based fault localization. Automated Software Engineering 15, 207–249 (2008)
9. Harman, M.: The relationship between search based software engineering and predictive modeling. In: Proceedings of the 6th International Conference on Predictive Models in Software Engineering, pp. 1–13. ACM Press, New York (2010)
10. Harman, M., Jones, B.F.: Search based software engineering. Information and Software Technology 43(14), 833–839 (2001)
11. Harrold, M.J., Rothermel, G., Wu, R., Yi, L.: An empirical investigation of program spectra. In: Proceedings of the ACM SIGPLAN-SIGSOFT Workshop on Program Analysis for Software Tools and Engineering (PASTE 1998), pp. 83–90. ACM, New York (1998)
12. Jaccard, P.: Étude comparative de la distribution florale dans une portion des Alpes et des Jura. Bulletin del la Société Vaudoise des Sciences Naturelles 37, 547–579 (1901)

13. Jones, J.A., Harrold, M.J.: Empirical evaluation of the tarantula automatic fault-localization technique. In: Proceedings of the 20th International Conference on Automated Software Engineering (ASE 2005), pp. 273–282. ACM Press (2005)

14. Jones, J.A., Harrold, M.J., Stasko, J.: Visualization of test information to assist fault localization. In: Proceedings of the 24th International Conference on Software Engineering, pp. 467–477. ACM, New York (2002)

15. Jones, J.A., Harrold, M.J., Stasko, J.T.: Visualization for fault localization. In: Proceedings of ICSE Workshop on Software Visualization, pp. 71–75 (2001)

16. Liblit, B., Naik, M., Zheng, A.X., Aiken, A., Jordan, M.I.: Scalable statistical bug isolation. In: Proceedings of the 2005 ACM SIGPLAN Conference on Programming Language Design and Implementation, PLDI 2005, pp. 15–26. ACM, New York (2005)

17. Naish, L., Lee, H.J., Ramamohanarao, K.: A model for spectra-based software diagnosis. ACM Transactions on Software Engineering Methodology 20(3), 11:1–11:32 (2011)

18. Ochiai, A.: Zoogeographic studies on the soleoid fishes found in Japan and its neighbouring regions. Bulletin of the Japanese Society of Scientific Fisheries 22(9), 526–530 (1957)

19. Park, S., Vuduc, R.W., Harrold, M.J.: Falcon: fault localization in concurrent programs. In: Proceedings of the 32nd ACM/IEEE International Conference on Software Engineering, ICSE 2010, vol. 1, pp. 245–254. ACM, New York (2010)

20. Perone, C.S.: PyEvolve, http://pyevolve.sourceforge.net

21. Renieres, M., Reiss, S.: Fault localization with nearest neighbor queries. In: Proceedings of the 18th International Conference on Automated Software Engineering, pp. 30–39 (October 2003)

22. Vargha, A., Delaney, H.D.: A critique and improvement of the "CL" common language effect size statistics of mcgraw and wong. Journal of Educational and Behavioral Statistics 25(2), 101–132 (2000)

23. Wong, E., Debroy, V.: A survey of software fault localization. Tech. Rep. UTDCS-45-09, Department of Computer Science, University of Texas at Dallas (November 2009)

24. Wong, W.E., Qi, Y., Zhao, L., Cai, K.Y.: Effective fault localization using code coverage. In: Proceedings of the 31st Annual International Computer Software and Applications Conference, COMPSAC 2007, vol. 01, pp. 449–456. IEEE Computer Society, Washington, DC (2007)

25. Yoo, S., Harman, M., Clark, D.: FLINT: Fault localisation using information theory. Tech. Rep. RN/11/09, Department of Computer Science, University College London (March 2011)

26. Yu, Y., Jones, J.A., Harrold, M.J.: An empirical study of the effects of test-suite reduction on fault localization. In: Proceedings of the International Conference on Software Engineering (ICSE 2008), pp. 201–210. ACM Press (May 2008)

27. Zeller, A.: Automated debugging: Are we close? IEEE Computer 34(11), 26–31 (2001)

28. Zeller, A.: Why Programs Fail: A Guide to Systematic Debugging. Morgan Kaufmann Publishers Inc., San Francisco (2005)

Applying Search Based Optimization to Software Product Line Architectures: Lessons Learned

Thelma Elita Colanzi[1,2] and Silvia Regina Vergilio[2,*]

[1] DIN - State University of Maringa, Av. Colombo, 5790, Maringá, Brazil
[2] DInf - Federal University of Parana, CP: 19081, CEP 19031-970, Curitiba, Brazil
thelma@din.uem.br, silvia@inf.ufpr.br

Abstract. The Product-Line Architecture (PLA) is a fundamental SPL artifact. However, PLA design is a people-intensive and non-trivial task, and to find the best architecture can be formulated as an optimization problem with many objectives. We found several approaches that address search-based design of software architectures by using multi-objective evolutionary algorithms. However, such approaches have not been applied to PLAs. Considering such fact, in this work, we explore the use of these approaches to optimize PLAs. An extension of existing approaches is investigated, which uses specific metrics to evaluate the PLA characteristics. Then, we performed a case study involving one SPL. From the experience acquired during this study, we can relate some lessons learned, which are discussed in this work. Furthermore, the results point out that, in the case of PLAs, it is necessary to use SPL specific measures and evolutionary operators more sensitive to the SPL context.

Keywords: SPL, MOEAs, software architecture optimization.

1 Introduction

Software Product Line (SPL) is a systematic approach for software reuse applied to a family of specific products within a well-defined domain. SPL approach allows to shift from the reuse of individual components to the large-scale reuse of a product-line architecture (PLA) [4]. A PLA defines the architectural elements required to realize feature commonality and variation across similar products.

The PLA design and its evolution is even more complex than traditional software architecting since, in order to maximize the reuse, some quality requirements such as modularity, stability and reusability play an important role. PLA design is a people-intensive, non-trivial and demanding task for software engineers to perform [12]. Software architecture design can be formulated as an optimization problem and has been investigated in related work [1,9,12]. These works have used Multi-Objective Evolutionary Algorithms (MOEAs) to optimize the design of software architectures and have achieved promising results. However, they have not been applied in the SPL context and do not consider specific characteristics of the PLA design.

---

* The authors would like to thank to CNPq for financial support.

G. Fraser (Ed.): SSBSE 2012, LNCS 7515, pp. 259–266, 2012.

Conventional architectural metrics [5] can be used to analyse the stability and modularity of any kind of software architecture, including PLAs, but it is also crucial to consider SPL metrics to evaluate PLAs due to its specific characteristics, such as variabilities, variation points, features assigned to elements, etc. In this sense, the research goal of this work is to investigate, through an exploratory study, how suitable is the use of existing search-based design approaches to the SPL context. To do this, we extend the existing approaches to PLA design by adding two specific metrics conceived to evaluate the cohesion of PLAs: CRSU (Compound Required Service Utilization) and CPSU (Compound Provided Service Utilization) [3]. Our extended approach is evaluated in a case study with the SPL Arcade Game Maker (AGM) [11]. Results of this study are presented, as well as, the main lessons learned.

The paper is organized as follows. Sections 2 and 3 respectively present related work and AGM. Section 4 describes how the case study was conducted. In the sequence, Section 5 relates the results and addresses the lessons learned during the study conduction. Finally, Section 6 concludes the paper.

2   Search-Based Software Design

Räihä published a survey of works on search-based software design [8]. Among them, the works most related to ours address the architectural design outset. They use search-based techniques in order to optimize software architectures under development and consider two problems: the Class Responsibility Assignment (CRA) problem [1,12] and the application of design patterns to design [9]. All of them use coupling and cohesion measures to evaluate the fitness of the solutions. There are two works based on the class diagram and MOEAs (SPEA2 and NSGA-II). They aim at improving the design by moving methods, attributes and relationships from one class to another, and adding and deleting classes [1,12]. Another work [9] applies architectural design patterns for mutations and two quality metrics for improving modifiability and efficiency of the design. Empirical studies [10,12] observed that the crossover operator does not offer significant advantage in the design optimization, when implemented in a random fashion. We can observe that the mentioned works successfully optimize software architectures, however, they have not been applied to optimize PLAs.

3   Arcade Game Maker (AGM)

Our study used the SPL AGM [11], which encompasses three arcade games: Brickles, Bowling, and Pong. It was created by the Software Engineering Institute (SEI) and conceived proactively in [2]. Its main variations are [7]: (i) specific rules of those games; (ii) kind, number and behavior of elements; and (iii) physical environment where the games take place. All variabilities of AGM were identified and represented rely on the use of the stereotype ≪variable≫ in the architecture models. This stereotype indicates that the component is formed by a set of classes with variabilities. For the sake of illustration, the model in Figure 1

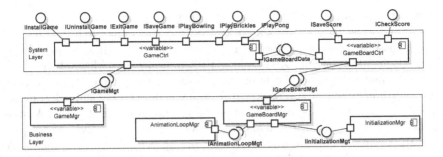

Fig. 1. AGM Component-based Architecture

represents a partial, high-level view of the PLA. Features are explicitly assigned to architectural elements using stereotypes, such as ≪Rule≫[1].

This architecture is in an intermediate stage since it is component-based, follows the layered architectural style and has manager classes that provide interfaces to enable the communication between layers. In order to improve the architectural quality, PLAs usually are conceived by using architectural styles, design patterns and frameworks. The AGM PLA has the system and the business layers. The system components have the suffix Ctrl and provide operations to the interface and dialog layers. The business components have the suffix Mgr and provide services to the system layer. The PLA is composed by 6 components and 45 classes.

The AGM PLA used in this work was analised in [2] by three specialists. The `GameCtrl` component was considered to accumulate too many responsibilities and, as a consequence, a candidate to manifest an architectural design anomaly.

4   Study Setting

Definition of the evaluation model: Aiming at obtaining highly cohesive, low coupling between components, stable and reusable PLAs, we used as objectives five metrics, supported by SDMetrics [13]: 3 conventional metrics, used in related work, and 2 SPL metrics. The metric Distance from the Main Sequence (D) [5] establishes a relation between abstraction and instability of a package. D values near to 0 represent components near the main sequence, i.e., there is a good balance between stability and abstraction [5]. The coupling metric Dependency of Packages (DepPack) [13] measures the number of packages on which classes and interfaces of this package depend. And, the size metric Number of Operations by Interface (NumOps) [13] counts the number of operations in each interface. CPSU and CRSU [3] allow to assess the internal cohesion of a PLA. CPSU is the rate of the number of services provided by components that is actually used by other PLA components. CRSU is the rate of the number of services required

---

[1] The detailed class diagram and all the solutions of our study case can be found at http://www.inf.ufpr.br/gres/apoio_en.html

by components that are actually used by other PLA components. CPSU and CRSU values close to 1 signify self-contained, fully functional architectures.

From these metrics, we defined five objectives in our evaluation model: (a) Sum of D; (b) Mean of NumOps of all the PLA interfaces; (c) Mean of DepPack; (d) CPSU; and (e) CRSU. We intended to minimize the three first objectives and to maximize the last two ones. CPSU and CRSU, D and DepPack provide, respectively, indicators about cohesion, stability and coupling of the PLA components. We believe that the size metric NumOps allows to observe some aspect of the PLA reusability since small interfaces are, in general, easier to reuse. We worked with models of the PLA in two levels: high-level and detailed level. Dep-Pack, D and NumOps were applied on the detailed level of the PLA, and CPSU and CRSU were applied on the high-level PLA.

Evolutionary and Multi-objective Approach: The approach is based on the evaluation model described before that considers specific SPL metrics in addition to the conventional ones. Each individual consists on a PLA, represented as a class diagram. Such diagram, converted to a XMI file, serves as input to SDMetrics. The MOEA used is inspired on NSGA-II. Each population is formed by 5 individuals (alternative design to the AGM PLA). One mutation operator is applied on each individual from parent population in order to obtain the offspring. For each individual, some mutation operator is chosen randomly, as well as, the source and the target architectural elements to be mutated. The mutations are based on [1,12]: *Move Method, Move Attribute, Add Class*. As the AGM PLA has manager classes with interfaces, we use two additional operators: *Move Operation* and *Add Manager Class*. The first operator moves operations between interfaces belonging to manager classes. The second one moves an attribute/operation to a new manager class. Architectural elements must maintain the assigned feature even if it is moved in the PLA. We did not apply the crossover operator since there is not a consensus on the benefits of such operator in this context [10,12]. Although, it is an interesting further work to perform.

The initial population is composed by the AGM PLA [2] and other four mutated individuals [9]. After achieving an offspring population, the algorithm selects the best 5 individuals (from parent and offspring populations) to obtain the next generation. The fitness of all individuals is obtained by using the metrics of the evaluation model with five objectives. As a result, the next generation is composed by non-dominated solutions from the first front, plus the solutions from the second front, and so on until that the new population is completed. The evolutionary process consists on five generations.

5   Results and Analysis

Table 1 contains the fitness of the solutions obtained at the end of the evolutionary process. All these non-dominated solutions have the same number of classes. They have the same CPSU and CRSU values and the best values for each other metric are in bold. Their fitness are slightly better than the original PLA's fitness (2.562, 3.57, 1.8, 0.579, 1). So, according to the metrics, we can state that

Table 1. Fitness of the non-dominated solutions found

Solution	Sum of D	Mean of NumOps	Mean of DepPack	CPSU	CRSU
4b	**2.528**	3.26	**1.75**	0.579	1
5b	2.736	**3.20**	2.50	0.579	1
6d	2.562	**3.20**	2.50	0.579	1
6e	**2.528**	3.26	**1.75**	0.579	1

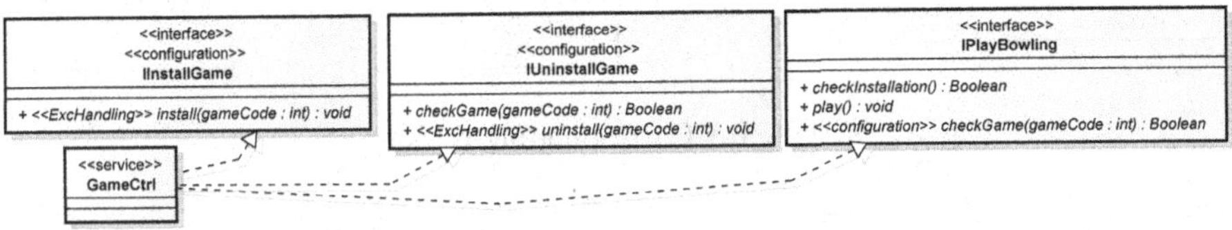

Fig. 2. Partial PLA Corresponding to Solution 4c

our solutions are more stable and more reusable than the original PLA. One responsibility of the component GameCtrl in Solutions **5b** and **6e** was delegated to a new manager class in only 5 generations. Therefore, in solutions found after many generations would be possible that some responsibilities assigned to this component were assigned to another one. As a result, GameCtrl would be not a candidate to manifest an architectural design anomaly.

From this study, the following lessons were learned:

Features Modularization: The used metrics are not sensitive to features modularization. The mutations can generate better fitness for some objectives, but in some cases they are not good changes since some SPL features became more scattered or tangled. To illustrate this, Figure 2 presents a fragment of Solution **4c**, which was obtained after the application of the mutation operator *Move Operation* for moving checkGame from IInstallGame to IPlayBowling in GameCtrl. Before the mutation, the feature *configuration* was assigned only to interfaces IInstallGame and IUninstallGame of GameCtrl. After that, this feature becames more scattered because there are 3 interfaces assigned to it in Solution **4c**. Due to their definitions, none of the used metrics was sensitive to this kind of scattering. On the other hand, in Solution **6c** (Figure 3) the feature *configuration* became more modularized and less tangled with other features after the mutation that moved the operation checkInstallation from the interface IGameMgt of GameMgr to the interface Interface_1 of Classe_2. Despite of these two facts, none metric was sensitive to this benefit for the PLA. The solution that better modularizes the feature *configuration*, is a dominated solution, and due to this, it is not included in the MOEA output.

PLA Cohesion: In our case study, CRSU and CPSU were not so sensitive to changes performed in the PLAs. In some cases, there are significant changes in the PLA, not captured by these metrics. It occurs because their definition are based on the satisfaction and utilization of the component services. So, changes inside a component that improve its internal cohesion are not detected. Many times changes inside components do not change the provided or required services.

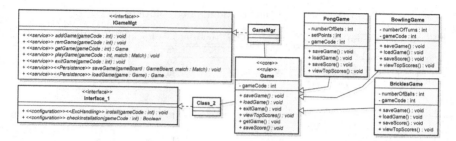

Fig. 3. Partial PLA Corresponding to Solution 6c

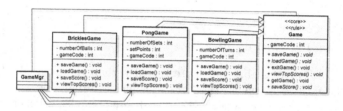

Fig. 4. Partial PLA Corresponding to Solution 5b

In addition to the feature modularization, in the SPL context it is interesting to have a PLA where the features are more cohesive as possible. However, CRSU and CPSU are not sensitive to feature cohesion. For example, Solution **6c** (Figure 3) has better feature cohesion than Solution **6d** regarding to the feature *configuration*, but this fact is not captured by these three metrics.

PLA Design Stage: The design stage of the PLA provided as the input to the search-based optimization impacts on the obtained results. With the AGM PLA the mutation operators could not obtain design solutions much better than it. In addition, the values of some metrics (CPSU and CRSU) did not change after mutations. For sure if we have performed a lot of iterations on the evolutionary process, we could obtain better design solutions. For instance, in some solutions the attribute **menu** was moved from **GameBoard** to a new class. This mutation could carry to a solution where all responsibilities regarding to the menu of the game were moved to that new class, so **GameBoard** would become a less bloated class. However, we could observe that the used mutation operators and metrics were not enough to achieve satisfactory results. So, considering that AGM PLA is in an intermediate stage, we can state that other mutation operators and other metrics more suitable to SPL context are necessary, for instance, mutation operators associated to design patterns or architectural styles.

Restrictions for the genetic operators: As mentioned before, the original PLA follows the layered style, where the business layer should not require services from the system layer. Despite of this, one solution (Solution **2d**) does not follow this rule. However, we have not established restrictions concerning to the architectural styles to apply the mutation operators. So, we can conclude that the genetic operators must be aware of the rules defined by the adopted architecture style. This can be done by defining restrictions to apply the operators.

There are some threats to our study: the sample size since our study focused on only one SPL; small populations; and few generations. However, this reduced number allowed us to perform an in-depth qualitative analysis that took a wide range of perspectives into account. The construct validity of the study with respect to the used metrics and operators is guaranteed by their previous successful application and evaluation by other researchers [1,3,5,6,12].

6   Concluding Remarks

The main contributions of this work are: the first case study where an search-based design approach was applied to SPL context and the lessons learned about such study. We also extended an existing approach by using two metrics defined to evaluate PLAs. The results point out that existing search-based approaches can be used to achieve better PLA design. Although, the results would be better if some points discussed in the lessons learned were improved. The two SPL metrics used were not sensitive to the points highlighted in the lessons learned.

Our ongoing work encompasses: a) the use of metrics to assess cohesion and modularization of SPL features; b) evolutionary operators sensitive to the SPL context, including modularization of features; and c) extensibility of the PLA. We also intend to investigate the crossover operator feasibility in the SPL context.

References

1. Bowman, M., Briand, L.C., Labiche, Y.: Solving the class responsibility assignment problem in object-oriented analysis with multi-objective genetic algorithms. IEEE Transactions on Software Engineering 36(6), 817–837 (2010)
2. Contieri Jr, A.C., Correia, G.G., Colanzi, T.E., Gimenes, I.M.S., Oliveira Jr, E.A., Ferrari, S., Masiero, P.C., Garcia, A.F.: Extending UML Components to Develop Software Product-Line Architectures: Lessons Learned. In: Crnkovic, I., Gruhn, V., Book, M. (eds.) ECSA 2011. LNCS, vol. 6903, pp. 130–138. Springer, Heidelberg (2011)
3. van der Hoek, A., Dincel, E., Medvidovic, N.: Using service utilization metrics to assess the structure of product line architectures. In: Proceedings of the 9th Int. Symposium on Software Metrics, pp. 298–308. IEEE, Washington, DC (2003)
4. van der Linden, F., Schmid, F., Rommes, E.: Software Product Lines in Action - The Best Industrial Practice in Product Line Engineering. Springer (2007)
5. Martin, R.: Stability – C++ Report. Tech. rep. (1997),
   http://www.objectmentor.com/resources/articles/stability.pdf
6. Martin, R.: Agile Software Development: Principles, Patterns, and Practices. Prentice Hall (2003)
7. Oliveira Jr., E.A., et al.: Systematic Management of Variability in UML-based Software Product Lines. J. of Universal Computer Science 16(17), 2374–2393 (2010)
8. Räihä, O.: A survey on search-based software design. Computer Science Review 4(4), 203–249 (2010)
9. Räihä, O.: Genetic Algorithms in Software Architecture Synthesis. Ph.D. thesis, School of Information Sciences, University of Tampere, Tampere, Finland (2011)

10. Räihä, O., Koskimies, K., Mäkinen, E.: Empirical study on the effect of crossover in genetic software architecture synthesis. In: Proceedings of World Congress on Nature and Biologically Inspired Computing (NaBIC), pp. 619–625. IEEE (2009)
11. SEI: Arcade Game Maker pedagogical product line (2012), http://www.sei.cmu.edu/productlines/ppl/
12. Simons, C.L.: Interactive Evolutionary Computing in Early Lifecycle Software Engineering Design. Ph.D. thesis, University of the West of England, Bristol (2011)
13. Wüst, J.: SDMetrics (2012), http://www.sdmetrics.com/

Problem-Specific Search Operators for Metaheuristic Software Architecture Design

Ramin Etemaadi, Michael T.M. Emmerich, and Michel R.V. Chaudron

Leiden Institute of Advanced Computer Science, Leiden University, Netherlands
{etemaadi,emmerich,chaudron}@liacs.nl

Abstract. A large number of quality properties needs to be addressed in nowadays complex software systems by architects. These quality properties are mostly conflicting and make the problem very complex. This paper proposes a hybridization process about the problem of optimization of system architecture, in which it uses quality improvement heuristics within an evolutionary algorithm. The solution can be represented in a systems model representation (instead of genotype-phenotype mapping approach) and then it is manipulated by specific and customizable transformations of system architecture. These transformations are based on patterns, for instance *Replicating-Component-Instant, Caching-Data*. In this case, various system quality improvement patterns such as known performance or security improvement patterns can be easily used for exploration in multiobjective evolutionary search.

Keywords: System Architecture Design Optimization, Genotype-Phenotype Mapping, Model-Driven Software Development (MDSD), Evolutionary Multiobjective Optimization (EMO).

1 Introduction

The architecture has deep impact on non-functional properties of a system such as performance, safety, reliability, security, energy consumption and cost. To construct a system that optimizes all its requirements simultaneously modern methods and techniques are needed. For the problem of automatically finding architectural designs researchers have in recent years proposed two main approaches: (1) Rule-based approaches and (2) Metaheuristic-based approaches. Rule-based approaches try to detect weak points (e.g. bottlenecks) in the architectural model based on predefined rules and apply predefined solutions (tactics, patterns) to alleviate these weak points. Metaheuristic approaches frame the challenge of designing architectures as an optimization problem and iteratively try to improve a candidate solution with regard to a given measure of quality. In this way these algorithms explore very large design spaces to find optimal architectural solutions.

Many metaheuristic approaches, such as evolutionary algorithms (EA), have standard representations (e.g. bit strings or real-valued vectors) and to apply them to a complex search space a mapping between the standard representation

G. Fraser (Ed.): SSBSE 2012, LNCS 7515, pp. 267–272, 2012.

and the represented solution needs to be defined, in the context of EA this is also referred to as genotype-phenotype mapping. Instead of working on standard representations, it is also possible to apply search operators directly on the natural representation of a solution, e.g. a systems model in case of software architectures. This approach supports the definition of problem specific heuristic transformations or crossover operators and will be applied here in the domain of multiobjective software architecture design. So our proposed approach uses the direct representation compared to previous approaches [5] [7] which use genotype-phenotype mapping.

The contribution of the paper is to define a new set of EA search operators that do not require a genotype-phenotype mapping and work on a direct representation of the problem. These operators are problem specific and motivated by software architectural patterns or antipatterns. They can be implemented as model-to-model transformation on the architectural model like UML model.

The paper is organized as follows: Firstly, Section 2 will discuss related work on direct representations and motivate its potential benefits in the software architecture domain. Then, Section 3 reviews state of the art techniques for software architecture optimization. The novel approach, that is the direct representation and problem-specific search operators, are introduced in Section 4. The paper concludes in Section 5.

2 Solution Representation and Operator Design in Evolutionary Algorithms

Evolutionary algorithms are often the method of choice for solving optimization problems with non-standard representations of the candidate solutions (e.g. special types of graphs). When applying EA to such non-standard domains, in general two approaches can be distinguished: (i) Introduce a genotype-phenotype mapping to a canonical representation, e.g. bit-string or continuous vectors on which standard operations, such as one-point crossover or bit mutation can be performed. (ii) Perform the search directly on the phenotype space and formulate problem-specific mutation and recombination operators as transformations of solutions in the phenotype space. While for the first approach out-of-the-box implementations of EA can be used after the genotype-phenotype mapping has been established, the second approach requires the formulation of new initialization, mutation and recombination operators. Often this effort is rewarded by a (much) higher performance of the EA as comparative studies in various domains show, e.g. in chemical process design [11] and decision diagram design [3].

In [4] mutation and recombination operators on graphs that represent chemical engineering process flowsheets were introduced in the form of graph rewriting rules. These rules define patterns in the flowsheet and how these patterns can be replaced by alternative patterns with a similar function. As opposed to operators that work with standard representations, this problem specific approach makes it easy to define causal transitions that lead from feasible structures to new feasible structures. Such tailored operators have a relatively high probability of finding improvements. Similar graph-based EA were successfully applied in analog circuit design [9] and multiobjective optimization of water distribution networks [1], among others. In software architecture design such direct representations with tailored operators have not yet been applied.

3   State-of-the-Art in Architecture Optimization

Koziolek et al. in [5] introduced for the first time the innovative way of using tactics as evolutionary algorithms operators. The approach employs tactic operators in addition to the general mutation and crossover operators. It still works on standard representations (bit strings) with a genotype-phenotype mapping and does not introduce any problem specific mutation or recombination operator. Since the solution does not support any initialization operator it should get an initial architecture as input. Also, it uses a fixed-length genotype to represent architectural solutions which can only be used for problems with few degrees of freedom. Moreover, it might suffer from *degeneracy*, which means that one phenotype is assigned to more than one genotype.

Martens at el. in [8] proposed another way of hybridization for optimization of system architectures. They propose two-step optimization: first analytic optimization and then a subsequent evolutionary optimization based on the Pareto candidates from the first step. In this paper, however, we use the domain knowledge in the search operators rather than defining a relay hybrid as in [8].

4   The Approach

4.1   Software Architecture Representation

In model-driven software engineering, model-based languages such as the Unified Modeling Language (UML) and the Architecture Analysis and Design Language (AADL) facilitate the modeling of software architecture and provide elements to understand it. Moreover, Architecture Description Languages (ADLs) and their accompanying toolsets have been proposed for modeling architectures, both within a particular domain and as general-purpose architecture modeling languages. For instance, EAST-ADL2 is an architecture description language which is designed specially for automotive embedded systems. So, these model-based languages make it possible to choose a proper representation model according to the domain problem. In this case, it is also needed to customize the search operator based on chosen representation. In the following we shortly discuss two popular modeling languages:

UML/MARTE Profile. The Unified Modeling Language standard (UML) is probably the most representative of model-based software development and has had significant success in the software industry. UML provides a set of diagrams to depict software structures graphically. The diagrams are developed as separate entities that express different aspects of the software. UML Profile provides a mechanism for extending UML models for different specific domains. MARTE (Modeling and Analysis of Real Time and Embedded systems) is a UML profile for model-based development of real-time and embedded systems. Because of it characteristics it is suitable for performance and schedulability analysis.

Architecture Analysis and Design Language (AADL). AADL comes from a computer language tradition, rather than a diagrams tradition. AADL is developed like a programming language not only the textual representation of the architecture. So it formally defines the syntax and semantics. However, it is possible to depict the textual representation graphically. Any software description in AADL is analyzable and has an unambiguous interpretation.

4.2   Problem-Specific Search Operators

Architectural and design patterns capture expert knowledge about best practices in software design by documenting general solutions that may be customized for a particular context. They make it possible to reuse the knowledge of software design and to focus on quality attributes such as performance. Software antipatterns are conceptually similar to patterns in that they document recurring solutions to common design problems (i.e. the bad practices) as well as their solutions: what to avoid and how to solve the problems [2]. Choosing a modeling language as solution representation make it possible to implement search operators in a standard model-to-model transformation language such as ATL or QVT. Moreover, by using our approach existing patterns or antipatterns (such as ECS in [2]) can be potentially implemented as a search operator. Following we give only few examples of possible patterns as operator:

Caching for improving performance. Basically, there are two main caching strategies that can be described as patterns [10], besides the other caching patterns: Primed Cache and Demand Cache. If the data required performing a certain usecase is known from the beginning, then the system can definitely cache it before the run starts, which is called Primed Cache. However, in case the data required by a usecase run can vary, the system can bring the data into the memory whenever required and keep it for future use. This situation corresponds to the Demand Cache pattern.

Figure 1a depicts caching search operator transformation defined by model rewriting rules. In this pattern *CompB* is replaced by a combination of *Cache* component and *CompB*. So, instead of directly calling of *CompB* by *CompA*, *CompA* calls the *Cache* component and then (if needed) it calls *CompB*.

Voter pattern for improving safety. Fault masking is one of the primary approaches to improve the normal behaviour of a system in a faulty environment. N-modular redundancy and N-version programming are the well-known fault masking methods. These approaches use redundant modules and a voting unit to hide the occurrence of errors. The voter arbitrates between the achieved results and produces a single output [6].

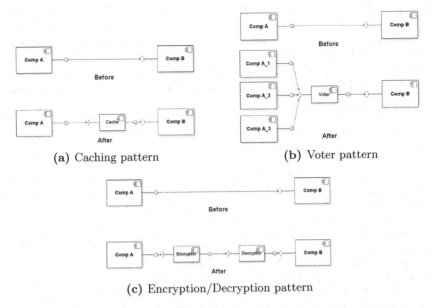

(a) Caching pattern

(b) Voter pattern

(c) Encryption/Decryption pattern

Fig. 1. Samples of Problem-specific Search Operators

Figure 1b depicts voter as a search operator transformation with three software replication. It shows that *CompA* is replaced by a combination of three replicated *CompA* and a *Voter* component. So, instead of directly calling of *CompB* by *CompA*, *Voter* calls the *CompB* and send a message from *CompA* depends on the voting algorithm.

Encryption/Decryption for improving security. Encryption provides message confidentiality by transforming readable data (plain text) into an unreadable format (cipher text) that can be understood only by the intended receiver after a process called Decryption, the inverse function that makes the encrypted information readable again.

Figure 1c shows this search operator transformation. In which *CompA* is replaced by a combination of *Encryptor* component and *CompA*. Also *CompB* is replaced by a combination of *Decryptor* component and *CompB*. So, *CompA* and *CompB* can communicate securely.

5   Conclusion

In this paper we introduce a new hybrid approach which combines two class of rule-based and metaheuristic-based approaches for solving automatic software architecture design problem. This approach uses evolutionary algorithms for global search and architecture design heuristics based on domain specific knowledge as local improvement. Hence, we could integrate knowledge of architecture design patterns and antipatterns to the automatic architecture design process in a systematic manner. The distinguishing features of our approach are direct representation within a modeling language and problem specific search operators that can be implemented by a model-to-model transformation language. Future work will have to investigate the performance of instantiations of the suggested approach since it has not been applied on a real case study yet. Also it would be interesting to derive designs for recombination operators for direct representations.

References

1. Baos, R., Fonseca, C., Gil, C., Mrquez, A.L., Vila Melgar, E.Y., Montoya, F.G.: Design and evaluation of evolutionary operators for water distribution network optimisation. In: META (2010)
2. Cortellessa, V., Marco, A.D., Trubiani, C.: Performance antipatterns as logical predicates. In: Calinescu, R., Paige, R.F., Kwiatkowska, M.Z. (eds.) ICECCS, pp. 146–156. IEEE Computer Society (2010)
3. Droste, S., Wiesmann, D.: Metric Based Evolutionary Algorithms. In: Poli, R., Banzhaf, W., Langdon, W.B., Miller, J., Nordin, P., Fogarty, T.C. (eds.) EuroGP 2000. LNCS, vol. 1802, pp. 29–43. Springer, Heidelberg (2000)
4. Emmerich, M., Grötzner, M., Schütz, M.: Design of graph-based evolutionary algorithms: A case study for chemical process networks. Evolutionary Computation 9(3), 329–354 (2001)
5. Koziolek, A., Koziolek, H., Reussner, R.: Peropteryx: automated application of tactics in multi-objective software architecture optimization. In: Crnkovic, I., Stafford, J., Petriu, D., Happe, J., Inverardi, P. (eds.) QoSA/ISARCS, pp. 33–42. ACM (2011)
6. Latif-Shabgahi, G., Bennett, S., Bass, J.: Smoothing voter: a novel voting algorithm for handling multiple errors in fault-tolerant control systems. Microprocessors and Microsystems 27(7), 303–313 (2003),
http://www.sciencedirect.com/science/article/pii/S0141933103000401
7. Li, R., Etemaadi, R., Emmerich, M.T.M., Chaudron, M.R.V.: An Evolutionary Multiobjective Optimization Approach to Component-Based Software Architecture Design. In: IEEE CEC, pp. 432–439. IEEE (2011)
8. Martens, A., Ardagna, D., Koziolek, H., Mirandola, R., Reussner, R.: A Hybrid Approach for Multi-attribute QoS Optimisation in Component Based Software Systems. In: Heineman, G.T., Kofron, J., Plasil, F. (eds.) QoSA 2010. LNCS, vol. 6093, pp. 84–101. Springer, Heidelberg (2010)
9. Natsui, M., Homma, N., Aoki, T., Higuchi, T.: Topology-Oriented Design of Analog Circuits Based on Evolutionary Graph Generation. In: Yao, X., Burke, E.K., Lozano, J.A., Smith, J., Merelo-Guervós, J.J., Bullinaria, J.A., Rowe, J.E., Tiňo, P., Kabán, A., Schwefel, H.-P. (eds.) PPSN 2004. LNCS, vol. 3242, pp. 342–351. Springer, Heidelberg (2004)
10. Rotaru, O.P.: Caching patterns and implementation. Leonardo Journal of Sciences (8), 61–76 (January-June 2006)
11. Sand, G., Till, J., Tometzki, T., Urselmann, M., Engell, S., Emmerich, M.: Engineered versus standard evolutionary algorithms: A case study in batch scheduling with recourse. Computers & Chemical Engineering 32(11), 2706–2722 (2008)

A Concept for an Interactive Search-Based Software Testing System

Bogdan Marculescu, Robert Feldt, and Richard Torkar

Blekinge Institute of Technology,
School of Computing,
Karlskrona, Sweden
bogdan.marculescu@bth.se

Abstract. Software is an increasingly important part of various products, although not always the dominant component. For these software-intensive systems it is common that the software is assembled, and sometimes even developed, by domain specialists rather than by software engineers. To leverage the domain specialists' knowledge while maintaining quality we need testing tools that require only limited knowledge of software testing.

Since each domain has unique quality criteria and trade-offs and there is a large variation in both software modeling and implementation syntax as well as semantics it is not easy to envisage general software engineering support for testing tasks. Particularly not since such support must allow interaction between the domain specialists and the testing system for iterative development.

In this paper we argue that search-based software testing can provide this type of general and interactive testing support and describe a proof of concept system to support this argument. The system separates the software engineering concerns from the domain concerns and allows domain specialists to interact with the system in order to select the quality criteria being used to determine the fitness of potential solutions.

Keywords: search-based software testing, interactive search-based software engineering, user centered.

1 Introduction

There is an increasing integration of software into many products. This makes software quality a relevant factor in the overall quality of all such products. However, systems engineers and integrators are not software engineering experts and, in particular, have little or no software testing experience.

One option to ensure proper quality of the software being integrated is to involve dedicated software engineers to handle software development and testing. There are, however, drawbacks to this approach. First, this approach is quite costly. This is all the more valid for smaller companies that do not have the resources to accommodate this expense. Another drawback is that software

G. Fraser (Ed.): SSBSE 2012, LNCS 7515, pp. 273–278, 2012.
© Springer-Verlag Berlin Heidelberg 2012

engineers do no have the domain expertise required to test a software component for the environment in which it will have to operate. Thus, if not adapted to the context, the software in question may have a lower quality as a system component, in spite of having high quality as a software component.

An alternative option is to package general testing solutions to be usable by non-experts in software engineering, i.e. systems engineers and integrators. This would allow the domain expertise of the systems engineers to be fully used, whilst still applying proven solutions for software testing.

Search-Based Software Testing (SBST) is an excellent fit for the latter option. It has been shown to be a good approach for many different types of testing [13,1]. It consists of a very generic search component, while those components that are domain-specific are those that systems engineers and domain specialists have their expertise in. These domain-specific components are the representation of the problem and the software as well as the quality criteria used for evaluation.

The contribution of this paper, therefore, is to propose a search-based software testing system that allows domain-specialist users to create test cases for the software they produce, without the need for specialized knowledge of software testing or search-based techniques.

2   Background

Search-Based Software Testing (SBST) is the application of search techniques to the problem of software testing. SBST has been applied to a variety of testing problems [13,1], from object-oriented containers [2] to dynamic programming languages [12]. SBST is a part of the wider area of Search-Based Software Engineering (SBSE), a term coined by Harman and Jones in 2001 [8], but a concept used earlier, see e.g. [19,4,9]. Since search-based approaches have been applied across the software development life-cycle [7], it is reasonable to expect that any conclusions referring to the general area of SBSE can be applied to the particular case of SBST.

The context of the system proposed in this work is that of software testing performed by domain specialists with relatively little knowledge of software testing or search-based techniques. The domain specialists would have extensive knowledge of the capabilities of the system-under-development, its own context and the limitations placed upon it, as well as the quality foci that they would have to pursue for each component of that system.

The combination of non-specialist software developers and the importance of domain knowledge and limitations makes it impractical to develop a fitness function up-front. To develop an appropriate fitness function for the component under test, the domain specialist would have to interact with the system and make adjustments to the criteria being used.

3   Interactive Search-Based Testing System

3.1   Running Example

To better illustrate the concepts discussed in this section, we will present an anonymized industrial example. The application we will use is that of a controller enabling a joystick or set of joysticks to handle a mechanical arm. The inputs for the controller software are the joystick signals and sensors that indicate the speed of the basket at the end of the crane. The outputs are the signals to the hydraulic pumps that drive the arm.

The System Under Test (SUT), in this case, is the software for the controller component. The goal of the Search-Based Software Testing System is to generate test cases that ensure the system's compliance to quality standards, ensure that no constraints are broken and discover any additional faults or unexpected behavior.

The Search-Based Software Testing System is the result of applying the concept presented developed using the proposed methodology in this work in the company in question. The system is meant to be tailored for the specific context and company it is expected to function in, yet be general enough to enable domain specialists to test new applications within the confines of that context.

3.2   Overview and Components

The figure 1 shows the structure of a complex Interactive Search-Based Software Testing system developed using the proposed approach.

Outer Cycle. The outer cycle is an interactive search-based system that uses the human domain specialist as a fitness function. It mediates the interaction between the domain specialist and the system by means of a component called Interaction Handler. For the purpose of this discussion we call a potential solution, or a solution candidate, any individual that is part of the population the human user is expected to evaluate.

The purpose of the Interaction Handler is to display the potential solutions shown to the human domain specialist and to collect their feedback. Feedback, in the type of system being proposed can refer to three separate issues:

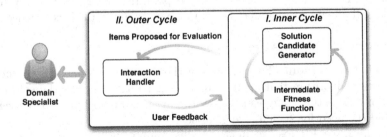

Fig. 1. Overview of an ISBSE system

- **Solution Candidate Feedback.** This describes feedback related to the solution candidates. In addition to selecting potential solutions for the next generation, the human domain specialist may assign values to each solution candidate they select for the next generation, giving them an evolutionary advantage.
- **Display Feedback.** This describes feedback related to the way solution candidates are displayed, the number of candidates displayed, and any additional information that is available or can be made available. Considering our running example discussed above, a domain specialist may choose to see memory required, response times for the output signals or discrepancies between expected output signals and actual output signals, in addition to the pass or fail status of each suite.
- **Quality Focus Feedback.** Since a search-based system can generate more solution candidates that a human can be expected to evaluate, some internal mechanism exists to enable a preliminary selection of potential solutions. This type of feedback allows the domain specialist to set or change the criteria by which this preliminary selection is performed. As the search for appropriate test cases goes on, it may become necessary to adjust the quality foci that set the selection criteria. As an example, an initial requirement of the joystick controller in our example may concern appropriate timing or accuracy of the output signal. Once the module is considered satisfactory from that perspective, searching for large variations or undesired behaviors may become more important. This type of feedback would allow the domain specialist to alter the focus of the search without restarting the search, and thus preserving the characteristics of the solution candidates already in the population.

The replacement of the fitness function with a human domain specialist restricts the number of potential solutions that the system can process in this manner. The additional information that the human can provide is an attempt to compensate for the lower number of solution candidates being processed by improving the selection mechanisms internal to the system.

Inner Cycle. The inner cycle is a search-based software testing system that uses a flexible fitness function. The purpose of this system is to generate and select the best solution candidates for the human domain expert to evaluate. This is meant to fully benefit from exploratory capabilities of search-based systems [5], while still allowing the human domain expert to apply their experience and insight.

The inner cycle itself has two components:

- **Search Component.** The purpose of the Search Component is to encapsulate the algorithm that creates the new generation of potential solutions. Encapsulating this component allow the existing algorithm to be changed, should the need for such a change arise.
- **Intermediate Fitness Function.** This component serves the purpose of the fitness function in any search-based system: it assigns each potential

solution a fitness value. The difference consists of allowing changes to be made to this component during the search process. Such changes originate in the feedback the human domain specialist provides and allows them to influence the direction of the automated search as well as performing their own selection.

The purpose of this component is not to replace human input, but rather to provide an initial screening of solution candidates, so that only those solution that are most likely to be successful are analyzed by the human domain specialist.

The interaction between the inner and outer cycles is achieved through the populations of candidates, the evaluations made by the domain expert and the feedback that guide the Intermediate Fitness Function. The generation and selection of the population, as well as the internal workings of the inner cycle are hidden from the domain expert.

4 Validation and Discussion

Validation efforts are focused on the development of a proof of concept system. This system is being developed in cooperation with an industrial partner and the initial validation will take place in that context.

The system presented here is designed specifically to address situations where domain expertise is the deciding factor in successful testing. This can be due to the complexity of the system under test and the influence external factors may have in its operation, as well as limitations in terms of the resources available for testing.

5 Conlusions

This paper has proposed a search-based software testing system designed to allow domain specialists with little software testing expertise to develop test cases for their applications. The value of such systems would be especially relevant for contexts were software testing experts are not available or where domain knowledge is the deciding factor in the success of the testing process.

References

1. Afzal, W., Torkar, R., Feldt, R.: A systematic review of search-based testing for non-functional system properties. Information and Software Technology (2009)
2. Arcuri, A., Yao, X.: Search based software testing of object-oriented containers. Information Sciences 178(15), 3075–3095 (2008)
3. Farrington, J.: Seven plus or minus two. Performance Improvement Quarterly 24(4), 113–116 (2011)
4. Feldt, R.: Generating multiple diverse software versions with genetic programming - an experimental study. IEE Proceedings - Software 145(6), 228–236 (1998)

5. Feldt, R.: Genetic programming as an explorative tool in early software development phases. In: Proceedings of the 1st International Workshop on Soft Computing Applied to Software Engineering (SCASE 1999), April 12-14, pp. 11–20. Limerick University Press, University of Limerick (1999)

6. Feldt, R.: An interactive software development workbench based on biomimetic algorithms. Tech. Rep. 02-16, Gothenburg, Sweden (November 2002)

7. Harman, M.: The current state and future of search based software engineering. In: Future of Software Engineering (FOSE 2007) (2007)

8. Harman, M., Jones, B.F.: Search based software engineering. Information and Software Technology (43), 833–839 (2001)

9. Harman, M., Mansouri, S.A., Zhang, Y.: Search based software engineering: A comprehensive analysis and review of trends techniques and applications. Tech. Rep. TR-09-03 (April 2009)

10. Kamalian, R., Yeh, E., Zhang, Y., Agogino, A., Takagi, H.: Reducing human fatigue in interactive evolutionary computation through fuzzy systems and machine learning systems. In: 2006 IEEE International Conference on Fuzzy Systems, pp. 678–684 (2006)

11. Maguire, M.: Methods to support human-centred design. International Journal of Human-Computer Studies 55(4), 587–634 (2001)

12. Mairhofer, S., Feldt, R., Torkar, R.: Search-based software testing and test data generation for a dynamic programming language. In: Proceedings of the 13th Annual Conference on Genetic and Evolutionary Computation, GECCO 2011, pp. 1859–1866. ACM, New York (2011)

13. McMinn, P.: Search-based software testing: Past, present and future. In: Fourth International Conference on Software Testing, Verification and Validation Workshops, pp. 153–163 (2011)

14. Metzger, U., Parasuraman, R.: Automation in future air traffic management: Effects of decision aid reliability on controller performance and mental workload. Human Factors: The Journal of the Human Factors and Ergonomics Society 47(1), 35–49 (2005)

15. Simons, C.L., Parmee, I.C., Gwynllyw, R.: Interactive, evolutionary search in upstream object-oriented class design. IEEE Transactions on Software Engineering 36(6), 798–816 (2010)

16. Simons, C., Parmee, I.: User-centered, evolutionary search in conceptual software design. In: IEEE World Congress on Computational Intelligence. IEEE Congress on Evolutionary Computation, CEC 2008, pp. 869–876 (June 2008)

17. Takagi, H.: Interactive evolutionary computation: fusion of the capabilities of ec optimization and human evaluation. Proceedings of the IEEE 89(9), 1275–1296 (2001)

18. Tarnow, E.: There is no capacity limited buffer in the murdock (1962) free recall data. Cognitive Neurodynamics 4, 395–397 (2010), doi:10.1007/s11571-010-9108-y

19. Xanthakis, S., Ellis, C., Skourlas, C., Gall, A.L., Katsikas, S., Karapoulios, K.: Application of genetic algorithms to software testing. In: Proceedings of the 5th International Conference on Software Engineering and Applications, Toulouse, France, December 7-11, pp. 625–636 (1992)

A Search-Based Framework
for Failure Reproduction

Fitsum Meshesha Kifetew

Fondazione Bruno Kessler, Trento, Italy
kifetew@fbk.eu

Abstract. The task of debugging software failures is generally time consuming and involves substantial manual effort. A crucial part of this task lies in the reproduction of the reported failure at the developer's site. In this paper, we propose a novel framework that aims to address the problem of failure reproduction by employing an adaptive search-based approach in combination with a limited amount of instrumentation. In particular, we formulate the problem of reproducing failures as a search problem: reproducing a software failure can be viewed as the search for a set of inputs that lead its execution to the failing path. The search is guided by information obtained through instrumentation. Preliminary experiments on small-size programs show promising results in which the proposed approach outperforms random search.

Keywords: failure reproduction, search-based software engineering, automated test data generation, adaptive search, debugging, instrumentation.

1 Introduction

Software testing is one of the most widely used techniques for assuring the proper functioning of a piece of software before releasing it to production. However testing can not ensure a totally bug free piece software. Consequently, software bugs are often revealed after the software has been deployed in production environments causing a *failure* of the software. The failure could be a crash, a hang, unexpected output, etc. When such software failures are encountered, the end users usually send bug reports to the software producer.

In order to fix a bug, developers try to reproduce the failure caused by the bug on their machines. This task of re-creating field failures in house is usually cumbersome and time consuming as it involves a considerable amount of manual effort. This is mainly because important information, such as input and the context that caused the failure, is missing from bug reports received from end users [4].

Automated support in reproducing field failures in house would hence greatly reduce the effort put in by programmers in the overall process of fixing bugs. To this end, a number of research works [3,5,6,9,7,11] have tried to address the issue of automated failure reproduction, many of which are based on Symbolic

G. Fraser (Ed.): SSBSE 2012, LNCS 7515, pp. 279–284, 2012.

Execution (SE). While SE offers several appealing features, it suffers from problems such as scalability, loop unrolling, black box functions, etc. This limits the applicability of existing solutions to large-size real world programs.

To address the problem of failure reproduction in a scalable and applicable manner, we propose a framework in which failure reproduction is formulated as a search problem [10]. In particular, a search is performed in the input domain of the software looking for those inputs that could reveal the failure observed in the field. The search is guided by a limited amount of execution data gathered from the deployed software by means of a light-weight program instrumentation.

2   State of the Art

The primary objective of research work in this area is to find test cases that lead program executions into failures which are similar to those observed on the field.

One of the proposed approaches is to generate unit tests by storing method arguments in memory and replaying them in the event of a failure [3]. While these unit tests may reproduce the failure, they provide little insight into the actual cause of the problem as they only show the manifestation of the problem. For instance, a *null* parameter could have caused a crash in a method. However more investigation is required to understand why and how the parameter became *null*. In contrast, a failure reproduction technique producing a full program execution would identify the input sequence responsible for generating such a *null* value, not just the *null* value itself.

Reproducing failures at the client side has also been proposed [5,6]. In such schemes, an execution profile is collected while the program is running on the client's machine and in the event of a failure, a client side component uses the collected profile and tries to find test cases that reproduce the failure, but are significantly different from the original user input. While this approach limits the exposure of sensitive client data, it could, however, incur a high performance overhead on the client.

Gathering limited amounts of execution data from deployed software and using it to reproduce failures in house is yet another approach which has been employed in a number of research works. For instance, Jin and Orso propose a customized SE-based algorithm for reproducing failures by using execution data collected from the field via instrumentation [9]. According to an empirical study carried out by the authors on several C programs, tracing *function call sequences* from deployed software was found to be an effective guide for the failure reproduction algorithm. Other SE-based approaches have also been proposed (for e.g. [7,11]), however, they all suffer from the inherent limitations of SE with respect to scalability, loops and black box functions. On the other hand, search based approaches can scale easily to large programs, and can handle black box functions and loops as well.

3   Research Problem

This research work tries to address the following research problems:

Scalable and applicable failure reproduction: while there are existing approaches to address the problem, their applicability is limited. Hence there is still a need for a scalable failure reproduction scheme that is applicable to a wide variety of programs.

Reasonable execution data collection: the failure reproduction scheme needs to be supported by a reasonable amount of execution data (trajectory) from deployed programs while maintaining an acceptable balance among performance, privacy and reproduction of failures. Existing instrumentation techniques should be enhanced in light of scalable failure reproduction schemes.

Adaptive search scheme: to apply search-based techniques for failure reproduction, the search needs to be adaptive with respect to (1) *search parameters*: search parameters significantly affect the outcome of the search [2] and manual tuning is impractical for our purpose as we target a potentially wide range of programs with different search landscapes, and (2) *trajectory*: it is possible that the trajectory could contain information which leads the search in the wrong direction, reducing its effectiveness in guiding the search.

4   Proposed Approach

To address the problem of reproducing field failures in a scalable manner, we propose a framework composed of two main parts: Instrumentation and Test Case Generation as shown in Figure 1.

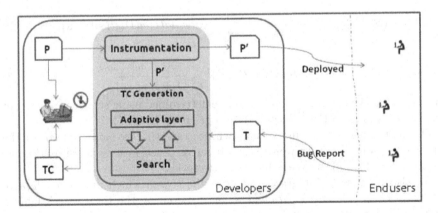

Fig. 1. Proposed Framework: *Program P', instrumented version of P, is deployed to end users. Trajectory (T), extracted from bug reports, is then used to generate Test Cases (TC) for reproducing a reported failure.*

4.1   Instrumentation

In this part of the framework, we thoroughly analyze the program applying a combination of static, dynamic as well as heuristic-based analysis techniques in order to study the nature of its control structure and the data elements it manipulates. Based on the analysis, we determine which part of the program state should be sampled (traced) in such a way that (i) it is sufficient to guide the search (ii) the performance overhead of the tracing is acceptable, and (iii) sensitive information is avoided. Then the program is instrumented accordingly and the desired program state is collected and included in the bug report which is sent back to the developers. We refer to the collected execution data as a *trajectory*.

More formally, a trajectory is defined as a sequence $T = \langle S_1, ..., S_n \rangle$, where each S_i is a variable-value mapping at observation point PC_i during the execution of the program. The mapping is a projection of the program state S_i at the execution point PC_i, in that not all program variables are necessarily traced.

4.2   Test Case Generation

The trajectory in the bug report is used to guide the search for potential sets of inputs that lead the program execution to the desired path (i.e failing path).

We propose an *objective function* based on the *distance* between the trajectory of the failing execution (obtained from the bug report) and the trajectory of a candidate individual. One way to compute the distance between two trajectories T_1 and T_2 could be defined as:

$$dist(T_1, T_2) = \sum_{S_i, S_j \in MatchedExecPoints(T_1, T_2)} (dist(S_i, S_j)) \tag{1}$$

$$dist(S_i, S_j) = \frac{1}{|Vars(Si)|} \sum_{v \in Vars(S_i)} (normalizedDist(s_i(v), s_j(v))) \tag{2}$$

$$normalizedDist(v, w)^1 = \frac{|v - w|}{(1 + |v - w|)} \tag{3}$$

where MatchedExecPoints (T_1, T_2) can be determined using sequence alignment algorithms, based on the values of the program counter PC.

The search encourages individuals whose trajectory is *more similar* to the target trajectory. In other words, the objective function tries to *minimize* the distance $dist(T_1, T_2)$ between the two trajectories.

We also propose to integrate a layer in the search which makes it adaptive with respect to (1) *search parameters*: by distributing the total search budget between the actual search and parameter control [8], and (2) *trajectory*: by automatically identifying the part of the trajectory that is more guiding to the search. One possibility could be to apply a binary-like search to effectively explore the trajectory while performing the actual search.

---

[1] Normalization technique proposed by Arcuri [1].

5   Preliminary Results

We have implemented a proof-of-concept level prototype of our proposed approach and we carried out experiments on *printtokens2*[2] (*size* $\approx 0.5KLOC$): a C program that accepts a stream of characters as input and parses it into tokens.

We developed a GA-based prototype that implements the first two parts of the proposed framework (see Section 4). The Trajectory contains traces of PC and values of arguments at function entry. Random search was used as a baseline.

To evaluate the performance of the GA-based approach, we computed the density of solutions in the final generation of the GA as *the number of solutions divided by the population size*. As can be seen in Figure 2, the GA-based approach produces a higher density of solutions as compared to random search.

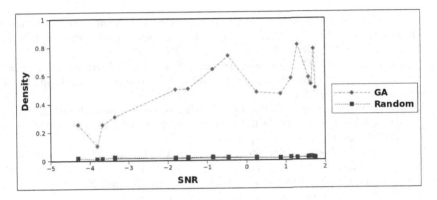

Fig. 2. Density vs SNR for GA and Random Search

To assess the effectiveness of the trajectory, we studied the subject program and the fault in detail so that we could decide which part of the trajectory is actually useful for the search (we call it *signal*) and which is not (we call it *noise*). After identifying and quantifying the signal and noise, we computed the *signal to noise ratio (SNR)* as *log(signal/noise)* for a number of trajectories. As can be seen from Figure 2, the higher the SNR, the higher the density of solutions obtained by the GA. It can also be seen that the GA produces higher densities of solutions even for trajectories with low SNR. The fluctuation in the values of density for GA seen in Figure 2 is due to the fact that there is a certain level of randomness in the GA search.

6   Plan and Conclusion

To address the problem of in-house failure reproduction, we propose an adaptive search-based framework supported by execution data from the field. An empirical

---

[2] Available from Software-artifact Infrastructure Repository (http://sir.unl.edu).

study with a proof-of-concept level prototype shows promising results in which the proposed search-based approach outperforms random search.

To complete the framework, we plan to address the following issues:

- devise a suitable instrumentation scheme that keeps performance overhead and exposure of sensitive data at a minimum
- assess parameter control mechanisms for implementing an adaptive search
- assess mechanisms for further improving the objective function
- evaluate the framework by conducting a series of empirical studies on real programs and real faults, and comparing the results with those of existing failure reproduction approaches.

References

1. Arcuri, A.: It does matter how you normalise the branch distance in search based software testing. In: 2010 Third International Conference on Software Testing, Verification and Validation (ICST), pp. 205–214 (April 2010)
2. Arcuri, A., Fraser, G.: On Parameter Tuning in Search Based Software Engineering. In: Cohen, M.B., Ó Cinnéide, M. (eds.) SSBSE 2011. LNCS, vol. 6956, pp. 33–47. Springer, Heidelberg (2011)
3. Artzi, S., Kim, S., Awasthi, P.: ReCrash: Making Software Failures Reproducible by Preserving Object States. In: Dell'Acqua, P. (ed.) ECOOP 2008. LNCS, vol. 5142, pp. 542–565. Springer, Heidelberg (2008)
4. Bettenburg, N., Just, S., Schröter, A., Weiss, C., Premraj, R., Zimmermann, T.: What makes a good bug report. In: Proceedings of the 16th ACM SIG-SOFT International Symposium on Foundations of Software Engineering, SIG-SOFT 2008/FSE-16, pp. 308–318. ACM, New York (2008)
5. Castro, M., Costa, M., Martin, J.-P.: Better bug reporting with better privacy. In: Proceedings of the 13th International Conference on Architectural Support for Programming Languages and Operating Systems, ASPLOS XIII, pp. 319–328. ACM, New York (2008)
6. Clause, J., Orso, A.: Camouflage: automated anonymization of field data. In: 2011 33rd International Conference on Software Engineering (ICSE), pp. 21–30 (May 2011)
7. Crameri, O., Bianchini, R., Zwaenepoel, W.: Striking a new balance between program instrumentation and debugging time. In: Proceedings of the Sixth Conference on Computer Systems, EuroSys 2011, pp. 199–214. ACM, New York (2011)
8. Eiben, A., Hinterding, R., Michalewicz, Z.: Parameter control in evolutionary algorithms. IEEE Transactions on Evolutionary Computation 3(2), 124–141 (1999)
9. Jin, W., Orso, A.: BugRedux: Reproducing Field Failures for In-house Debugging. In: 2012 34th IEEE and ACM SIGSOFT International Conference on Software Engineering (ICSE) (June 2012)
10. McMinn, P.: Search-based software test data generation: a survey: Research articles. Softw. Test. Verif. Reliab. 14(2), 105–156 (2004)
11. Zamfir, C., Candea, G.: Execution synthesis: a technique for automated software debugging. In: Proceedings of the 5th European Conference on Computer Systems, EuroSys 2010, pp. 321–334. ACM, New York (2010)

Evolutionary Testing of PHP
Web Applications with WETT

Francesco Bolis, Angelo Gargantini, Marco Guarnieri, and Eros Magri

Dip. di Ing. dell'Informazione e Metodi Matematici, Università di Bergamo, Italy
{francesco.bolis,angelo.gargantini,marco.guarnieri,eros.magri}@unibg.it

Abstract. One of the current core requirements of web applications is the continuity of the service, because loss in availability can lead to severe economic losses. This is the main reason behind the growing interest in web application testing that offers to researchers several challenges, due to the peculiar nature of these applications. Several classical testing techniques have been extended to deal with web testing. In this paper we propose to extend to web application testing a recent search-based approach that optimizes the generation of the whole test suite. This approach has several advantages over common approaches that optimize the generation of a single test case at a time. We show the technological challenges we have had to face, the architecture of the tool WETT we have developed, and some preliminary results of the experiments.

1 Introduction

The wide diffusion of Internet combined with mobile technologies has produced a significant growth in the demand of web applications with more and more strict requirements of reliability, usability, inter-operability and security [4]. Due to market pressure and very short time-to-market, the testing of web applications is often neglected by developers, although several works, such as [9], analyze the high costs of unavailability of web applications.

To address this problem, the testing community has tried to extend *traditional* testing methods in order to make them suitable for testing web applications. However, traditional testing theories, methods, and tools cannot be used in most cases just as they are, because of the peculiarities and complexities of web applications. For instance, web applications are almost always connected to databases and they are distributed with a client part (often a simple web browser) and a server part (a web server). The diversity of technologies and programming languages involved in the development of modern web applications represents a serious problem for traditional testing techniques. For instance unit tests may have limited efficacy, and thus acceptance testing techniques are preferred, since they try to capture the behavior of the entire web application [2]. However, the automated generation of oracles for acceptance testing is a hard task, because it requires a formal model from which the oracles can be extracted, and thus the manual definition of them is yet a concrete alternative. Due to this

G. Fraser (Ed.): SSBSE 2012, LNCS 7515, pp. 285–291, 2012.

fact the test suite must remain of manageable size to avoid the burden of introducing oracles in large test suites.

For these reasons, web application testing still offers many open research issues and challenges. One possible way to deal with it, is to adapt *search-based* techniques for test generation [8]. Note that there exist several approaches for web testing [2,4], however, only a few of them attempt to extend search-based approaches to web application [1,7]. In our paper we try to apply the approach presented in [6,5] and implemented in a tool called EVOSUITE, that generates and optimizes whole test suites towards satisfying a coverage criterion. That approach improves over the state of the art in search-based testing by keeping the size of the test suite under control and by maximizing the coverage of the whole test suite instead of that of single tests.

An evolutionary approach for test generation of web applications has already been presented in [1]. Whilst we share with it several concepts, our approach is different also because we do not assume any automated oracle. This has a strong impact in the design choices we have made, including the following ones. (1) Our tests are described in terms of user actions in a high level language (presented in Section 2) and not like [1] as arrays of parameters. Tests preserve an intelligible meaning and the user can better complete the oracle part and reproduce the actual scenario that causes a possible bug. (2) The size and the length of our tests is constantly under control. For example, in [1] test suites had around 160 tests, while our tests for the same case study contains only around 10 tests. (3) Our tool evolves the whole test suite and not single test cases, so coverage is maximized but overlapping in coverage among tests is also minimized.

Also Marchetto and Tonella [7] use a search-based approach in order to test Ajax web applications. A search-based algorithm is used for the exploration of the state space. Their approach shares with ours the goal of making the generated tests more efficient, i.e. of increasing the coverage with respect to the number of events in the tests. To this goal, they use a hill climbing algorithm in order to create test suites that maximize diversity.

In this paper we explain the basic algorithms and the architecture of the Web application Evolutionary Testing Tool - WETT- we have developed. We report also some initial experimental results.

2   Our Approach

Given the fact that an important part of web application behavior, and thus code, is tightly related to the interaction with the users, we have chosen to define our test suites in terms of the events that can be executed on the web application interface, and thus, in order to automate the testing phase, we have chosen to use Sahi[1], a *capture and replay tool* that lets users express test cases using scripts and then execute them. Each test case is a list of Sahi statements, chosen among the ones presented in Figure 1. We choose a random test suite TS as initial population. We call $|t|$ the length of the test case t (i.e. the number

---

[1] Sahi Web Automation and Testing Tool http://sahi.co.in/

of statements in t), AUT the application under test, and we have defined a threshold k that represents the maximum length of a test case.

The genetic algorithm used in our application is shown in Algorithm 1. The algorithm is adapted from [5]. It takes as input the initial test suite TS, the maximum number of iterations it, the fitness threshold ft, and the probability of a crossover cp. It evolves the initial test suite and returns as output the evolved test suite *population*. The algorithm evolves the test suite until the iteration threshold is reached or an adequate test suite is found. At each iteration the algorithm selects the subset of the current population with the best values of fitness using the function $elite(population)$, then it evolves the current elite. It selects two indi-

```
_navigateTo ()
_setValue ( textbox () )
_setValue ( textarea () )
_setSelected ()
_click ( radio () )
_click ( link () )
_click ( _image () )
_click ( _imageSubmit () )
_click ( _submit () )
_click ( _reset () )
```

Fig. 1. Sahi statements

viduals using tournament selection, with the function $tournament(population)$, and it modifies them using the crossover operator with a probability cp. Then the two individuals are mutated and the algorithm adds the two parents or the two new individuals to the current elite, depending on which are the best ones in terms of the fitness. The algorithm continues to evolve the elite until the test suite has grown more than the current population, then it starts another iteration considering the current elite as the population.

We have defined the fitness function in terms of the statement coverage achieved by the test suite. However, given the fact that a web application consists of several pages, in our fitness function we consider also the number of pages covered by the test suite, i.e. we increase the fitness function of a certain value $(n - l)r$ if the number n of covered pages is greater than a certain threshold l, where r represents the reward per page. In this way, we can reward with higher fitness values those test suites that exercise the AUT both in terms of covered statements and in terms of visited pages. Explicitely, the fitness achieved by the test suite T is $f(T) = \alpha * C(T) + \beta * r * (max(n, l) - l)$, where α and β are two parameters used to weight the influence of the statement coverage and the page coverage on the overall fitness value.

Our approach uses the *two points crossover* technique, i.e. given two individuals P_1 and P_2, we generate two new individuals O_1 and O_2 such that O_1 contains the firsts x statements of P_1, then the next $y - x$ statements of P_2, and the lasts $|P_1| - y$ statements of P_1 and the opposite for P_2, where $x, y \in [0, min(|P_1|, |P_2|)]$ are random values such that $x < y$.

In the mutation phase we apply three operators, each one applied with probability $\frac{1}{3}$: (a) *Remove*: Given a test case t, each statement is removed with a probability $\frac{1}{|t|}$, (b) *Change*: Given a test case t, each statement is modified with a probability $\frac{1}{|t|}$ (we change the parameters of the statement by choosing new parameters at random from configuration files), (c) *Insert*: We add a new statement to the test case t with a probability β^2, when a new statement is added

Algorithm 1. Genetic Algorithm

Input : *ts*, *it*, *ft*, *cp*
Output: *population*
begin

    population ← *TS*;
    iteration ← 0;
    while (*iteration* < *it*) ∧ (*fitness*(*population*) < *ft*) **do**
        E ← *elite*(*population*);
        while $|E|$ < $|population|$ **do**
            P_1, P_2 ← *tournament*(*population*);
            if *random*([0, 1]) < *cp* **then** O_1, O_2 ← *crossover*(P_1, P_2);
            else O_1, O_2 ← P_1, P_2;
            mutate(O_1, O_2);
            f_p ← *max*(*fitness*(P_1), *fitness*(P_2));
            f_o ← *max*(*fitness*(O_1), *fitness*(O_2));
            T_b ← *best individual of population*;
            if $f_p > f_o$ **then** E ← $E \bigcup \{P_1, P_2\}$;
            else
                for $O \in \{O_1, O_2\}$ **do**
                    if *length*(O) ≤ 2 * *length*(T_b) **then** E ← $E \bigcup \{O\}$;
                    else E ← $E \bigcup \{P_1 \text{ or } P_2\}$;

        population ← E;

---

we add another statement with probability β^{i2} (where i is the number of already inserted statements), until no new statements are added or the maximum length of the test case is reached. The statements to add are selected randomly among the ones presented in Listing 1, and the parameters are chosen at random from the configuration files. Configuration files contain the definitions of the elements of the AUT that can be used as arguments for the Sahi statements.

In order to prevent the undefined growth in length of the test cases, which could cause a *bloat*, we adopted several strategies: (i) we have defined a maximum length for the test cases, (ii) we have used the *tournament selection* method, that lets us choose only the individuals with high fitness values, (iii) our fitness function considers also the number of pages explored and, thus, we can reward better the individuals that explore several pages instead of the ones that explore exhaustively a single page.

2.1 Architecture

As already said, each WETT test case is a list of Sahi statements, which are executed by the Sahi tool that invokes the browser and executes the scripts. Figure 2(a) presents our process, whereas Figure 2(b) shows the architecture of the WETT tool. In order to generate the initial population, we execute manually a certain number of test cases using Sahi, and then we select at random the population among the statements used in these test cases. Then we parse the scripts in order to import them into the WETT application, where we evolve the current population. In order to compute the fitness function, we execute the evolved test suite by using Sahi, which communicates with the AUT using the

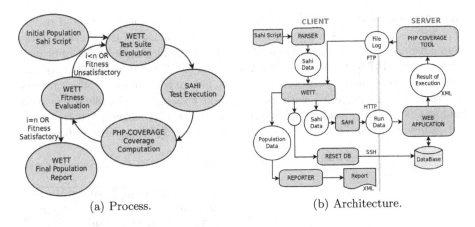

(a) Process.   (b) Architecture.

Fig. 2. Our Approach

HTTP protocol. We use *XDebug*[2] and *PHP-Coverage*[3], which are installed on the server that executes the AUT to measure the coverage achieved by the test suite. Given the fact that these tools produce coverage reports on the server, we import the coverage results in WETT via FTP. When WETT computes a test suite that satisfies the fitness requirement or it reaches the iteration threshold, we produce the final test suite and an XML report containing the information about the process. Given the fact that usually web applications use databases to store the current state of the application, we delete the content of the database after the execution of each iteration by executing an appropriate script (RESET DB) by SSH.

3   Initial Experimental Results

In order to evaluate the performance of WETT, we have selected as case study the *Schoolmate*[4] PHP application, which was already used in other works [1,3]. We report here the results of our preliminary experiments. We run WETT five times, each one with a different seed. The use of Sahi implies a considerable time overhead since the execution of a single candidate test case is around one minute. Each experiment started from a test suite with 10 randomly generated test cases and with the elitism that chooses at each iteration the 5 best test cases in terms of fitness. The results in Figure 3 show how the fitness evolves during the iterations. Note that the fitness of the test suite can also decrease from an iteration to the next one due to the elitism that selects each time only a subset of the current population and to the bloat control that limits the length of the test cases.

Table 1 reports the size of the best test suite for every iteration, the coverage achieved by the test suite, and the number of PHP interpreter warnings,

---

[2] XDebug, Debugger and Profiler Tool for PHP http://xdebug.org/

[3] PHPCoverage, code coverage tool for PHP http://phpcoverage.sourceforge.net/

[4] Schoolmate http://sourceforge.net/projects/schoolmate/

which are non-fatal errors. The table reports also the number of visited pages. In comparison with other approaches, our test suite achieves the average statement coverage of 21.7% with 13 test cases, while [1] achieves the 56.5% of branch coverage with 167 test cases and [3] achieves the 64.9% of line coverage with 724 tests. In comparison with a random approach [3], we achieve much better results in terms of coverage (8.3% with random) and test suite size (1396 tests with random). Note that although our approach achieves a low coverage, it keeps the test suite very small. Our approach visits, in average, 18.4 pages out of 63 pages.

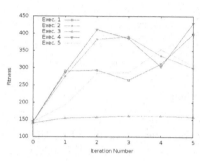

Fig. 3. Fitness Function

Table 1. Experimental results

#	Test Suite Size	Coverage	Warnings	Visited Pages
1	12	24.94%	36	19
2	17	24.44%	96	20
3	7	10.06%	132	15
4	13	26.87%	72	20
5	13	22.12%	144	18
Avg.	12.4	21.69%	96	18.4

4   Conclusion and Future Work

We found that extending the whole test suite generation approach to web application testing is feasible and that our approach has some advantages and drawbacks. We were able to automatically generate small tests suites achieving a discrete coverage. However, the achieved coverage remains low and the required time for test generation makes the number of iteration small. The former drawback is primarily due to the fact that our approach does not consider the state of the web application when mutating test cases and that the initial test suite is randomly chosen: this results in test scripts with a low quality and slows the evolution process. We plan to integrate a model-based/model discovery approach in order to solve this problem. The latter drawback is due to the use of Sahi as capture and replay tool. We plan to consider other alternative tools.

References

1. Alshahwan, N., Harman, M.: Automated web application testing using search based software engineering. In: Proc. of ASE 2011 (2011)
2. Andrews, A., Offutt, J., Alexander, R.: Testing web applications by modeling with FSMs. Software and Systems Modeling (2005)
3. Artzi, S., Kiezun, A., Dolby, J., Tip, F., Dig, D., Paradkar, A., Ernst, M.: Finding bugs in web applications using dynamic test generation and explicit-state model checking. IEEE Trans. Softw. Eng. 36(4) (2010)
4. Di Lucca, G., Fasolino, A.: Testing web-based applications: The state of the art and future trends. Information and Software Technology 48(12) (2006)

5. Fraser, G., Arcuri, A.: Evolutionary generation of whole test suites. In: Proc. of QSIC 2011 (2011)
6. Fraser, G., Arcuri, A.: Evosuite: Automatic test suite generation for object-oriented software. In: Proc. of ACM SIGSOFT ESEC/FSE, pp. 416–419 (2011)
7. Marchetto, A., Tonella, P.: Search-based testing of ajax web applications. In: Proc. of SSBSE 2009 (2009)
8. McMinn, P.: Search-based software test data generation: a survey. Softw. Test, Verif. Reliab. 14(2) (2004)
9. Pertet, S., Narasimhan, P.: Causes of failures in web applications. CMU Technical Report (2005)

Author Index